# THE GROWTH OF
SCIENTIFIC IDEAS

Title Page from Borelli's *De Motu Animalium* (Lugd, 1685)

# THE GROWTH OF SCIENTIFIC IDEAS

WILLIAM P. D. WIGHTMAN
M.Sc., Ph.D., F.R.S.E.

GREENWOOD PRESS, PUBLISHERS
WESTPORT, CONNECTICUT

Library of Congress Cataloging in Publication Data

Wightman, William Persehouse Delisle.
The growth of scientific ideas.

Reprint of the 1966 ed. published by Oliver and
Boyd, Edinburgh.
Bibliography: p.
1. Science--History. I. Title.
Q125.W66 1974        509        74-4732
ISBN 0-8371-7484-8

© *1950 W. P. D. Wightman*

Originally published in 1950 by Oliver and Boyd, Edinburgh

Reprinted with the permission of William P. D. Wightman

Reprinted by Greenwood Press, Inc.

First Greenwood reprinting 1974
Second Greenwood reprinting 1977

Library of Congress catalog card number 74-4732

ISBN 0-8371-7484-8

Printed in the United States of America

# PREFACE

THAT epoch-making work, *An Essay Concerning the Human Understanding*, has been described as being " not philosophy, but a preparation for philosophy ". If it be permissible to compare small matters with great, it is as a preparation for the study of the History of Science that I should wish this little work to be judged.

Any attempt to enlarge and clarify the study of the History of Science calls for no apology. For in the proposals for educational reform sponsored by many and various Public Bodies there has been a striking unanimity in the call for a greater emphasis on the historical aspects of science at every educational level. The manner in which this demand is to be met calls, however, for serious consideration. We shall stand the best chance of giving an effective response if we bear in mind that what has prompted it is the growing sense of frustration in the face of the modern dilemma. This dilemma may be expressed in some such terms as these: The crisis of the modern world, material and spiritual, is largely the result of the phenomenal growth of scientific ideas and of technological expansion; it is inevitable therefore that a scientific education is gradually displacing the traditional one based on the culture of classical antiquity. But the further results of this process have been far from happy: such an education, it seems, fails to give the sense of human continuity and all those intangible qualities which we express in the name *culture*. Thus, while the continued isolation of our future leaders in the classical twilight would maintain the intellectual gulf which still yawns between many of our governors and the community, yet on the other hand to deprive them of acquaintance with the triumphs and failures of the past would be to deny our parentage, to establish a society in a cultural vacuum, with all the possibilities of anarchy and frustration which that so evidently implies. The mere prescription of " some acquaintance with the history of science " has in the past often been held, and with some justice, to be a mere sop to contemporary modes of thought in which a " scientific " education could be in a small measure " liberalised " by accretion of historical fact. In the last twenty or thirty years we have seen a radical change in outlook. The history of science has been rewritten with a *purpose*. Sometimes this purpose, as for instance that associated with the political aims of proving the emergence of a master race or the dialectical evolution of a classless society, may be open to the objection of special pleading. But the point I am wishing to emphasise is that if the purpose has been bad, it has at any rate been a purpose. I believe that a sense of purpose, which means a principle of coherence, may be introduced into the historical study of science without giving way to any kind of special pleading. I do not for a moment claim any originality in the method adopted in this book. The experienced reader, if any such should happen to open it, will see that any merits it may have are due to the influence of two great thinkers of the recent past—R. G. Collingwood and A. N. Whitehead. In their several ways they have, I believe, emphasised the two-fold nature of historical fact, namely that every standpoint is what it is by virtue of its origin from the past and its urge towards the future. In its relevance to scientific ideas this means that no scientific entity— atom, organ, or wave—is *merely* itself, but is constantly evolving in a context of associated ideas. Hence the necessity for a knowledge of history; not to " liberalise " science, but to understand it.

This book therefore makes no claim to be another history of science, but a guide to the study of the development of scientific thought as illustrated by a few dominant ideas. An attempt is made to appreciate the problems as they appeared to the leading thinkers of each age, to describe in outline the nature of the solutions attempted, and to assess (in the light of later modification) the proportion of essential truth to psychological imagery which is inevitably interwoven therewith in the enunciation of this truth. In this way science becomes revealed as a *struggle*, a struggle no less charged with humanistic value than the struggle for political liberty or national expression; in a word, as a branch of *Litterae Humaniores*.

The method of treatment dictated by these considerations is to combine that of the scholar with that of the experimentalist; more emphasis is laid upon the question of sources and authorities than is customary in a book intended for general reading. While little actual documentation is provided, the reader is nevertheless by rather lengthy citations brought into more intimate contact with the minds of the great discoverers. Only in this way can that sense of historical actuality be acquired by which such apparently futile controversies as phlogiston, caloric, animal spirits, and the like, may be shown to be significant stages in the cultural advance. It will be evident that such a method is intensive as to problems, extensive as to background. This has its

# PREFACE

dangers, in that the ideas so treated will be such as appeal to the author and lie within his compass. To compensate for this, rather full bibliographies are appended. If the use of these by the reader is stimulated by the hope (alas too easily fulfilled!) of revealing the author's ignorance, one main purpose of the book will have been incidentally achieved.

Since all along the aim has been to indicate the interconnexion of concepts drawn from various contemporary fields, and of theory with technological demands and advances, the individual sciences have not been treated in isolation. Nor has there been any attempt at comprehensiveness. Thus some readers will doubtless be scandalised by the almost complete omission of any reference to the thoery of organic chemistry. My reason for this and other omissions is that this exceedingly involved and often confused development was in a sense a domestic affair—of great interest to the specialist, but demanding a book to itself to make it even significant to the general reader.

I have tried to avoid any special pleading. Thus while I admit wholeheartedly our debt to such writers as Crowther, Farrington, and Hogben, I feel that the balance has swung too far in the direction of regarding the socio-economic as the sole directive in the cultural advance. I have tried as far as possible to take the record as I have found it, neither failing to emphasise the connexions where they seem particularly instructive, nor seeking them where other factors seem more important. Lest the story should seem colourless I have enlivened its course with such biographical detail as seemed appropriate.

Who is the " general reader " to whom the book is addressed? It is, I hope, everyone, whether undergraduate at the University of the future, or member of that rapidly increasing public which has learnt enough general science at school to wish to continue the inquiry into the origin of our present-day scientific concepts and the part played by science in the creation of our culture. Moreover such a historical study undertaken in a critical spirit provides, I believe, a better introduction of " scientific method " than any abstract discussion of this difficult subject.

I have been very fortunate in the help I have received. Chapters XXIII and XXIV were read in manuscript by Professor Max Born, F.R.S.; Chapters IX, XIII, XIV, and XXI, by Sir Edmund Whittaker, F.R.S. Their kindness not only enabled me to remove errors in detail, but also to gain a deeper insight into several of the topics discussed. They are not of course responsible for the interpretation or final form of the chapters. My colleagues, Mr E. P. West and Mr S. Read, have in the course of discussion helped me more than

## PREFACE

they know; the former, in particular, revised the note on potential (p. 316). To my wife and daughter, Phyllis Ticciati, I am indebted for friendly criticism and much assistance with the preparation of the typescript and index.

Acknowledgment and thanks are due to the following publishers who have given me permission to reproduce the undernoted illustrations from their publications: Messrs George Allen & Unwin Ltd. for Figs. 6, 7, 12, 13, 22, 27, 29 and Plates II and IV from Wolf: *History of Science*; G. Bell & Son Ltd., for Fig. 23 from Newton: *Opticks*; Cambridge University Press for Fig. 32 and Plate V from Singer and Rabin: *Prelude to Modern Science*, and Plate VI from Arber: *Herbals*; Longmans Green & Co. for Plate VII from Bayliss: *Principles of General Physiology*; Macmillan & Co. Ltd. for Figs. 25 and 26 from Partington: *Short History of Chemistry*; Oxford University Press for Figs. 9 and 15 from Singer: *Short History of Science*, and Fig. 31 from Singer: *Short History of Biology*; Oliver & Boyd Ltd. for Figs. 11, 28, 33, 35, and Plates III and VIII from Wightman and Chesters: *A Modern Introduction to Science*.

Finally I should like to pay tribute to my publishers for their great patience and understanding.

WILLIAM P. D. WIGHTMAN

EDINBURGH, *November*, 1949


## PREFACE TO THIS REISSUE

During the fifteen years or so since the book was written I have come to appreciate more justly both the importance of the medieval contribution to the " growth of scientific ideas " and the character of the highly complex movements of thought concealed under the deceptively simple term " Renaissance ". Here and there (notably pp. 62, 208, 271, 335-36) the consequences of my former lack of insight are only too apparent; but to correct them would involve more drastic re-writing than is feasible for this edition. In my *Science and the Renaissance* (Edinburgh, Oliver & Boyd Ltd; New York, Hafner; 1962) I have undertaken a fundamental revaluation of both these topics.

WILLIAM P. D. WIGHTMAN

ABERDEEN, *June*, 1966

# CONTENTS

## PART I
## MATTER AND MOTION

| CHAPTER | | PAGE |
|---|---|---|
| I | THE BIRTH OF SCIENTIFIC MAN | 3 |
| II | ALL NATURE FOR THEIR PROVINCE | 9 |
| III | ORDER AND NUMBER | 19 |
| IV | THE GEOMETRY OF THE HEAVENS—THE GREEK CONTRIBUTION | 29 |
| V | THE GEOMETRY OF THE HEAVENS—THE EARTH DETHRONED | 44 |
| VI | THE TELESCOPE AND ITS REVELATIONS | 55 |
| VII | THE RISE OF TERRESTRIAL MECHANICS | 60 |
| VIII | THE DAWN OF UNIVERSAL MECHANICS | 72 |
| IX | THE MATHEMATICS OF MOVING BODIES | 82 |
| X | THE NEWTONIAN REVOLUTION | 99 |
| XI | IS THE LAW OF GRAVITATION TRUE? | 113 |
| XII | THE CELEBRATED PHENOMENA OF COLOURS | 120 |
| XIII | ON WAVES AND WAVINESS | 129 |
| XIV | "THE SUBJECT OF THE VERB TO UNDULATE" | 146 |
| XV | THE SEARCH FOR THE ELEMENTS OF MATTER | 158 |
| XVI | THE BURNING QUESTION | 176 |
| XVII | THE "SUBTLE FLUIDS"—I. CALORIC | 197 |
| XVIII | THE "SUBTLE FLUIDS"—II. MAGNETIC AND ELECTRIC | 206 |
| XIX | HOW THE STRANGE BEHAVIOUR OF A DEAD FROG LED TO THE DISCOVERY OF NEW ELEMENTS | 219 |
| XX | WHEN IS AN ATOM NOT AN ATOM? | 229 |

## CONTENTS

| | | |
|---|---|---:|
| XXI | THE NEWTON OF ELECTRICITY | 244 |
| XXII | FROM COLOURED FLAMES TO STAR STUFF | 255 |
| XXIII | THE "GO" OF THINGS | 268 |
| XXIV | THE "PARTICULAR GO"—FROM PARTICLES TO FIELDS | 298 |

### PART II
### NATURE AND LIFE

| | | |
|---|---|---:|
| XXV | WHAT IS LIFE? | 321 |
| XXVI | ORGANISMS AS MACHINES | 340 |
| XXVII | THE MICROSCOPE AND THE ORIGIN OF LIFE | 353 |
| XXVIII | WHAT'S IN A NAME? | 365 |
| XXIX | THE SEARCH FOR THE STRUCTURAL UNITS—THE STATIC THEORY OF FORMS | 379 |
| XXX | THE HISTORY OF HISTORY—THE DYNAMIC THEORY OF FORMS | 395 |
| XXXI | THE DARWINIAN REVOLUTION | 415 |
| XXXII | THE INSIDE STORY OF LIFE | 430 |
| XXXIII | CONTINUITY AND ORGANISATION | 459 |
| | GENERAL BIBLIOGRAPHY | 475 |
| | CHRONOLOGY | 477 |
| | INDEX OF NAMES | 481 |
| | INDEX OF SUBJECTS | 489 |

# ILLUSTRATIONS

PLATE

I  Title Page from Borelli's *De Motu Animalium*     *Frontispiece*

FACING PAGE

II  Greenwich Observatory in Flamsteed's Time     106

III  Apparatus used by Joseph Black, from a collection in the Royal Scottish Museum, Edinburgh     .     185

IV  Galvani's Experiments on "Animal Electricity"     .     219

V  Dissection Scene from the *Fasciculo di Medicina*, Venice, 1493     .     .     .     .     .     .     336

VI  *Sonchus sp.* (Sowthistle) as depicted in the *Codex Aniciae Julianae*     .     .     .     .     .     .     367

VII  Claude Bernard demonstrating an experiment to his friends     .     .     .     .     .     .     442

VIII  Drawings made by Koch in 1876 to illustrate development of Anthrax Spores     .     .     .     .     449

TO MY WIFE
MILDRED WIGHTMAN

"*Here and elsewhere we shall not obtain the best insight into things until we actually see them growing from the beginning . . .*"

ARISTOTLE: *Politics*

# PART ONE
# MATTER AND MOTION

CHAPTER I
# THE BIRTH OF SCIENTIFIC IDEAS

IDEAS like persons are born, have adventures, and die. But unlike most persons they do not disappear from this mortal stage; their ghosts walk, often to the confusion of new ideas. The birth of ideas, like that of persons, does not occur spontaneously, but is a culmination of travail following on, may be, centuries of gestation. Moreover their conception is often the result of a clash of cultures, if not of temperaments.

Until comparatively recent times it was believed that science, like almost every strand of Western civilisation, had its origin in the Eastern outposts of the Greek confederation. It was admitted that the Greeks had borrowed a good deal of assorted information of a practical nature from the older cultures of the East, principally Egypt and what is loosely known as Babylonia. But all this information was of a purely " rule of thumb " nature and hardly merited the name of science. Our views on this matter have recently undergone a profound change. Due in part to the enlargement of our knowledge of the quantity and quality of the eastern achievements, due perhaps equally to the enlargement of our conception of what constitutes *science*, we are in no doubt that *science* was born with infinite pains and in many places, neither by the happy inspiration of one man, nor at any clearly defined epoch. The flint knappers who *learnt by observation, imagination and trial* how to extend the range of the appliances obtainable from the single material at their disposal were assisting at the birth of science.

It is significant that all the early historians agreed that science had its birth at the *outposts* of Greece, namely where the vigorous young civilisation was in intimate contact with a rich, if declining, culture. There was thus no doubt that the first speculative natural philosophers of whom we have any record were not creating something out of a cultural vacuum. To deny the title of men of science to those ingenious workers who created the technique of multiplication and division; who made an error of only one inch in the 755¾ feet base lines of the Great Pyramid; who discovered how to mark out the passing of the seasons by taking as a unit the lapse of time between two heliacal risings of the star Sirius—would be to narrow down the meaning of the term beyond what in this industrial age we should be willing to do. Yet, although the heritage of natural knowledge upon which the Greeks were able to draw is proving year by year to be far richer than was first supposed, nevertheless most contemporary historians would agree that although the Ionians did not create science they undoubtedly transformed it. The evidence for this is stated by Professor Gordon Childe, a writer who has done much to correct the former unjust estimate. In relation to arithmetic, the science in which Babylonia made the most notable progress, he says of the clay tablets which constitute the " documents ": " It looks as if the research they disclose was really limited in its scope by consciously conceived possibilities of practical application. In any case no attempt was made to *generalise* the results " (italics mine). Herein lies the nature of the transformation effected by the Greeks. It was not merely that their problems were more " theoretical " in the sense of being more remote from the problems of daily activity—such a criterion for " science " would hardly be allowed in an age when the boundaries between " pure " and " applied " science are seen to be artificial and mere conveniences. What distinguishes the Greeks of the seventh century before our era from any previous thinkers *of whom we have any record* is that their concern is not with triangular *fields* but with *triangles*; their curiosity is not directed to the nature of the fire which is *baking bricks* but to the nature of *fire itself*. And they went further: they asked, if we can trust the testimony of Aristotle, writing three hundred years later, " What is the nature of Nature? " Lastly, with regard to mathematics, although in some respects they seem to have failed to garner the rich harvest of *methods* invented by their forerunners, they recognised, as the former did not, the necessity for deductive *proof*.

## THE BIRTH OF SCIENTIFIC IDEAS

In a word, Greek thought differed from all that had gone before in respect both of *generality* and *rigour*. In *general* thinking, which opens up possibilities undreamed of by the severely " practical man ", we pass from *percepts*—things to which we can point, like triangular fields—to *concepts*—creations of the process known as *abstraction*. If we ignore everything about a number of fields except the fact that they all happen to have three sides, we have arrived at the concept of triangularity. We cannot point to a triangle, because in order that triangularity may be manifested there must be something else—a hedge, a pencil line —and our abstraction has thereby become merged in the concrete instance. Moreover, since we cannot point to a triangle, we have to define it in words, that is, in the form of relations between other abstractions, such as straight lines and planes, none of which " exist " in nature. What makes this type of thought so immeasurably more *powerful* than thought (in so far as this is possible) about individual instances, is that we can worry out by deduction all the characteristics that *must* "go with" triangularity as such. Thence we return to the concrete individual, knowing that in so far as triangularity is manifested in it, so also is an infinite number of other possibilities which would never have suggested themselves by mere contemplation of a three-sided parcel of land. Science is, among other things, the generalisation of perceptual experience by means of adequate concepts; the concepts grow with the experience; history is the record of their growth; philosophy, among other things, the critique of their adequacy.

It must not be thought that the Greeks from the outset *knew* what they were doing; or at any rate could have expressed it in intelligible terms, because the terms themselves were created by their procedure. The conscious recognition came to birth painfully in the thought of Parmenides (*ca.* 500 B.C.) and especially of Plato.

In what follows we shall therefore take it as proved that *so far as the existing evidence goes* the birth of scientific IDEAS took place in Asia Minor in the seventh century B.C., on the ground that at that time and place the idea of nature as a whole first took shape, was explicitly stated, and systematically expounded. In our study of the growth of scientific ideas we shall therefore take as our starting point the same thinker as has traditionally been regarded as the founder of science, Thales of Miletos; but in the examination of the seminal ideas of science we shall have to take into consideration the hinterland of thought and practice which made possible the emergence of this idea into explicit form.

So far by way of justification of our starting point: a justification made in general terms without any reference to detailed fact. Before we set out on our long journey down the centuries it is important that we should be quite clear about what we hope to discover, what are our means of discovery, what reliance may be placed on their adequacy, and what we hope to gain as a result of the discovery. In a word, since this book, though not a history of science, is offered as a preparation for that study, it will be necessary to indicate in a very simple way the author's philosophy of history, especially in relation to method.

All history is based on documents; and it is an axiom of historical method to use as far as possible the records of eye-witnesses or at least those of writers contemporaneous with the events under discussion. Since in the history of scientific ideas we are concerned only with the productions—written or constructed—of individuals or small groups of individuals, it is axiomatic that we should use when possible the writings or actual constructions of the thinkers themselves. In the case of Egypt and Babylonia, although such material is meagre—frequent additions being made as new excavations are carried out—it is commonly, though by no means always, original in that it consists of baked clay tablets, mural decorations, coffin lids and the like, which describe the state of contemporary knowledge or depict contemporary practice. But of what we have agreed to call the birth of scientific ideas, and indeed of their early childhood, we have no contemporary records whatever. The reason for this is technological and geographical: there was in Ionia no clay such as was available in Babylonia; no papyrus as in Egypt. The teachings of the Ionian sages had to be highly condensed into pithy epigrams or verses suitable for memorising. Such " books " as they wrote must have been limited to very few copies. Consequently we are almost entirely dependent on the writings of

Aristotle, who lived three centuries later, for summaries of what was believed *in his day* to have been their teaching. Moreover, it must not be forgotten that not a single contemporary copy of even Aristotle's works is extant. The earliest presumed copy now in existence was compiled in the tenth century and is believed to be one of many copies taken from another of the sixth century. So however scrupulous the various copyists may have been (and we know that many of them filled in obscure passages in earlier copies without noting how much they had called on their own imagination) the ravages of time and the elements (many of Aristotle's works were literally buried for many years) have rendered impossible a completely authentic reproduction. It is the business of classical scholars to decipher and compare earlier texts and to produce what to their individual judgment is the most likely version. This version is still in fourth century B.C. Greek. For those like the author of this book whose knowledge of Greek is fragmentary, English translations must be prepared, and here again doubts arise as to the exact connotation *to the writer* of such key words as φύσις, ἄπειρον, and the like.[1]

It is only too easy to read into the words of the Ionians prophetic visions of latter-day physics; for this reason, if for none other, the history of science must be set in the context of the history of language and human institutions of every kind.

Having examined those texts or translations which have best stood up to the fire of eager critics, the student of the growth of scientific ideas must then make up his mind how he intends to interpret these often very naïve, sometimes very obscure, sayings of the pioneers. Until the recent past, writers on the history of science have usually contented themselves with the exposition of their own views or, worse still, somebody else's views, on the extent to which the pioneers had or had not anticipated the discoveries of " modern " science.

My aim throughout this book will, on the contrary, be first to set out the expressed views of each thinker—in his own words so far as space permits. In attempting an interpretation of each thinker, I shall try to answer the following questions: (*a*) What is he trying to do—that is, to what unspoken question is he trying

[1] For a fuller account of the historical path of the Greek texts, see below, p. 322.

to find an answer? (*b*) **Why has this question assumed a special importance for** *him*? (*c*) **What is the nature of his answer?** (*d*) **How far did it satisfy his contemporaries?** (*e*) **How far does it satisfy us**—that is, to what extent was his answer a *real* answer?

Our aim therefore in Collingwood's phrase is to *relive* the ideas of the past in order that, by watching them grow, we may the better understand what scientific ideas really *are*. Incidentally we shall come to see, as Professor Conant has pointed out, how completely false is the belief in the triumphal march of science down the ages. Progress has resembled much more closely the infiltration of modern war, in which tactical objectives have been seized by bold leaps, and held precariously while the main body has been slowly consolidating the ground in the rear. This explains perhaps the frequent and disconcerting inconsistencies of great thinkers.

CHAPTER II
## ALL NATURE FOR THEIR PROVINCE

THERE is no doubt that sometime in the seventh century before our era, probably between 630 and 620,[1] there was born at Miletos the man Thales, whom later Greeks counted as one of the Seven Sages. We can place the date of his birth fairly accurately because of the tradition reported by the historian Herodotos that he foretold an eclipse of the sun which put an end to the war between the Lydians and the Medes. If this is true, it was a lucky shot; but even if it is no more than a fable, it gives us a good idea of the sort of problems people were interested in at that time. It also suggests that he was acquainted with Babylonian astronomy, in particular of the cycle of relationships between the sun and the moon as viewed from the earth, known as the Saros. Of his writings—if there were any—we know nothing; though Diogenes Laertius, a writer of the second century A.D., reports a belief that Thales wrote a book on astronomy for sailors—a significant suggestion even if unfounded. Of his *reported* sayings Aristotle mentions three: (a) All things are made of water on which the earth floats; (b) All things are full of " gods "; (c) The lodestone and amber have souls.

To anyone whose knowledge of science was restricted to the standpoint of the present day these sayings would appear as childish nonsense. It is only because we are aware of the gradual development and modification of these views by his contemporaries, and of the further fact that no one before Thales seems to have put forward views of a like kind, that we revere him as the founder of natural philosophy, and through that, though at a considerable remove, of natural science. If we are to appreciate the greatness of Thales we must divest our minds at once of any idea that he was trying to found the science of *chemistry* on the

[1] All dates of Pre-Socratic thinkers are to be regarded as approximate only.

orderly but ungeneralised observations of the copper smelters, glassmakers, and dyers of Egypt and Babylonia. Chemistry as we now know it is concerned with the composition and interaction of " pure " substances, that is, artificially refined " units of thought ", to use that valuable term introduced by Mr. H. T. Pledge (see General Bibliography). To take an example: pure salt is composed exclusively of the elements sodium and chlorine and contains no water; but *pure salt* is a product of modern laboratory technique and virtually did not exist in Thales' day. Thales on the other hand was dealing with things as they are, and not with things neatly sorted and cleaned up by chemists. His dictum then, though certainly not wholly true, was, at its face value, very far from being nonsense. The greater part of the earth's surface is water; water pervades every region of our atmosphere; life as we know it is impossible without water; water is the nearest approach to the alchemists' dream of a universal solvent; water disappears when fanned by the wind, and falls again from the clouds as rain; ice turns into water as does the snow that falls from the skies; and a whole country surrounded by a barren desert is fertile, rich, and populous because a huge mass of water sweeps through it annually. All these facts—the last particularly, for we know that he had spent much time in Egypt—gave Thales good grounds for thinking that *if there is any one thing at the basis of all nature*, that thing must be water. If there is any one thing! This supposition, that is to say, the asking as it were of this question, constitutes Thales' claim to immortality. The fact that he made a guess at the answer, and a pretty good guess at that, is of minor importance. If he had championed the cause of treacle as the sole " element " he would still have been rightly honoured as the father of speculative science. True, others before him (such as Homer and Hesiod) had sketched the origin of the world from one substance, but they were not content to deal with *verae causae*, that is with things whose existence can be verified by observation. To attempt to explain the origin and process of the world by having recourse to gods and spirits endowed with special powers, is merely to beg the question, since the existence of such beings can never be proved (nor of course disproved) by the means wherewith we know that world. In a word, it was Thales who first attempted to explain the variety of nature as the modifications of something *in nature*.

What of the other sayings of Thales? Here we are on more debatable ground. Most historians are agreed that they not only imply a theory as to the " go " of things, but also that the " go " is part of the stuff. (I have used the words the " go of things " because I have never discovered any more expressive of the dawn of natural wonder than this constantly repeated phrase of the youthful James Clerk Maxwell.) In the absence of any evidence provided by Thales himself (even the remark about there being gods in all things is not actually *quoted* by Aristotle) we are compelled to fall back upon the later commentators on the plausible assumption that what they have to say will at least give a fairly just idea of the general intellectual outlook at the time when Thales was living; and this after all is what chiefly concerns us here. As usual Aristotle seems to be our safest guide, and he states that Thales " was led to this probably by the observation that the nutriment of all things was moist, and that heat itself is generated by moisture, and living beings live by it . . .". Now if we take this in conjunction with the saying about the magnet we cannot doubt but that Thales regarded nature as a single immense organism composed of lesser organisms of which the earth was one, and also every tree and stone upon it.

At a superficial glance this attempt to explain the " go " of things is merely a bolder and yet more fantastic form of the myths in which the earlier poets, Homer and Hesiod, saw the " creation " of the earth out of the ocean, and filled every tree and brook with a god or nymph. But there is a subtle change in the emphasis; the gods and nymphs of mythology were human persons writ large—the products of the poetic fancy. The " gods " and " souls " of Thales were conceived, so it seems, as *part of the nature of the natural object* regarded as an organism. This is a vital step towards objectivity, that is towards the placing of man *within* the wider system of nature. It is not yet pure objectivity; that lies far ahead.

Why was this all-important step taken in such a way as to make the new world-view so fantastic? Here we have an example of the necessity to place ourselves as far as possible at the standpoint of the thinker we are considering. Divesting ourselves of all preoccupations with airborne microbes and the technique of sterilisation we see at once that the boundary between the living and the non-living fades out. The cattle graze upon the grass; the grass feeds upon the earth; slime feeds upon the rocks; maggots grow out of slime. Everything grows out of something else. What is man that he should draw a line between the grass and the earth, between the slime and the rock? But everywhere for growth there must be water; and for the fishes of the sea this alone seems to suffice. So the whole world grows out of water.

Seen in this light, is the picture any more perverted than that in which the soil with its teeming population is regarded merely as an anchor and a sponge for the roots of the trees we plant?

There is another reason—more scientifically orthodox—for the hylozoism ($ὕλη$ = stuff, $ζωή$ = life) of Thales and his followers. Mathematics and astronomy were not the only forms of knowledge learnt by the Ionians from Egypt; there is little doubt that a knowledge of the human body was also taken over. This " knowledge " was of very uneven quality. Owing to the close association of the physicians with the powerful priesthood the practice of medicine was confused and darkened by superstition. The surgeons on the other hand seem to have been regarded as an " inferior " social caste, hence were left to their own devices, with good results for the accuracy of their observations. It is doubtful how the great Greek tradition of medicine had its origin. From a very early date—probably before Thales—there existed a closed guild of practitioners known as the Asclepiads. Whether their name betokened the *fact* that their forbears had actually been associated with the priests in the temple of Asklepios (Lat. *Aesculapius*) or merely their *hope* that they were the descendants of the god himself (a not uncommon trait in the Greek character of this and even later periods), it is probably impossible to say. But we have it on the authority of Hippokrates, the greatest physician of antiquity, writing in the fifth century B.C., that there had existed a body of physicians in " olden times " whose practice was based on careful observation of signs and symptoms in health and disease. It is not unlikely that this freeing of medical practice from the shackles of ecclesiastical authority was in part due to the *local* character of Greek religion, each city state having its own gods and priesthood, which would therefore never have attained the wealth and consequently the power of the priesthood of the oriental countries. In any case the careful attention paid to the relations between diet, climate, and other physical factors and health cannot have escaped the notice of Thales. And we may reasonably conjecture that it may have given his nature-philosophy a biological bias.

We may agree then that Thales asked one of the most important questions in human history; that the answer was framed in terms intelligible to his contemporaries, and in a form perhaps fantastic but certainly not nonsensical to us. Criticism may however be levelled at it on two quite different points. First, that the answer, even if a *real* answer, was barren, in that no useful consequences could be drawn from it for the improvement of man's lot. Second, that in any case it was in fact no real answer. The first criticism would never have occurred to the Greeks of his day. Speculation on the nature of things was restricted to the rich and powerful for whom an adequate slave population provided those amenities which were likely to be desired by thinkers of such temper. The second point of weakness was recognised implicitly by his immediate successor, Anaximandros, namely that Thales had given no sort of clue as to *how* water could give rise to all the variety of natural beings unless it contained within itself something else with which it could react in some way or other; but in that case water as such could not be *the* primordial stuff.

Anaximandros was a more subtle thinker than Thales; with astonishing anticipation of modern views on motion he saw that there was no need to attempt to explain how the earth is supported (Thales' statement that it floats on water is of course little in advance of the Indian myth of a cosmic tortoise to bear the burden) in the absence of any reason for motion in one direction rather than another. What concerns us here is his statement that " all things must in equity again decline into that whence they had their origin; for they must give satisfaction and atonement for injustice each in the order of time." This[1] is generally interpreted as meaning that if the stuff out of which nature " grows " were endowed with a *definite* quality such as the " moistness " of Thales' " water " it is inconceivable how its *opposite* (dryness) could ever be produced. He therefore calls the material cause of the universe (he introduced the word ἀρχή— " principle " rather than " stuff ") ἄπειρον—the " boundless ". This word has given scholars a lot of trouble; literally it seems to mean no more than " infinite " which would probably apply to Thales' water. But Anaximandros goes on to say that the naturally occurring opposites—hot and cold, wet and dry—are separated out from it. It therefore seems likely that he had in mind some primordial stuff, in itself quite indeterminate, yet capable of producing determinate qualities. A crude analogy might be the chargeless interior of a hollow charged conductor. *How* this came about is of course unexplained. But from our standpoint it was a capital advance to recognise, if indeed he really did so, the *kind* of conceptual scheme which has played such an important part in the generalisation of science.

That Anaximandros was still thinking in terms of life and growth is shown by his opinion that living creatures are evolved out of the elemental moisture by the action of heat—an observation which must have been only too familiar, though misinterpreted, in a warm climate.

The subtlety of Anaximandros—so admirable to us—must have been a stumbling-block to his contemporaries, for his disciple, Anaximenes, reverted to the more naïve standpoint of Thales: "As our soul, which is air, holds us together, so breath and air encompass the universe." It is perfectly clear that by " air " Anaximenes meant all aeriform stuff—steam, mist, vapours, even fire. The Greek word is ἀήρ, which is used indiscriminately by the poets; it was only later (p. 15) that the distinct material nature of atmospheric air was demonstrated.

Anaximenes did however make one essential step forward.

[1] A most interesting alternative interpretation has been developed by Dr. Kelsen to the effect that the modern scientific concept of causality has grown out of the almost universal sentiment of retribution for wrong done. In the primitive form the lightning stroke is "caused" by the angered god: no other explanation is sought. Anaximandros' view is thus transitional between animism and naturalism.

## ALL NATURE FOR THEIR PROVINCE

It is clear to us that a simple stuff can give rise to any form of variety only if it is either mixed with something else (in which case it would cease to be " the " material cause of nature) or capable of varying its relationships with respect to the space it occupies. The latter device was adopted by Anaximenes when he spoke of the formation of actual clouds, water, wind, earth as being formed from the ἀηρ by *condensation and rarefaction*. Though, as is inevitable with the pioneer, he is claiming for his principle of change more than it could achieve, yet when we think of the variation of form and texture which can be superinduced on substances held to be chemically (i.e. materially) identical (e.g. water, steam, and ice; diamond and graphite) by variation of the spatial relationships of identical units, we must concede to Anaximenes a permanent niche in the gallery of natural philosophers.

The first of the above alternatives, namely that there must be more than one primary substance (though the number may nevertheless be incomparably smaller than the huge variety of things which nature displays) was adopted by Empedokles of Akragas, who declared that four primary substances—air, fire, earth and water—are sufficient to account for all else. Empedokles lived about fifty years later than Anaximenes, and he had profited by the speculations of many other thinkers, some of whom we are shortly to consider in another connexion. In fact he is often regarded as an " eclectic ", in that he brought forward no original views of his own, but welded together what he considered to be the best parts of earlier teachings.[1] To us the adoption of four " roots " as he called them, seems largely arbitrary; but not wholly so, as each of them contains one pair of the " opposites ", warm and cold, wet and dry. He recognised the hopelessness of trying to explain all things as proceeding from one stuff; but others had shown how very much might be explained from these two pairs of opposites; let them suffice then. It was really the

---

[1] That this was not entirely so is evidenced by the fact that when he included " air " among the four elements he called it not ἀήρ but αἰθήρ since by actual experiments with closed vessels under water he had shown the existence of a distinct corporeal substance (αἰθήρ) different from the " misty " ill-defined ἀήρ of Anaximenes. We can agree with Burnet in regarding this as " one of the most important discoveries in the history of science ".

tacit admission that further progress along this road was impossible; the Greek world, which shunned experiment, had shot its bolt as far as the discovery of " elements " was concerned; nevertheless this eclectic view of Empedokles (adopted in a modified form by Aristotle) served mankind for two thousand years. We shall see later that the fact that modern chemistry has discovered about ninety elements is not in *direct* contradiction to this view of the ancients; it is all a matter of definition; and the progress of the last two hundred years has been made possible by more accurate (and, as we shall see, narrower) definition of the concept of an element.

If we look back at what we have learnt about the attempts of the early nature philosophers to find the " roots " of things, we shall see that it is in effect a search for that which has the greatest power to produce other things, and at the same time is the most permanent among all things. While the search was proceeding, as related above, another completely opposed point of view was established by Herakleitos. To the query, what is most permanent, most real, among things? he replied in effect, "impermanence". Change alone is real; the eternal hills, the solid ground, are so only to our mortal eyes. "All is fleeting "; from moment to moment the whole world is changing: " You cannot step twice into the same rivers; for fresh waters are ever flowing in upon you."

This constitutes a remarkable transformation of the early Ionian views on the nature of the primordial stuff. We should probably be crediting him with an outlook even further in advance of his times if we were to attribute to him a belief that change or " becoming " is *by itself* the most real thing;[1] on the contrary one of the extant fragments of his poem clearly states that " This world . . . was ever, is now, and ever shall be an ever-living fire . . . The transformations of fire are, first of all, sea . . ." It is clear that it is some stuff akin to the " air " of Anaximenes. On the other hand it is not the *permanence* of the stuff that endows

[1] The nature of " being " itself, as being more real than individual " beings " was developed by Parmenides. His argument may be roughly summarised as follows: If a thing changes it must both " be " and " not be " which is absurd. But nature displays to our senses a panorama of ceaseless change; very well, our senses must deceive us and nature is an illusion. On this view science is illusory; but the value of Parmenides' thought is that it shows up a certain confusion in that of his predecessors. Discussion of this would carry us too far into pure philosophy.

it with primordial power, but its endless motion and change like the ever-changing shapes of visible flame.

To the discovery of the pre-eminence of change Herakleitos added a view if possible even more revolutionary and more fertile. We can do no better than hear his own famous words: " Homer was wrong in saying ' Would that strife might perish from among gods and men '. He did not see that he was praying for the destruction of the universe; for if his prayer was heard, all things would pass away." In this passage, Herakleitos, probably unconsciously, struck the keynote of modern science; in nearly all of the most fruitful theories, such as the theory of electrical potential and the theory of survival of the fittest, we have exemplified that belief in tension, which among organisms becomes a gradually more conscious struggle, and which alone makes possible change and progress. Nor is it mere chaos and friction; there is a higher harmony within the apparent discord, as Herakleitos so admirably expressed in the words, " What is at variance agrees with itself. It is an attunement of opposite tensions, like that of the bow and the lyre."

By his immediate successors Herakleitos was known as the " obscure " and the " riddler ", and Socrates is reported to have said that it needed a Delian diver to sound the meaning of his work; but from our point of view his " dark sayings " are paradoxically the most illuminating of all those of the Ionian sages. With his recognition of the fact that whatever lies at the base of things must possess at once a " restlessness " and " an attunement of opposite tensions ", their speculation reaches its high-water mark. To our minds it may appear superfluous to call in fire as the bearer of these qualities; but we must remember that the conception of the conditions of existence apart from any existing thing, namely the process called by philosophers " abstraction ", is a very difficult matter involving a range of intellectual power beyond that of these pioneers.

We have already seen that Empedokles unconsciously exhibited· the perfection of the Ionian solution *within its inevitable limitations*; and the impossibility of attaining to completeness without breaking through those limitations by denying the validity of the problem, namely of reducing nature to *one* stuff.

But even before Herakleitos was writing his " dark sayings " there had appeared a man whose intellect was a brilliant example of the other side of the Greek genius, and whose teaching was to influence mankind almost as much as that of any man who has ever lived. We shall study this and its consequences in another chapter.

## CHAPTER III
## ORDER AND NUMBER

There is a picturesque story that Pythagoras was led to his view of nature by listening to the ring of the smith's hammer on the anvil. He was struck, so the story goes, by the differences in the pitch of the sounds emitted by hammers of different weights, and, returning home, experimented to such good purpose that he was able to discover that the pitch of the note was simply related to the weight of the hammer, and that strings stretched by weights, one of which was double the other, sounded an octave apart. Unfortunately none of these results is true; so we must dismiss the story also.

Another story no better authenticated than the first is however much more plausible. This was, that he found that two similar strings equally stretched will sound an octave apart if one is exactly twice the *length* of the other; and further that when the strings are sounding the harmonies known to musicians as the " fifth " and " fourth " the lengths of the strings are in the ratio of 3:2 and 4:3 respectively. These results are not only perfectly true, but we know that the Greeks had sung their epics to the accompaniment of a stringed instrument for centuries before Pythagoras' time.

Whether the story be true or no, there is little doubt, according to the testimony of commentators living at a time but little removed from the lifetime of Pythagoras, that he summed up his views on the nature of the world in the paradoxical saying, " Things are numbers." Now Pythagoras is probably most widely known as the first to prove the already familiar property common to right-angled triangles; yet it is undoubtedly true that, great and fertile as that discovery was, it is not to that but to his teaching that " things are numbers " that Pythagoras owes his secure position among the immortals; for this view marked a radical departure from that of the Babylonian men of science, whose remarkable advances in the handling of numbers sprang almost entirely from the contemporary importance of accounting.

Once again the hasty will cry out against any reverence for such fantastic doctrines; but a moment's consideration will show us that the whole course of man's history was changed for good or ill by this strange utterance of a man born in the little island of Samos more than five hundred years before the commencement of our era.

We must first recall that what these pioneers sought was that which is most fruitful, and most permanent, in the scheme of nature; and what, taught Pythagoras, more so than the *numerical relations* between things? We do not know how Pythagoras arrived at this conclusion, but we should probably not be far wrong to reconstruct his argument somewhat on these lines: " It matters not ", he might have said, " what the string be made of,[1] and whether plucked or struck, and by what man played; its length alone, that is a property which can be completely described by a number, determines what will be the pitch of its note; and those combinations of notes which are pleasing to our ears arise from strings whose lengths stand to one another as simple *numbers*. If music then be based on numbers, why not all else? When all chance and fleeting qualities have been sifted away, *numbers* alone remain. Things, in fact, *are* numbers."

There are two charges which may be brought against this argumentation which deserve our attention. First, that the idea of numerical *proportion* would never have occurred to a pioneer in natural philosophy; and second, that numbers, even though permanent, bear no sort of resemblance to things. The interest of these objections is that the replies to them both strengthen the argument and illuminate the nature of the thought.

The first objection is easily answered by pointing out that, according to Pliny, before Pythagoras was born Thales had already used the constancy of proportion between corresponding sides of equiangular triangles to measure the height of a pyramid by comparing the length of its shadow with the length of shadow of

[1] This is admittedly only approximately true; but the deviations can themselves be described in terms of numerical quantities, e.g. density.

a rod of known height (Fig. 1). Indeed the rapid progress in Geometry from the moment when Thales first began to show that the practical " recipes " for mensuration used by the Egyptians depended upon certain general truths such as the fact that the angles at the base of an isosceles triangle are equal, was one of the greatest marvels of Greek civilisation; though it would be dangerous to assume that Thales " proved " the theorems upon which such measurements are made.

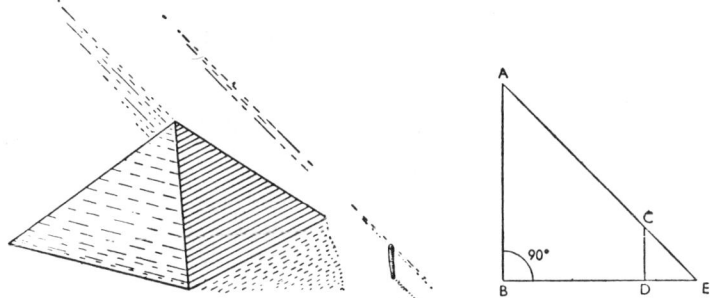

FIG. 1. Measuring height by length of shadow cast by (parallel) solar rays.

Since Pythagoras left nothing in writing, any attempt to reconstruct his contribution to the growth of the *idea* of order and number must be largely conjectural; but some fragments of Aristoxenos, a fourth-century author of a work on *Harmony*, claim that Pythagoras was the first to carry the study of arithmetic beyond the needs of commerce. Moreover from our wide knowledge of the sort of things the Greeks were writing and talking about from near the beginning of the fifth century onwards, we can say without fear of contradiction that by then the use of number as a tool for accounting (Babylon) and mensuration (Egypt) had been largely superseded by an interest in the nature of number itself. Now if Pythagoras and his school (actually a semi-religious brotherhood) were not responsible for this change, who was? No other claim has been put forward by any near-contemporary writer.

The second argument which might be urged against our serious acceptance of the dictum, " Things are Numbers," opens up the whole question of the idea of number. The idea or concept of a thing is bound up with its mode of representation; the more abstract the "thing", the more is this the case. Now we have literally no *image* of *a* number as such: we may picture to ourselves three blind mice or five fishes; but to picture " three " or " five " by themselves is a sheer impossibility. We are therefore driven to the use of symbols. Symbols may be pictographic, such as ○ for circle, or ideographic, as in fact are our " numbers " to-day, though the ideogram 3 has probably evolved from the pictogram ≡, and so less obviously for the others (except 1!). Now there is more than a little evidence for the belief that the founders of the Pythagorean sect (which lingered on until about the beginning of the fourth century B.C.) had no such ideograms, but represented numbers by groups of dots. Tradition takes the matter even further back by alleging that the dots represented pebbles (Gk. ψηφοι; Lat. *calculus*—hence " calculation "). Now though the absence of a straightforward numerical notation, such as we have, made it difficult for the Greeks to carry out arithmetical calculations which a modern preparatory school boy or girl could do without effort, nevertheless it gave an insight into geometry regarded as an understanding of the relation between numbers, represented by lines of various lengths, rather than in the proof of a variety of equalities and inequalities such as our modern academic geometry starts with. This achieved its final development in the *Elements* of Euclid. An example will make this clear.

Suppose a long line into which a short line is to be fitted. Two cases only can arise: that in which the short line " goes into " the long one an exact number of times, and that in which a piece of the longer line is left over. In the latter case the remainder must be fitted into the smaller line, and so on until a line is found sufficiently small to divide without remainder; this will be the common measure—the " greatest common measure " —between the two original lines. Now that we apply this process to *numbers* instead of lines, we commonly call it the " highest common factor " (H.C.F.).

This was not the path taken by the early Pythagoreans, who seem to have amused themselves by arranging dots in various patterns and so discovering the relationship between the sums of successive integers and triangles, the series of odd integers and

## ORDER AND NUMBER

squares, the series of even integers and oblongs, and so on, doubtless including solid " heaps " as well.

All this was excellent, in that it revealed mathematics as essentially an organic unity. What was not so excellent was its becoming entangled with the religio-mystical side of Pythagoreanism, with the consequence that to everything was assigned a definite number, from the holy tetraktys to the earth and justice! When a powerful intellectual revolution is accompanied by monstrous excrescences such as this, the latter are apt to have an associated influence for centuries. How many lesser men were strangled by this thicket of nonsense we shall never know; only indomitable courage, honesty, and singleness of purpose saved one of the greatest, Johannes Kepler, from a like fate two thousand years later (see below, p. 51).

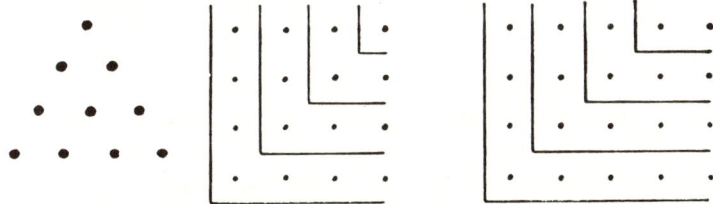

FIG. 2.   Pythagorean number-diagrams.

The discovery of the property possessed by all right-angled triangles is Pythagoras' most familiar memorial; but it was not an unmixed blessing for his own *entourage*, for it follows that if the length of each side of a square is called *one*, the length of the diagonal cannot be represented by any *number*. It is no use saying that it is " root two " or 1.414 . . . ; neither of these is a number in the original Greek sense. A contemporary rumour had it that Hippasos of Metapontion, one of the brotherhood, was excommunicated for divulging the secret rites, but it was darkly hinted that he was drowned at sea for letting the irrational cat out of the bag. But worse was to come.

It has been noted above that the identification of " things " with " numbers " is possible only by reducing the former to assemblages of measurable relations in space. Bearing this in mind we may applaud Pythagoras for seizing upon the one factor in nature which is universal, essential, and eternal. So long as we are satisfied with a so to speak petrified world, the basis of nature seems to be geometry; but a difficulty arises as soon as we try to understand the nature of motion and change. This difficulty was laid bare by the famous philosopher, Zeno, who was born in southern Italy about half a century after Pythagoras had settled there. Only a few fragments of his works have been preserved as quotations in later commentators, but Aristotle thought so highly of his critical insight that he discussed his famous paradoxes at length. We shall consider only the simplest one— that in which the athlete Achilles is trying to overtake a tortoise which has received a " start " in the race. "Achilles will never overtake the tortoise. He must first reach the place from which the tortoise started. By that time the tortoise will have got some way ahead. Achilles must then make up that, and again the tortoise will be ahead. He is always coming nearer but he never makes up to it." And yet, as we know, Achilles *would* undoubtedly have overtaken the tortoise. Now when an apparently sound argument yields us a result in flat contradiction to the delivery of our senses, it is necessary, for our peace of mind, to find out what false assumptions have been made at the start. If we examine the argument closely we shall see that the paradox is arrived at only by *tacitly* introducing an artificial condition as to the flow of time, that is, that we are justified in considering the mutual relations in space of two objects during intervals of time which are continuously *decreasing*; for if it be assumed that their speeds though different are nevertheless constant, it follows that, since the space between the objects is admittedly decreasing, the times allowed for Achilles to reach the tortoise are also progressively diminishing. Perhaps if " time " behaved in this way Achilles never would catch up a tortoise; but experience shows us the contrary result, so we must conclude that whatever Zeno's paradox *does* teach us, it certainly cannot be called upon in defence of the view held by the school of philosophers to which he belonged, namely that nature is illusory.

There is much more in Zeno's acute critique of the teachings of the Pythagoreans that is of great interest; another of his paradoxes cannot be refuted as simply as the above, but it is also of great difficulty, so we cannot discuss it here. The one thing however that we must bear in mind is that the belief that objects are capable of actual indefinite division without limit, in the way that numbers (which are inventions of our minds) are, leads to contradictory consequences. The importance of this lies in the fact that it was largely to overcome this difficulty that Leukippos framed the central postulate of what has for long been known as the atomic theory.

Now when progress in science is blocked by apparent inconsistencies in theory there are two ways of dealing with the matter. One is to pretend that they do not exist, as the early Pythagoreans did in regard to " incommensurables "; the other is to examine the basic concepts, which are in any case creatures of the mind even if derived from " things " by observation, and by reshaping them remove the cause of the trouble. If we can accept the conversation in Plato's dialogue *Theaetetus* as a true account, the contradiction of the irrational was removed by Theodorus and his pupil Theaetetus by the simple if daring expedient of defining *number* in more general terms. The passage is worth quoting for its own sake. The main purpose of the dialogue is to determine the nature of knowledge in general, and after a false start Theaetetus is telling Socrates how Theodorus has been teaching him and other young men geometry and arithmetic—his purpose being to give an answer to Socrates' question as to the meaning of knowledge. Theodorus had told them that the sides (or roots) of squares representing three, etc., square feet are not commensurable in length with the line representing one foot.

The idea occurred to us [he continues], seeing that these square roots are evidently infinite in number, to try to arrive at a single collective term by which we could designate all these roots. We divided number in general into two classes. Any number which is the product of a number multiplied by itself we likened to the square figure, and we called such a number " square " or " equilateral ". Any intermediate number such as three or five or any number that cannot be obtained by multiplying a number by itself, but has one factor either greater or less than the other, so that the sides containing the corresponding figure are always unequal, we likened to the oblong figure, and we called it an oblong number. All the lines which form the four equal sides of the plane figure representing the equilateral number we defined as *length*, while those which form the sides of squares equal in area to the oblongs we called *roots*, as not being commensurable with the others in length, but only in the plane areas to which their squares are equal. And there is another distinction of the same sort in the case of solids. [I have omitted the encouraging remarks of Socrates interjected from time to time.]

The other contradiction was removed by Eudoxos of Knidos, perhaps the greatest original genius among mathematicians before Archimedes. We have it on the authority of Archimedes, and of a nameless commentator on the Fifth Book of Euclid's *Elements*, that Eudoxos gave a definition of *ratio*, which was applicable to incommensurables as well as to " rational " numbers. Since this at the same time showed how to define *equal* numbers it did away with the need for assuming " infinitesimals " such as Zeno's paradoxes laid bare. In Euclid's words, " Magnitudes are said to have a ratio [i.e. are commensurable] to one another which are capable when multiplied of exceeding one another." From this it follows that the " magnitudes " may be of any kind—whether commensurable in *rational* numbers or not —providing they are of the *same* kind, and that either of them when multiplied a sufficient number of times will exceed the other. Now 0, however many times multiplied, can never exceed any number however small, hence $\frac{1}{0}$ is *not* a " ratio ", consequently its value is indeterminate. We shall have occasion to revert to this matter in Ch. IX; here we merely quote the late Sir Thomas Heath's comment on it: " The intrinsic greatness of the theory needs no further proof when it is remembered that the definition of equal ratios in Euclid Bk. V [based on Eudoxos' definition] corresponds exactly to the theory of irrationals due to Dedekind, and that it is word for word the same as Weierstrass' definition of equal numbers."

Growing partly out of the Pythagorean number theory, and partly perhaps in response to a weakness in the philosophical implications of the Milesian pioneers pointed out by Parmenides, was the theory attributed to Leukippos that matter, as was *later* shown by Eudoxos, cannot be *infinitely* divided. Leukippos himself seems to have been of a very retiring nature. Epikuros indeed, who was teaching similar views about 150 years after the presumed date of the former's birth, even questions whether he ever existed; but Epikuros always liked to think he was really first in the field! We must once more be content to accept the testimony of Aristotle that it was indeed Leukippos who, with his pupil Demokritos, originated the view which was not only the most effective reply to the paradoxes of Zeno, but in the development which it afterwards underwent was ultimately handed down to the modern world as the most fruitful of the Greek speculations about the nature of things. The effectiveness of this theory consisted in its unconditional acceptance of the reality of motion, change, and a multiplicity of existent objects, combined with the denial of the infinite divisibility of the stuff of which these objects are made. The ultimate basis of nature, it was held, is an infinity of *atoms*, that is material particles indivisible because *full*, moving in space which is mere emptiness. The creation of objects is simply the aggregation of these particles according to definite spatial patterns, the nature of these patterns being determined by the shape of the atoms.

We shall at once agree that this system is the summit and perfection of Greek physical speculation. It adds to the partial truths discovered by the earliest Ionian philosophers and Herakleitos the precise mathematical relationships discovered by Pythagoras. Of course its truth cannot possibly be *proved*, since the magnitude of the atoms is supposed to be far less than the smallest sensible particle; indeed the Greeks themselves failed to make any appreciable use of it in the unravelling of nature's tangled skein. But when after a lapse of nearly two thousand years it was revived, there was hardly any branch of science of whose foundations it did not form a part. It is only after these have been discussed in later chapters that we shall be in a position to realise the admirable intuition which, in the absence of any adequate detailed evidence, guided these early thinkers to the discovery of an essential feature in the structure of nature.

On the other hand there is a danger, amply exemplified in the history of thought, that its seductive simplicity and comprehensiveness may blind men to its essential one-sidedness. There can be little doubt that matter, as we experience it, is characterised by atomicity, but it is only too easy to assume with Demokritos that, because the concept of atoms helps us to understand the behaviour of *matter*, therefore nothing exists except the atoms and the blind forces which bring them together, and later drive them apart again. This belief forms the basis of mechanistic materialism which is held by some people to be forced upon us as a consequence of the astonishing success of the atomic theory in science; we shall be better able to discuss this matter at a later stage.

Although the atomic theory persisted as a philosophical doctrine throughout classical times the most influential schools of thought gave to Greek science a determined twist in a quite different direction.

## CHAPTER IV
# THE GEOMETRY OF THE HEAVENS
### *The Greek Contribution*

THE change that came over Greek science at the time when Demokritos was giving its final answer to the question, " What is the stuff of nature and how is it ordered? " had many causes; but none was more important than the comparatively sudden realisation that the possibility of knowledge is itself a problem. As is so often the case when some great light is about to dawn in men's minds, it came at first as a faint glimmer through a mist of doubt. The first man to voice a doubt as to the possibility of any real knowledge was Protagoras, an early contemporary and fellow-citizen of Demokritos, who realising that all our so-called knowledge is gleaned from fleeting impressions of sights and sounds and the like, announced his discovery in the famous apophthegm, " Man is the measure of all things." By this he meant that philosophise as we may about nature and eternity, we are each and all of us bound captive within the circle of our own sense impressions, which, as we know from the existence of colour-blindness, optical and acoustical illusions and the like, we can never trust implicitly. The only facts I *know* to be true are those impressions I am receiving *now*. The green patch (of my blotting paper), the swishing sound (of the wind) are indisputable facts; that the green patch also will absorb ink is not, strictly speaking, a fact at all; it is a hypothesis based upon another hypothesis, namely memory. It certainly has a very high degree of probable truth, as I verify for myself every time I turn the damp writing upon the paper; at that moment it is a fact. But to assume as we do, and quite rightly so for the purposes of everyday affairs, that every relation stored in our memory is of the same degree of truth as are our immediate sense-impressions, is to let in at once the unrecognised possibility of error, as we admit every time we say, " Did that really happen or did I dream it ? "

All this sounds very hard, and it is hard. Knowledge and truth are nearly always hard, but their pursuit is often pleasant. Unemployment, payment of heavy income tax, war, and revolution are also hard, and they are never pleasant; but they are the price the world is paying for its failure to distinguish between " real " facts, " probable " facts, and mere delusions decked out in fanciful language to masquerade as facts. " Most people would rather die than think," as Lord Russell once reminded us; to this he added in characteristic manner, " Many do."

However, to make a hard thing harder it has to be admitted that this great truth discovered by Protagoras is itself so easily made into a half-truth that to many of his contemporaries it behaved like an *ignis fatuus*, leading them to proclaim that the individual man's selfish desires and interests are alone real and should therefore be his only concern. This frivolous and unreasonable inference was corrected by Socrates who, by an ingenious method of question and answer, compelled the Sophists, as they were called, to contradict themselves and reveal the essential shallowness of their thought.

With the wisdom, humour, and beauty of Socrates' discussions of the problems of knowledge and behaviour, as handed down to us by the poetic genius of his pupil Plato, we have nothing to do. For the study of nature, except in so far as it contributed immediately to man's spiritual well-being, he had a hearty contempt; here he was wrong as we shall see; but we have had to refer to him, because he first gave expression to the necessity for a critical examination of the basis of all views on the nature of things; and because it was his pupil Plato who turned the attention of Greek nature philosophers from the immense variety, struggle, and confusion which characterise the earth, to the apparently simple ordered harmony which was the sensible heavens.

The second major problem which Greek natural science set itself was to reduce the motions of the heavenly bodies to order. This problem, first posited by the followers of Pythagoras, was given, as we might say, a much wider publicity by Plato. We shall not attempt to review the progress of astronomical discovery

THE GEOMETRY OF THE HEAVENS 31

among the Greeks; but in order that we may assess the value of their systems it is absolutely necessary that we should have before us the facts as they were presented to the Greeks of Plato's time and later. Without these it is impossible to avoid the conclusion that since the Greek astronomical systems were " wrong " and their speculations often fanciful and absurd, it is a waste of time to study them. If on the other hand we place ourselves as nearly as may be in the position of the Greeks, that is with no instruments save roughly vertical rods and graduated circles, and half-blinded by mythological and intellectual prejudices, we shall come to admit that their procedure was at first thoroughly scientific and their results " true " within the limits of their experience. That they did not carry their investigations right through to the ultimate test which modern science demands, we must also admit; the reason for this we shall later enquire.

As a result of two centuries of development of the knowledge acquired from Egypt and Babylon, the Greek astronomers of Plato's day were in possession of the following important facts:

(1) The earth is surrounded by a dome-shaped vault covered with points of light, the vast majority of which keep their positions so closely fixed *relative to each other* that the various groups have been called Orion, Cassiopeia, Boötes, etc. These are called " stars " to distinguish them from the " planets " (or wanderers) which change their positions relatively to each other and to the " fixed " stars.

(2) The celestial vault or sphere revolves round the earth once daily from east to west, carrying with it all the heavenly bodies; moreover the point which does not revolve (as shown by the very small circle described by a star near it) is not immediately overhead.

(3) Observations on the height of the sun on 21st June at noon (i.e. its highest point during the day) and on successive days throughout the year, show that it not only swings once daily with the celestial sphere, but also crawls down to a minimum height and back to the maximum in one year, after which it repeats the cycle. The maximum and minimum positions are found to be $23\frac{1}{2}°$ above and below respectively the mean positions. Moreover the sun is observed to pass through different constella-

tions at different times of the year, but to repeat its passage each year.

(4) The remaining planets similarly move through different constellations at different times, but the cycle of motion differs for each planet; moreover the passage is not a regular progression like that of the sun, but each planet from time to time apparently stops in its career, turns back for a while, and then proceeds once again in the original direction.

The hasty will of course deny that it is a " fact " that the celestial sphere revolves once daily round the earth. We must admit that the " sphere " or " vault " is not a fact, but the revolution of the stars *is*. If any one deny it he must equally deny that a cricket ball thrown " straight up " does travel in a straight line and not in a curve. To which he will probably reply that " of course it travels in a straight line relative to the earth, but to anyone fixed in space its path would be a curve." Well, the Greeks *were* on the earth and *not* fixed in space, so we come back to what we started with, namely that it is a fact that the stars revolve round the earth![1]

Now we come to the correlation of these facts; and this involves the framing of a hypothesis, that is a supposition suggested by the facts which will help to explain them in a simple way. The hypothesis may or may not be true, and it can *never* be *proved* to be true; but the more facts it helps to explain, the more truth it probably contains. It cannot be too strongly emphasised that people who say that " Science must be true because it deals only with facts " are under the gravest misapprehension. The exclusive concern with facts is merely the orderly collection and recording of data; this becomes science only when hypothesis, that is the departure from fact, is resorted to.

Let us examine the possible hypotheses to account for the first " fact " mentioned above, the one which we have already admitted to be in reality a hypothesis, that is the " celestial sphere ". The actual " facts " are the points of light keeping their mutual distances constant. Now these points of light either

[1] Note that " facts " are no longer " true " when the conditions of their observation are changed. That is what Protagoras was hinting at and what Einstein has abundantly proved. There are, as far as we can see, no " absolute facts " but only relative ones.

## THE GEOMETRY OF THE HEAVENS 33

reside on a common surface or they do not; but in the face of their constancy of motion with respect to each other, the hypothesis that they are all entirely disconnected appears to stretch probability too far, and the Greeks, *in default of any knowledge as to the motion of the earth*, rightly rejected it.

Next we have to consider that motion itself. Unless it can be assumed that the earth makes a complete diurnal revolution in a contrary sense, the revolution of the celestial sphere is not a hypothesis but a " fact ". The same is true of the motions of the sun and other planets. The real trouble begins only when we take note of the fact that each of these planets has *two* sets of motions, one with the celestial sphere, and one " on its own ". Here is where the " facts " must be left as uncorrelated events or *some* hypothesis must be brought forward to account for them. This step was taken by Plato who, obsessed by the beauty and orderliness of the geometrical harmonies which the Pythagoreans had seen in nature, felt himself compelled to see in the starry heavens their most perfect revelation. The hypothesis which he propounded was that the heavenly bodies " really " revolve eternally in circles, whose centre is the earth. He proposed to the geometers the problem of explaining the apparent anomalies.

Before we discuss their response to this appeal it would be well to consider Plato's hypothesis a little more carefully; the more so, since, as some would say, it retarded the progress of astronomy for nearly two thousand years. At the risk of being wearisome I must reaffirm that the essence of science is to frame hypotheses—good hypotheses if possible—but better bad ones than none at all. Now Plato's hypothesis was not a good one since it was dictated by no facts at all save the apparent existence of a sphere rotating about the earth as a centre. On the other hand, once the fixed position of the earth is assumed, it is difficult to see what hypothesis could have been suggested that would have been better. The business of science is to reduce apparent confusion into relations between simple " invariants ": Plato's triumph consisted in recognising that an " invariant " must be chosen; his failure lay in choosing the wrong one.

As has nearly always been the case with bold, if at the same time rather ill-founded, hypotheses, that of Plato was the stimulus to more refined observation and the exercise of mathematical ingenuity of a very high order. The first, and perhaps the greatest of these mathematical devices was due to Eudoxos, a contemporary of Plato; here is the account which Aristotle, Plato's greatest pupil, gives of it:

> Eudoxus assumed that the sun and moon are moved by three spheres in each case; the first of these is that of the fixed stars, the second moves about the circle which passes through the middle of the signs of the zodiac, while the third moves about a circle latitudinally inclined to the zodiac[1] circle; and of the oblique circles, that in which the moon moves, has a greater latitudinal inclination than that in which the sun moves. The planets are moved by four spheres in each case; the first and second of these are the same as for the sun and moon . . .; the third is, in all cases, a sphere with its poles on the circle through the middle of the signs; the fourth moves about a circle inclined to the middle circle (the equator) of the third sphere; the poles of the third sphere are different for all the planets except Aphrodite and Hermes, but for these two the poles are the same.

This is evidently an exceedingly complicated business; but it is important to gain at least a general idea of what it all means. Since all the planets share the daily motion of the stars and at the same time move at a uniform rate near the ecliptic (though the rate is not the same for all the planets), their motion may be resolved into two components, one the motion of a point on the equator of a sphere identical with the celestial sphere, the other that of a similar point on a sphere concentric with the celestial sphere but having its axis at right angles to the plane of the ecliptic, and a period of rotation equal to the time taken by the planet to describe the " zodiacal circle " (ecliptic). The third and fourth spheres were introduced to account for the " stations " or " retrogradations " of the planets, and in this case the poles of the third sphere were actually carried round by the revolution of the second sphere, the poles of the fourth sphere standing in a similar relation to the third sphere. Thus while the motion of the fourth sphere is causing the planet to describe one circular path, the motion of the *axis* of this sphere in a circle impresses another circular motion upon the planet; but the motion of the axis of that sphere, whose revolution is causing the motion of the

---

[1] i.e. The ecliptic or apparent path of the sun in the heavens. It is a great circle inclined at a constant angle to the equator of the celestial sphere.

axis of the fourth sphere upon whose equator the planet is actually revolving, impresses a further component of circular motion upon the planet, which, as a result, does not move in a circle at all, but in a sort of flattened figure of eight, called by Eudoxos the " hippopede " (" horse fetter ").

From the point of view of his contemporaries, Eudoxos had achieved a perfect solution of Plato's problem; for he had shown that the heavenly bodies, though " appearing " to move in complex paths, " really " move in curves, which are the results of combining several circular motions. Their attitude is naïvely expressed by the following words of Simplicius who, though he lived several hundred years later, shared Plato's views on the function of astronomy: " Eudoxos of Knidos was the first of the Greeks to concern himself with hypotheses of this sort, Plato having, as Sosigenes says, set it as a problem to all earnest students of this subject to find what are the uniform and ordered movements by the assumption of which *the phenomena in relation to the movements of the planets can be saved* " (italics my own). This expression, the " saving of the phenomena ", gives us a clue to the only real error, if we may call it such, in the scientific method of Plato, namely that in the framing of his hypothesis he laid far too much stress upon a preconceived form of harmony and beauty upon which he felt any " perfect " universe must " really " be planned; and if to mortal eyes the things of sense (phenomena) seemed different, then it was the business of geometry, which alone has insight into the eternal forms, to show wherein the deception lay. It is all very well to censure the Greeks for this tendency to twist the facts, but it is questionable whether any science worthy of the name could have developed without this transitional stage in which the insight of the human intellect (particularly the Greek intellect!) was much exaggerated. Moreover if there was twisting, there was no falsification, of the facts; for as soon as new ones were discovered, which Eudoxos' brilliant device failed to " save ", then, although these were regarded as further deceptions which had to be removed, nevertheless they were neither ignored nor denied.

Before we pass on to consider what these facts were, we should note that Eudoxos, in the service of a rather questionable hypothesis, had truly, if imperfectly, anticipated one of the greatest discoveries of modern mathematical physics, namely harmonic analysis. First fully worked out by Fourier in the early years of the nineteenth century, in connexion with the flow of heat by conduction, it has made possible great advances in the theory of sound, and the modulation of a wireless carrier wave. So when we hear the " swish " of our loud speaker break suddenly into a graceful melody of Mozart,·we may recall the Greek geometer Eudoxos seeking to resolve an irregular curve into regular components. But we must remember that whereas we regard the irregular as the " real " and the resolution a mathematical device for simplification, he thought just the opposite. Perhaps we are both right!

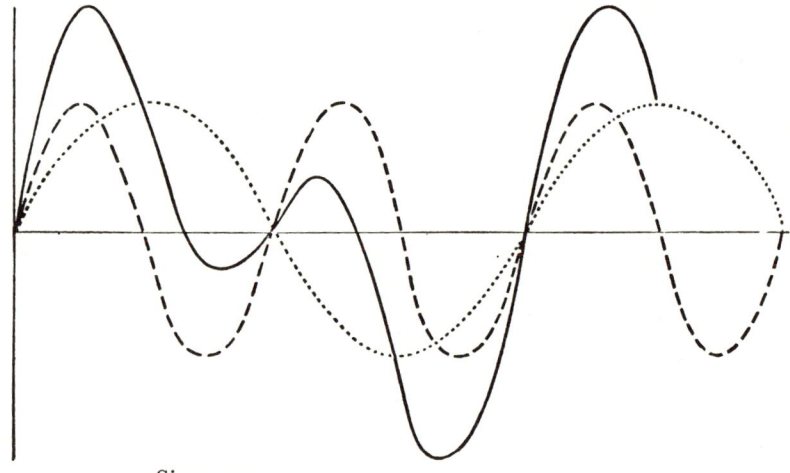

.......... Sine curve.
_ _ _ Similar curve with double frequency.
―――― Wave formed by compounding
FIG. 3. Resolution of a complex wave into simple components.

During the lifetime of Aristotle, and for some time afterwards, no advance was made on the device of Eudoxos. Kalippos, it is true, added a number of spheres to those of Eudoxos to account for minor discrepancies in the motions of some of the planets. Aristotle himself made no alteration in the geometrical scheme,

## THE GEOMETRY OF THE HEAVENS

since he had neither great ability nor great reverence for geometry; but in order to bring it into harmony with his view that the motion of the celestial sphere (set in motion by God) caused the motion of all the remainder, introduced reacting spheres between those of Eudoxos (and Kalippos) which, while leaving the motions of the planets unchanged, welded the whole into one mechanical system. This concept of " cause " was quite foreign to Plato's scheme of thought; but it marks the beginning, though rather a confused one[1] of scientific method in the modern sense.

The shift of the centre of Greek culture from Athens to Alexandria, which took place towards the end of the fourth century B.C., was accompanied by a great enlargement of man's vision, accompanied at the same time by an increase of specialism and the rise of a professional class. A huge library and museum were collected under the supervision of Ptolemy Philadelphos, where men of learning could converse together and use instruments provided at the public expense. Books, copied by an army of scribes, became articles of commerce, maps of a considerable part of the earth's surface were drafted on the evidence supplied by such explorers as Hanno, who journeyed down the west coast of Africa, and Pytheas who sailed beyond Britain into the polar seas. It is not surprising then that some bold spirits tried to " save the phenomena " by the simple but heretical expedient of admitting that the earth perhaps might not be the centre of the whole universe and, as such, endowed with eternal immobility. The later Pythagoreans had hinted as much, and Plutarch states that in his last years Plato had favourably considered this view; but it was not till about 350 B.C. that Herakleides of Pontos began that progressive simplification which, if it had not been prevented by prejudice and the weight of authority, might have altered the course of the history of astronomy. He taught that precisely the same geometrical result will follow whether the whole universe is made to swing round the earth daily from east to west, or the earth alone is allowed to revolve with the same angular velocity in a contrary sense. In addition to which he pointed out the great simplifi-

[1] Confused because scientific method aims at explaining natural phenomena as due to natural causes. Aristotle, by including God as " prime mover ", took the whole question out of that sphere in which science, from the nature of its method, is applicable.

cation which followed the assumption that Mercury and Venus revolve round the sun during the latter's revolution round the earth. His teachings made little impression; about seventy years later however Aristarchos of Samos, having shown by ingenious geometrical demonstrations that the sun's diameter is between six and seven times that of the earth, made the literally revolutionary statement that the earth is not stationary in any sense, but in addition to its diurnal rotation about its own axis possesses an annual revolution about the fixed sun. Although no trace of Aristarchos' hypothesis is to be found in his extant work, we have undoubted testimony to the authenticity of the hypothesis in the following quotation from a book by his late contemporary Archimedes: " His hypotheses are that the fixed stars and the sun remain motionless, that the earth revolves about the sun in the circumference of a circle, the sun lying in the middle of the orbit . . ." The only reply to this wonderful anticipation of modern astronomy was, so far as we know, that " Cleanthes thought it was the duty of Greeks to indict Aristarchos of Samos on the charge of impiety for putting in motion the Hearth of the Universe." Only the Chaldaean, Seleukos, seems to have championed this view, but in vain.

So the splendid opportunity passed; and when a century later the greatest astronomer of antiquity, Hipparchos, applied himself to the " saving of the phenomena " he returned to the device of Eudoxos, which in the meanwhile had been simplified by Apollonios of Perga acting probably on the hint contained in the teaching of Herakleides. Instead of spheres they used the concept of jointed rods by which a planet moved ever in a circle, whose centre moved on another circle centred at the earth; such a device is known as an epicycle, and accounts, it will be readily seen, for the looped paths of the planets. But Hipparchos was faced with a further set of phenomena to " save ", namely that the sun travels from vernal to autumnal equinox in 186 days, but does the return journey (during our winter) in only 179 days. Any hypothesis concerned with the motion of the heavenly bodies must of course ultimately account for this; but to a Greek it was not an " effect " whose " cause " must be sought, but, as it were, a delusion to be explained away. The words of Geminus, a later commentator, cast a flood of light on this most interesting characteristic of the Greek intellect: " They (the Pythagoreans) could not brook the idea of such disorder in things divine and eternal as that they should move at one time more swiftly, at another time more slowly . . ." " Here we shall show, in the case of the sun, for what reason, *though moving at uniform speed, it nevertheless traverses equal arcs in unequal times* " (italics mine). Here it is openly assumed that the sun *is* moving at a uniform speed

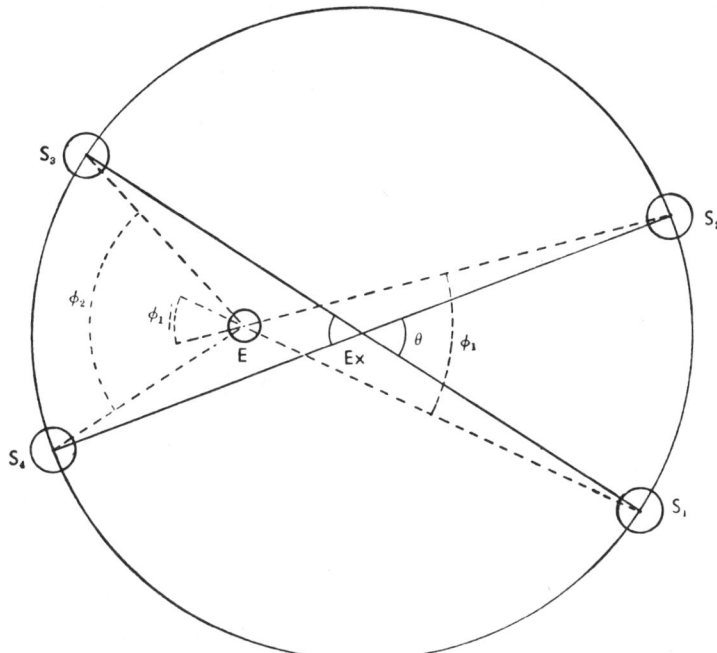

FIG. 4. Hipparchos' eccentric system for the apparent path of the sun.

(i.e. goes equal distances in equal times) and in the sequel it is shown how Hipparchos was able to reconcile the two conflicting ideas, namely by supposing the sun to travel at a uniform velocity on a circle whose centre was slightly removed from the centre of the earth. He then proceeded to combine the epicyclic representation of the paths of the planets with this eccentric device for accounting for their unequal rates. He never completed the system, which was ultimately perfected and published by Claudius Ptolemy in a work generally referred to as " The Syntaxis ", or "Almagest " to give it its contracted Arabic title. The so-called " Ptolemaic system " was the triumph of Platonic metaphysics; with it Greek astronomy, having unscientifically rejected the hope of a renewed vitality held out by Aristarchos, reached completion but also extinction. For 1400 years the Ptolemaic system served astronomers as a means for predicting the positions of the heavenly bodies with surprising accuracy; it is beside the point to say that such a theory is " wrong " and foolish. Wrong-headed it certainly was; for in the absence of any possible means of deciding whether it or the theory of Aristarchus was " true ", it behoved astronomers to adopt the simpler and more elegant of

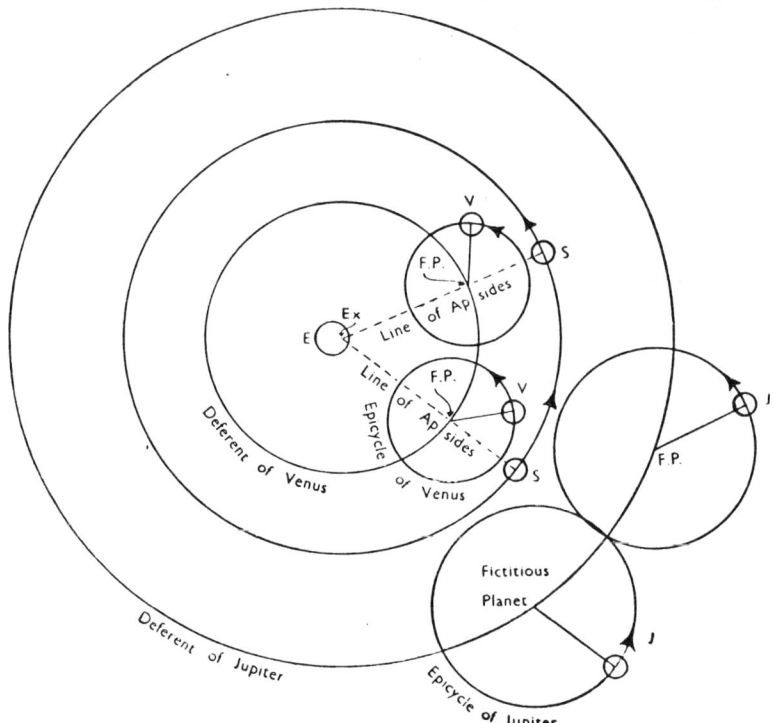

FIG. 5. Simplified diagram of the motions of Venus and Jupiter according to the Ptolemaic system.

## THE GEOMETRY OF THE HEAVENS 41

the two. We shall close this account of Greek astronomy with a brief (and simplified!) reference to the Ptolemaic system, not to stress its importance, but to indicate the folly of choosing it in preference to its rival.

The diagram (Fig. 5, which is not to scale) indicates, in a form much simplified as compared with the developed forms of the Ptolemaic system, the movements of the sun, Venus, and Jupiter relative to the fixed earth. The sun moves on a circle which is eccentric to the earth at a uniform rate. On two circles eccentric to the earth move at a uniform rate what is sometimes called the " fictitious planets " corresponding to Venus and Jupiter respectively; these fictitious planets were conceived to be the moving centres (or, in later forms, ex-centres) of two epicycles on which moved the planets themselves at a uniform rate; but whereas the centre of the epicycle of Venus keeps always on the line joining the sun to the earth, and the planet describes its epicycle in its own characteristic time (about $7\frac{1}{2}$ months), that of Jupiter makes a complete revolution on the deferent in nearly 12 years, the planet itself describing its epicycle in one year; this it does by keeping the line joining the planet to the centre of the epicycle always parallel to the line joining the sun to the earth. In this way and with the addition of further epicycles, by which the system became unthinkably complex, the Ptolemaic system accounted with a fair degree of accuracy for the apparent motions of the heavenly bodies as seen from the earth, assumed to be fixed in the centre of the universe.

Before leaving the Greeks we must briefly notice a discovery of the first magnitude made by Hipparchos, which, though at the time thought to be unrelated to the other heavenly motions, was shown 1800 years later to be an inevitable consequence of the physical relationships of the earth and sun. According to Ptolemy, Hipparchos found that the star Spica occupied a position 6° E. of the autumnal equinoctial point in the heavens, whereas Timocharis about 160 years earlier had recorded it as being 8° E. Now the equinoctial point is near that occupied by the sun at noon at the equinox, i.e. it is one of the two points where the celestial equator cuts the ecliptic; it was used by Hipparchos as the origin of co-ordinates by which the positions of the stars could be fixed. On finding this deviation in " celestial longitude " of Spica he compared his measurements of this co-ordinate with as many as possible of those recorded by Aristillus and Timocharis; he found the same deviation to hold good for every star thus compared. Being thus confronted with the dilemma of supposing either that a large number of widely separated stars had all changed their position by the same amount and in the same direction, or that the equinox, from which all the measurements were taken, had itself moved, he wisely chose the latter alternative. This discovery, which is marked by the exercise of true scientific patience, judgment, and caution, also illustrates the fact, common enough in the history of science, that sound and fertile discoveries may be made by investigators cramped by adhesion to faulty and misleading hypotheses. Incidentally evidence has recently been adduced for the belief that precession was already known to the Babylonian astronomers two centuries earlier.

We have now come to the end of our study of Greek astronomy; and those readers who have managed to scramble through this maze of perverted ingenuity will doubtless express a sigh of relief. Indeed they may even go as far as Alphonso X who in A.D. 1252, when the Ptolemaic system was explained to him, exclaimed that " Had God consulted him at the creation, the universe would have been on a better and simpler plan! "

## NOTE ON BIBLIOGRAPHY

Since this work makes no attempt to give a balanced account of the history of science, it seems desirable to suggest works from which such a view can be gained. These are collected in the General Bibliography on p. 475. The list is of course by no means exhaustive but should assist the reader to continue his studies according to his inclinations and the time at his disposal.

At the end of each chapter or group of chapters are collected the sources which the author has found especially valuable. The reader who wishes to explore further any points of special interest, and to form his own interpretation will thereby be enabled to do so with the minimum of trouble. Many of the works cited contain further detailed references.

# THE GEOMETRY OF THE HEAVENS

## SOURCES FOR CHAPTERS I-IV

The quotations of the fragments of the Ionian nature philosophers are drawn mainly from J. Burnet's *Early Greek Philosophy* (3rd ed. London, 1920). A complete translation of Diel's collection of the Pre-Socratic fragments is now available: K. Freeman, *Ancilla to Pre-Socratic Philosophy* (Oxford, 1948). References to many less well-known authors are to be found in K. Freeman's *The Pre-Socratic Philosophers* (Oxford, 1946). For the interpretation of this period I have also drawn on R. G. Collingwood's *The Idea of Nature* (Oxford, 1945).

The oriental background of Greek mathematics and astronomy is most interestingly told in V. Gordon Childe's *Man Makes Himself* (3rd imp., London, 1941). A simple account of Greek Mathematics is given in Turnbull's *The Great Mathematicians* (London, 1929). This may be supplemented (also the Egypto-Babylonian background) by E. T. Bell's *The Development of Mathematics* (2nd ed., New York, 1945). Sir Thomas Heath's *Manual of Greek Mathematics* (Oxford, 1921) is more detailed.

Sir Thomas Heath's *Greek Astronomy* (Library of Greek Thought, London, 1932) gives adequate translations of the original sources and a valuable outline history of the period.

The reader interested in the author's view of the function of history should read A. N. Whitehead's *Science and the Modern World* (Cambridge, 1926, and Pelican), and *Adventures of Ideas* (Cambridge, 1933); also R. G. Collingwood's *Autobiography* (London, 1939), and *The Idea of History* (Oxford, 1946). H. A. Hodges' *Wilhelm Dilthey* (London, 1944) is also useful. Of special interest is the section on History in M. Oakeshott's *Experience and its Modes* (Cambridge, 1933). Finally, J. B. Conant's *On Understanding Science* (Oxford, 1947) should on no account be missed.

CHAPTER V

# THE GEOMETRY OF THE HEAVENS

*The Earth Dethroned*

WITH the decline of the Alexandrian schools of Hellenistic culture Europe relapsed into a state of barbarism usually called the " Dark Ages ", though evidence has recently come to light that throughout this period and the ensuing Middle Ages isolated discoveries were made, but for want of a favourable social environment seem to have been forgotten again. The love of learning was preserved in the Byzantine Empire which was the last relic of the classical world; at this source the Arabs drank freely, and it is one of the ironies of history that at least one of the many causes which finally brought about the re-awakening of Western Europe from its centuries of uneasy dreams was the discovery by the Crusaders of the comparatively high state of culture reigning in the lands of those very infidels whose presence was " defiling " the Holy Sepulchre. Thus when the Church had gained sufficient wealth and power to establish seats of learning in England and France, it was no uncommon thing for their leading scholars to journey to Granada and the other Moorish universities of southern Spain.

If to the Arabs we owe the preservation of Greek culture, we owe at the same time little advance thereon. Thus it came about that although the Christian Church ultimately co-opted Aristotle as the orthodox revealer of secular knowledge, beyond whose words, except by way of commentary and elucidation, it was considered heretical to pass, it was nevertheless by the intellectual audacity of two of its own monks that the Ptolemaic system, which had served the world for fourteen centuries, was called in question.

When in 1543, at the request of Pope Clement VII, Nicolai Koppernigk (better known by his Latinised name of Copernicus),

## THE GEOMETRY OF THE HEAVENS

being then on his deathbed, received the first printed copy of his work, *De Revolutionibus Orbium Coelestium*, Europe was already in a state of intellectual ferment. As is indicated above, printing was now an accomplished fact, so whatever was being taught as new doctrine would rapidly reach a far larger public than ever before. In 1492 the islands of the American continent were discovered; in 1516 the world was at last proved to be spherical by circumnavigation. From these discoveries followed not only the enlargement of men's vision, heretofore restricted to the cloister or the walled town, but also a great increase in wealth and a rise in the standard of life. These factors doubtless accelerated enormously the effects of the sack of Constantinople in 1453, whereby Greek manuscripts and scholars capable of elucidating them were driven westwards into Italy.

In these circumstances it might have been expected that the hypothesis of Copernicus, that the sun, and not the earth is the centre of the universe, and that the earth and planets revolve round it in paths which can be represented by mathematical figures much more simple than those of the Ptolemaic system, would have been eagerly discussed and speedily adopted. But it was not so. The reasons for this were doubtless many, but they are typified in the criticism of Tycho Brahe (born three years after Copernicus' death) who expressed himself as unable to accept the heliocentric system because the earth is too sluggish to move; because the consequences which may be deduced from such a movement could not be observed; and moreover, the whole doctrine is contrary to scripture. These criticisms, it should be remembered, are not those of a monk, but of an astronomer freed by royal patronage from fear of persecution. In these criticisms we may see the mental confusion which during the Renaissance still characterised the outlook of the learned. The reference to the " sluggish " earth is, we may believe, a tacit admission of a residue of the Aristotelian opinion that motion is according to essence; it is the " nature " of the earth to stay where it is. Why the sun, which Aristarchos had shown to be at least two hundred times the bulk of the earth, should be whirled through space to accommodate a " sluggish " earth, Tycho did not explain, let alone how the whole universe was able to rotate without flying to pieces. The third objection, that the doctrine is contrary to scripture, need not cause in us to-day so much scorn as we might at first be inclined to give way to. If the state of Tennessee in the most progressive country of the modern world could but a few years ago persecute its citizens for teaching the doctrine of evolution, we can forgive, for this lapse into prejudice and muddle-headedness, this man born into a world yet hardly freed from a millenium of continuous domination by ecclesiastical authority.

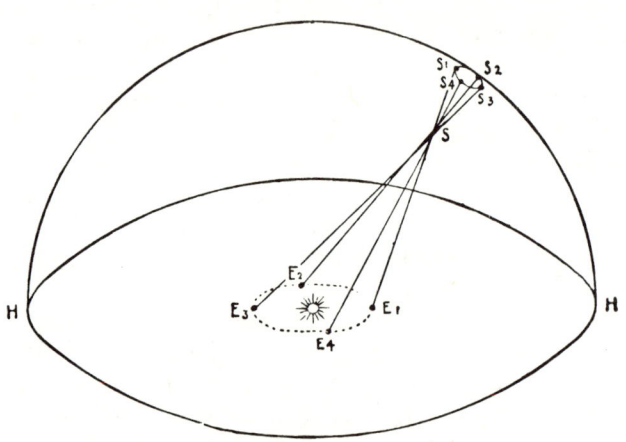

Fig. 6. Annual stellar parallax—the apparent movement of a star on the celestial sphere due to the annual orbital motion of the earth.

The second objection must give us pause. If the world is itself spinning, why do we not feel it, why are objects not swept off its surface? Above all, if the earth is sweeping round the sun, why do the stars not describe annual parallactic circles with respect to the sun? Since none of these questions could be answered, Tycho was justified in denying to the Copernican hypothesis the status of a scientific *theory*, i.e. a hypothesis whose consequences are verified by observation. Now the Ptolemaic system was backed by no evidence, save that of the " common sense " of the ancient world, and the fact that it worked. The Copernican likewise was backed by no evidence, but it worked much more elegantly and simply. Moreover, the standpoint of common sense had shifted with the spread of mechanical inventions. So we must conclude that it was not its scientific insufficiency that was its stumbling block to Tycho, but its unorthodoxy and its conflict with what *he* regarded as common sense. In point of fact Tycho made a compromise and adopted an hypothesis which was an extension of that of Herakleides of Pontos, and also attributed by Cicero to the Egyptians. This was to the effect that the planets revolve round the sun, which in turn revolves round the fixed earth. That the Tychonic and Copernican systems were geometrically equivalent, was shown by Kepler in a most ingenious argument involving the idea of relative motion.

An insight into Tycho's mind may best be gained by his reaction to his own discovery of a new star in November 1572. Of its newness he had no doubt, for he had " almost from boyhood known all the stars of the heaven perfectly (there is no great difficulty in attaining that knowledge). A miracle indeed, either the greatest of all that have occurred in the whole range of nature since the beginning of the world, or one certainly that is to be classed with those attested by the Holy Oracles, the staying of the sun in its course in answer to the prayer of Joshua and the darkening of the sun's face at the time of the Crucifixion. For all the philosophers agree, and facts clearly prove it to be the case, that in the etherial region of the celestial world no change, in the way either of generation or of corruption, takes place." Thus far the dogmatist, looking back to the Authority of Holy Writ even in matters of natural fact, and beyond that to the Aristotelian cosmos in which change and chance are suffered only in the " sublunary sphere "! But Tycho, good son of the Renaissance, also looked forward. He measured the position of the star relative to a known star in Cassiopeia, but could detect no change within 1′ of arc in circumstances in which, were the star below the orbit of the moon, or even in the spheres of the planets, a change must have been detectable. He concluded therefore, despite Aristotle (though he makes no mention of this), that " this star is not some kind of comet or of fiery meteor ... but that it is a star shining in the firmament itself, one that has never been seen before our time, or in any age since the beginning of the world."

Fig. 7. Tycho Brahe directing observations. Sights, mural quadrants, and clocks are in use.

## THE GEOMETRY OF THE HEAVENS

Before leaving Tycho, we must point out that he was doing something far better than giving his blessing to the Copernican system, which in retaining circular orbits was itself a bar to progress, namely spending his nights and days in accumulating observations which were ultimately the means of overthrowing its rival and itself—observations which enabled Kepler to produce the Rudolphine Tables, and were not surpassed in accuracy for over a century.

Meanwhile the Copernican system was being praised and popularised in a most unexpected quarter. Wearied by the aridity of monkish learning, Giordano Bruno rebelled against his Order and travelled over Europe extolling the joy of free enquiry and the innocent delight of learning from nature herself. He went much further than Copernicus, who had retained the Aristotelian concept of a " sphere " of the fixed stars;[1] on the contrary taught Bruno, the universe is infinite, and the stars, scattered indiscriminately throughout it, are worlds like ours. For this as for many of his other views (for instance that the porcupine plucks out its quills and hurls them with deadly accuracy at its foes!) he had no evidence; he was plainly an undisciplined enthusiast rather than an exact thinker who, nevertheless, by his imaginative genius and subsequent martyrdom, did much to prevent the Copernican hypothesis from fading out of notice.

It is important to remember that the Copernican system did no more than reduce the number of epicycles by altering the geometrical point of reference. It did not reject the epicycles as a device, nor did it call in question the truth of Plato's insistence on the circularity of the orbits. Above all, by stating that the sun was not co-planar with these orbits, it tacitly denied any physical interaction between the sun and the planets. In 1609 in a book entitled *Astronomia Nova, seu Physica Coelestis tradita commentariis de Motibus Stellae Martis*, the whole Greek conception of the universe was completely overthrown and, by the irony of fate, by a man whose ambition it was to display it in a yet more glorious light.

[1] The Englishman, Thomas Digges, was one of the first to deny the existence of this (1576).

Johannes Kepler, to whom this great step was due, is a figure almost unique in the history of science. One might almost say that he was a weapon, a beautifully balanced and keen-edged weapon, in the hands of fate. That he regarded himself in this light is evidenced by his own words announcing his last work in 1618: " The book is written, the die is cast. Let it be read now or by posterity, I care not which. It may well wait a century for a reader as God had waited six thousand years for an observer." The irony of the situation lay in the fact that, despite his mystical vision, his intellectual integrity compelled him to enunciate laws which were destined to destroy for ever that mystical view of the universe so dear to him, and pave the way for a thorough-going determinism, which has more or less dominated scientific thought ever since. Having joined Tycho Brahe as an assistant shortly after the publication of his *Mysterium Cosmographicum*, he ultimately gained his master's respect so completely that the great mass of observations which Tycho had gathered night by night on the planet Mars, was placed in his hands to make of it what he might.

Kepler went about his business in the sure conviction that the true orbits of the planets would express in simple numerical relationships the harmony of a perfect universe; an essential factor in this harmony was to be the circular, or *pseudo*-circular, form of the orbits. He quickly discovered what Copernicus had failed to realise—that the sun lies in the *plane* of the orbits. This confirmed what had always been in his mind, and which marks him out as standing as it were on a ledge between the old order and the new, namely that the sun is not merely the most convenient geometrical point of reference for the construction of the orbits, but that it exercises some physical *control* over the planets. It is true that he did not envisage this concept of cause as being in the nature of a merely natural sequence; on the contrary he seems to have identified the sun with the very power of God Himself.

His prejudice on behalf of the circular nature of planetary orbits seems to have been supported likewise by an almost religious ardour. It was only after he had with immense labour tried and discarded one variant after another of the epicyclic hypothesis—many of them fanciful to a degree—that he at last

# THE GEOMETRY OF THE HEAVENS

began to realise that he was approaching nearer and nearer to a form of orbit which in the absence of this prejudice he might have found much earlier. The description of his failures occupies thirty-nine chapters of his book, and he admits himself that this " opinion was a more mischievous thief of my time in proportion as it was supported by the authority of all philosophers, and apparently agreeable to metaphysics ". When at last he discovered the law, it was to the effect that each planet describes an *elliptical* orbit, at one of whose foci is the sun. It must be clearly understood that this form of the orbit is that which an observer on the sun would see, provided that he assumed his own station (the sun) to be at rest; since we now know that this assumption is untrue, we realise that the law has no *absolute* status; that is to say, the planets do not describe ellipses " in space " relative to the fixed stars for, since the sun is moving, the path of the planet must to an observer fixed in space[1] be an " open " curve.

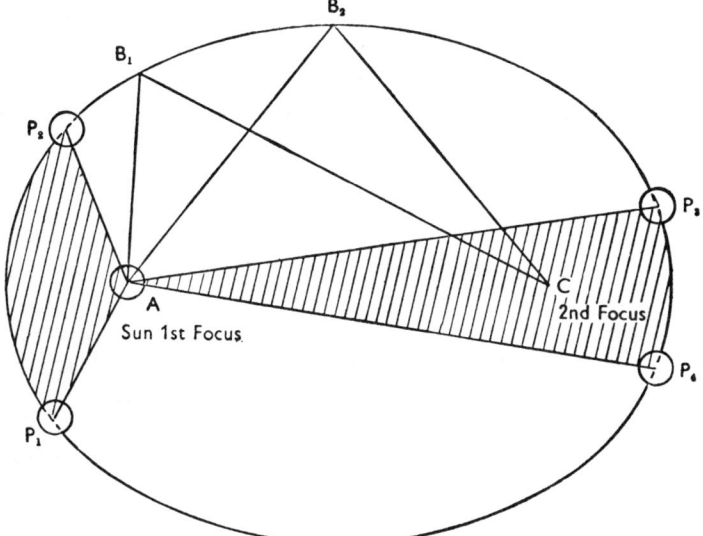

FIG. 8. Kepler's Law of Equal Areas.

[1] Modern mathematical physics has shown that in all probability no such observer could ever, even in theory, exist; but this does not affect the argument given here.

The discovery of his second law rewarded Kepler for any sorrow he may have had for being the means of laying once for all the epicyclic ghost which had haunted astronomers for two thousand years; this law was concerned with orbital velocity.

It will be recalled (p. 38) how Geminus ascribed to Hipparchos the credit for having " saved the phenomena " when the latter had shown that the sun could proceed with a constant velocity and appear not to do so. In the same way we find Kepler equally delighted to have " saved the face " of Mars by showing that although the actual orbital velocity varies from moment to moment, there is nevertheless *something* which is described at a uniform rate, namely the area swept out by the radius vector, that is, the line joining the planet at any moment to that focus of the ellipse at which the sun is situated. But his joy was greatest when as a result of immense mathematical labour he was able in 1619 to publish a third law in a book bearing the suggestive title of *Harmonices Mundi*.

Kepler's pleasure at the nature of this third law was due to the fact that not only to Plato's notions of spatial harmony did he pay homage, but also to the Pythagorean belief that since the pitch of notes depends only on their numerical ratios, the planets in their courses must fill the ether with celestial music. It was actually a search for these numerical ratios that led Kepler to enunciate his so-called " harmonic law ", namely that the squares of the times which the several planets take to complete their orbits relative to the stars are proportional to the cubes of the respective mean distances of the planets from the sun. Anyone might have been proud of so wonderful a discovery, but to Kepler it gave especial pleasure, not only on account of the actual musical harmonies he was able to deduce therefrom, but because in the course of his laborious calculations he discovered what he considered to be the crowning glory of his planetary theory, namely a complete vindication of his early prophecy (set forth in his *Mysterium Cosmographicum* 1596), which was to the effect that the planetary orbits can be fixed with reference to a point in space by inscribing in the sphere containing Saturn's orbit a cube which will be found to circumscribe the sphere of Jupiter's orbit; and so on through the five regular polyhedra. Plato was

## THE GEOMETRY OF THE HEAVENS

right then, thought Kepler, when he said that " God for ever geometrises." But it was in the treatment of this matter so dear to his heart, that Kepler failed to achieve the ideal which became in his later contemporaries the aim of scientific discovery, that is to be cautious in the use of analogy, and not to endow the argument with a dogmatic finality beyond what the facts warrant. Now it is difficult to see what real analogy there can be between the radius of the planetary orbits and the circumscribed and inscribed regular polyhedra. Because there happens to be a

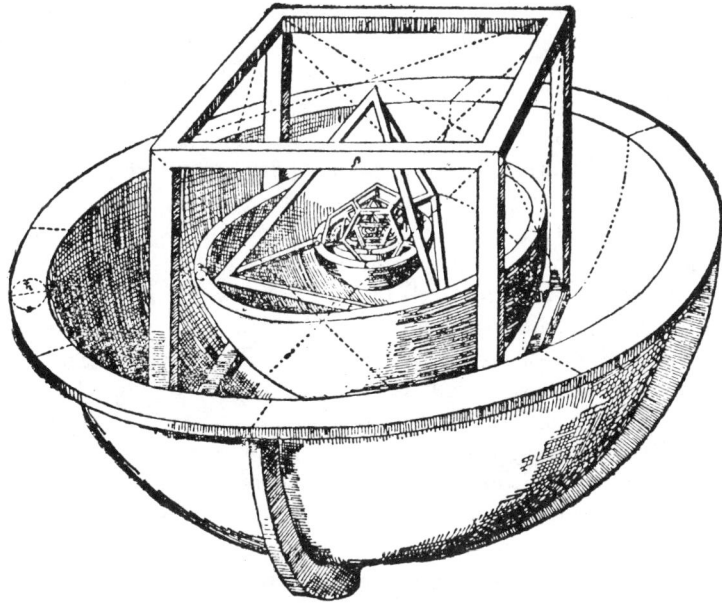

FIG. 9. Kepler's figure of the (supposed) relations between the planetary orbits and the regular polyhedra.

rough correspondence in fact, it would be foolish to assume that since there can only be five such polyhedra—the cube, tetrahedron, octahedron, icosahedron and dodecahedron—therefore there are only six planets. But this is precisely what Kepler did say: " The number of planets, or orbits about the sun, was derived by the most wise Creator from the five solid figures." Fortunately for his peace of mind it was not till more than a century after his death that a succession of new planets began to appear.

We must now leave Kepler in his solitary ecstasy over these fictitious harmonies and divine geometry, which had enabled him to overcome the great difficulties of determining the planetary orbits, and of showing at last the complete wrong-headedness of the Platonic hypothesis and its cumbrous superstructure from any point of view save that of an arbitrarily fixed earth. This orb no longer, after Kepler, ruled the heavens, but was shown up as one of the lesser planets—" of a most contemptible smallness ", as Kepler put it. To Kepler is the glory of having expressed in the simplest and most elegant way the geometry of the heavens as it would appear to an observer fixed in space; a geometry which contains within itself something more than mere geometry, namely a hint of the underlying cause to which these orderly relationships in space and time are due. This Kepler dimly glimpsed; but, being yet obsessed by Aristotelian notions of God as the sustaining cause of motion, failed to grasp. The way to such a discovery was clear; but before it could be made, four other conditions had to be fulfilled: (1) Some striking evidence was needed by which the truth of the Keplerian system could be directly observed or easily inferred; (2) The method and character of the " new science " had to be displayed; (3) A basis of conversion between geometry and algebra had to be effected, whereby the facts of the former might be rapidly correlated by the simpler and swifter operations of the latter; (4) The nature of motion must be investigated, and its laws, if any, enunciated.

Kepler died in 1630; by that time the first three conditions had been realised and the ground prepared for the realisation of the fourth.

CHAPTER VI

# THE TELESCOPE AND ITS REVELATIONS

THE actual discovery of the telescope is wrapt in mystery. There is a legend that on the Greek promontory known as the Pharos was a glass through which ships passing at sea appeared several times their normal size. There is another story, which is probably true, that Hans Lipperhey, a Dutch spectacle maker, chancing to look through two convex lenses in line with one another, saw to his amazement the church steeple enormously enlarged and standing upside down; he communicated the discovery to his friends, but did not develop it. Our only certain information is that an entry in the archives of the States General at the Hague shows that an award was made to Lipperhey without an exclusive patent, since other claimants had made good their claim to have made such an instrument. This was in October 1608. To the Dutch the invention seems to have been regarded as merely a toy. But Galileo, writing in 1610 in his *Sidereal Messenger*, says:

About ten months ago a report reached me that a Dutchman had constructed a telescope, by the aid of which visible objects, although at a great distance from the eye of the observer, were seen distinctly as if near; and some proofs of its wonderful performances were reported, which some gave credence to but others contradicted. A few days after I received confirmation of the report in a letter written from Paris by a noble Frenchman, Jaques Badovere, which finally determined me to give myself up first to inquire into the principle of the telescope, and then to consider the means by which I might compass the invention of a similar instrument, which a little while after I succeeded in doing, through deep study of the theory of Refraction; and I prepared a tube, at first of lead, in the ends of which I fitted two glass lenses, both plane on one side, but on the other side one spherically convex, and the other concave.

It would appear from this account that Galileo had to work out for himself the kinds of lenses to use; and he suggests that he did this, not by the simple process of trial and error, but " through deep study of the theory of Refraction ". Now the laws of refraction were not enunciated with mathematical precision until

1621 (Snell); but the phenomenon, which consists merely in the apparent displacement of objects when viewed through various media, or varying thicknesses of the same medium, had been very fully studied from early times. To understand the theory of the matter it is necessary to return to the Greeks, whose studies of light put the subject almost at once on the level of an exact science.

It was probably Euclid who first realised that to make any advance in formal optics, i.e. the department of science which frames general principles to explain the varying appearances of bodies in different circumstances, it would be necessary to isolate, in imagination, an axis along which the " visibility " is propagated in any particular direction: this was, and is, called a " ray ". In actual experiments, of course, the most we can do is to isolate a narrow beam, or " pencil ", of rays, which are mathematical fictions only. With respect to these rays Euclid made the same mistake as Plato, supposing that the action proceeds, like a tentacle, from the eye of the observer to the object; this is evidently a prejudice of exactly the same kind as that which made the earth the centre of the universe. And it had similar results, namely that being *geometrically* identical with the " true " state of affairs which demands the presence of some external luminous body before sight of any kind is possible to us, it did not prevent the discovery of important *geometrical* properties of light rays. In the first place Euclid gives adequate reasons, such as the sharpness of shadows and points of light from small orifices, for the law that the rays of light travel in lines which are straight. Applying this discovery to the reflection of light by mirrors he was able to show, firstly that the incident ray, the reflected ray, and the normal to the surface lie in one plane; and secondly that the angle of incidence is equal to the angle of reflection. It was probably the knowledge of these laws which enabled Archimedes to focus the rays of the sun on the Roman ships by means of spherical mirrors at the siege of Syracuse. We cannot be absolutely certain of the authenticity of this report, but we know that by his ingenuity the town held up the Roman siege for a far longer time than they had anticipated.

That the positions and sizes of heavenly bodies are distorted when they approach the horizon had been known for some time; but it was Ptolemy who first made a systematic investigation of the problem, on the supposition that the incoming ray is bent out of its course when it reaches the earth's atmosphere, just as a stick appears bent when half immersed in water. The conclusion he arrived at was that the incident angle is always proportional to the refracted angle, which is true for small angles.

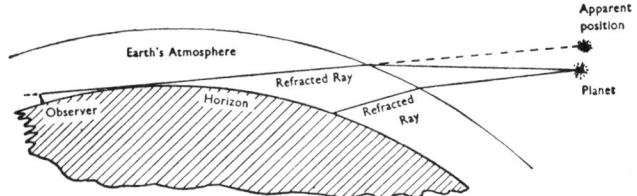

FIG. 10. Apparent position due to atmospheric refraction.

Geometrical optics was one of the branches of science in which the Arabs made appreciable advances beyond their Greek masters; and by the time that Galileo felt impelled to a " deep study of refraction ", the theory of the simple magnifying glass, and of convergent and divergent lenses was well understood; it would indeed have been quite possible for him to have calculated what are known as the focal lengths necessary for the lenses of his telescope, but in the absence of an exact knowledge of the law of refraction, the curvature of the lenses necessary to give this focal length must have been determined by simple trial and error.

In the *Siderius Nuncius* (Starry Messenger), having told us how he made his telescope and determined directly its magnifying power, Galileo describes his observations on the moon, now seen for the first time to be, like our own planet, covered with mountains, valleys, and volcanic craters. Moreover that its light is due wholly to reflection of light from other bodies, mainly the sun, is put beyond dispute, because the faint illumination of the dark part of the crescent moon is shown to be due to the reflection of moonlight back from the earth to the moon.

When turned on the stars the telescope does not magnify the individual but enormously increases their number; moreover the milky way, always regarded as something of a mystery, now appears as an enormous mass of distinct stars; the so-called " nebulae ", or cloudy stars, are found to be merely clusters of small stars.

But much as this huge enlargement of the universe casts doubt upon the finite universe of Aristotle, there was another discovery, which was the best answer to those who objected either to the possibility of any body other than the earth being the centre of a planetary system, or on the other hand of the possibility of the earth going round the sun without " losing " the moon. This discovery was made on 7th January 1610 when, viewing the planet Jupiter, Galileo " noticed a circumstance which I had never been able to notice before, owing to want of power in my other telescope, namely that three little stars, small but very bright, were near the planet ". Two lay to the east of the planet, one to the west. On the following night " led by some fatality " he turned again to look at Jupiter and " found a very different state of things, for there were three little stars all west of Jupiter ". Not daring yet to accept the belief he so much wished to accept, namely that here was a miniature planetary system of three moons revolving round Jupiter, in case the planet had itself moved beyond the little stars, he " waited for the next night with intense longing "; but the sky was covered with clouds. On 10th January however two little stars only were visible, the third he believed was hidden by Jupiter. On the 12th there were three again. " I therefore concluded," he says, " and decided unhesitatingly, that there are three stars in the heavens moving about Jupiter, as Venus and Mercury round the Sun." The next day he discovered a fourth.

When Kepler heard of these discoveries he was so delighted that he corresponded with Galileo, and it was in a letter quoted in the introduction to his *Dioptrics* that Galileo announced his discovery which was the most direct " proof " of the falsity of the Ptolemaic system, namely that the planet Venus passes through the same phases as the moon. Since on the Ptolemaic view the sun never lies between the earth and Venus, a full cycle of phases such as the moon displays would be impossible. On the Copernican and Keplerian view however such phases should be observable.

Here at last then was evidence appealing directly to the senses and not being based merely on geometrical elegance and simplicity, that the earth together with the other planets revolves

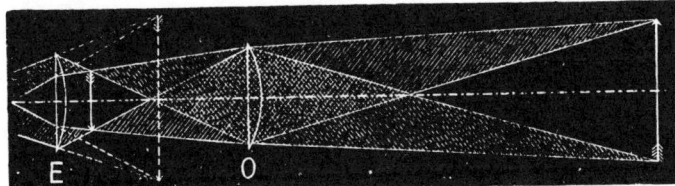

Fig. 11. The optical system of the (Keplerian) astronomical telescope.

round the sun, some being eternally accompanied by satellites. We cannot doubt that this evidence, and the publicity given to the problem by Galileo's subsequent persecution for supporting the theory, were responsible for hastening its general acceptance. It is important to note that the Church had the support of a considerable body of " scientific opinion "; indeed Galileo in one of his letters to Kepler says that " Kepler and the rest of the school of Copernicus . . . (are) regarded by the philosophers of our times, who philosophise on paper, with an universal agreement, as men of no intellect, and little better than absolute fools."

Just as progress in astronomy was thus enormously accelerated by the progress in the science of optics, so its crowning glory, that of displaying the motions of the heavenly bodies as due to one underlying cause, could not have been achieved without those advances in mathematics and mechanics which we have already mentioned and to which we shall now turn.

### SOURCES FOR CHAPTERS V-VI

Shapley and Howarth, *op. cit.* in Gen. Bibl., Arts., Tycho Brahe, Kepler.

Galilei, *The Sidereal Messenger*, trans. E. S. Carlos (London, 1880).

The reader is advised to study the background of medieval and Renaissance Europe in order to assess these periods as justly as possible. Many of the earlier works under-rate the positive contribution of the Middle Ages, and over-rate the importance to the progress of *science* of the Renaissance.

G. G. Coulton's *Studies in Medieval Thought* (Edinburgh, 1940), provides a short and very readable account of the former period; while Sir Charles Sherrington's *Jean Fernel* (Cambridge, 1946) gives a clear picture of the transition in Paris, and delineates Fernel as a typical transitional figure in biology.

See also Gen. Bibl.

## CHAPTER VII
# THE RISE OF TERRESTRIAL MECHANICS

THE Keplerian system was perfect within the limits of accuracy set by the degree of instrumental precision existing in his time. The motions of the heavenly bodies were summarised in the form of concise statements of fact, which were independent of any preconceived views as to what they " ought " to be. But it was the genius of the seventeenth century, what marked it out as different from any previous century of human culture, to feel dissatisfied until the behaviour of things had been traced back to the operation of *known* causes. To this view no one gave more eloquent expression than Francis Bacon in the following words of the Father of Solomon's House to one of the party storm-driven to the *New Atlantis*: " The end of our foundation is the knowledge of causes, and secret motions of things; and the enlarging of the bounds of human empire, to the effecting of all things possible." The recognition that the causes of phenomena may lie deep hidden in the nature of things; the shift of emphasis from the preoccupation with the after-life to the enlargement of the empire of man; an unbounded optimism: these are the marks of the opening of the Scientific Age. No contemporary work gives a better idea of the temper of the age than Thomas Sprat's *History of the Royal Society of London* (1667). " It was therefore ", he writes, " some space after the end of the Civil Wars at Oxford,[1] in Dr. Wilkins' his lodgings, in Wadham College, which was then the place of resort for vertuous and learned men, that the first meetings were made, which laid the foundation of all this that follow'd. The University had, at that time, many members of its own, who had begun a *free way* of reasoning; and was also frequented by some gentlemen, of

[1] In a more detailed work (*History of the Royal Society of London*, 1756) Birch stated that the first meetings (referred to by Boyle as the *Invisible College*) took place in London in 1646, and that later the members divided, some going to Oxford, others remaining in London.

philosophical minds whom the misfortunes of the kingdom, and the security and ease of a retirement amongst gown-men, had drawn thither." About 1658 most of the original members, being then in London, joined with several others to meet at Gresham College. Their design was to " make faithful records of all the works of nature, or art, which can come within their reach . . . They have studi'd to make it, not only an enterprise of one season, or of some lucky opportunity; but a business of time; a steddy, a lasting, a popular, an uninterrupted work. They have attempted to free it from the artifice, and humors, and passions of sects; to render it an instrument whereby mankind may obtain a dominion over *Things*, and not only over one another's judgements." This shift of emphasis from preoccupation with *words* to the observation of *things* is nowhere better brought out than in the famous passage describing how " they have exacted from all their members a close, naked, natural way of speaking; positive expressions; clear senses; a native easiness: bringing all things as near the mathematicall plainness, as they can: and preferring the language of artizans, countrymen, and merchants, before that, of wits and scholars ".

The empirical approach, concern with the observed properties of things, is illustrated by a passage from *The Advancement of Learning*, which, although referring to subject matter we shall take up at a later stage, is nevertheless worth quoting here:

> For it is a thing more probable that he that knoweth well the natures of weight, of colour, of pliant and fragile with respect to the hammer, of volatile and fixed in respect of the fire, and the rest, may superinduce upon some metal the nature and form of gold by such mechanique as belongeth to the production of the natures afore rehearsed, than that some grains of the medicine projected should in a few moments of time turn a sea of quicksilver or other material into gold [*op. cit.*, p. 109].

Here Bacon is speaking of the futility of the so-called " Philosopher's Stone ", but his comparisons are applicable to all the problems of nature. The gist of the message is that we must seek out the common natures of the things we are comparing, and thus, remembering that " all true and fruitful natural philosophy hath a double scale or ladder, ascendent and descendent, ascending from experiments to the invention of causes, and descending from causes to the invention of new experiments ", we may knit all nature together into one coherent assemblage of events.

Now what is the " nature " of the heavenly bodies that the astronomers from all times sought to know? What common property characterises them and is shared with them by bodies that we on earth can handle? There may be many, but one only is certain, that is the power of *motion*. To the seventeenth-century thinker therefore it was " a thing more probable " that he who studied the common characteristics of moving bodies would be in a better position to discover the nature of motion itself than one who, like Aristotle, explained all motion as the internal urge of bodies to seek the places " proper " to them in the scheme of things—a hypothesis not growing out of observation but dictated by human reason and purpose. To this extent the " New Science " was, as the late Professor Whitehead pointed out, *anti*-rational, resembling rather the remorseless working of fate (ἀνάγκη) portrayed in Greek tragedy. The man who more than any other both realised the necessity for this changed method, and by the skilful use he put it to, clearly revealed the nature of motion, was Galileo. Throughout the sixteenth century the growing contact with the greatest minds of Ancient Greece had directed the stirring impulse to a renewed effort to understand the workings of nature—an impulse which had lain dormant for over a thousand years. Indeed that great artist and engineer, Leonardo da Vinci, had anticipated Galileo by a century, but his colossal mind had so teemed with ideas that time nor patience sufficed to set them in order for the guidance of posterity. Of this we shall treat later; but in respect of the nature of *motion*, it was only in 1638 that the publication of *Mathematical Discourses Concerning Two New Sciences Relating to Mechanics and Local Motion* marked the foundation of dynamics and at the same time established the principles of what became the dominant form of modern scientific method.

In seeking to reduce to order the phenomena displayed by moving bodies, Galileo departs from the classical method in a number of ways, but none is as important as the recognition that the solution of the problem of *how* bodies move under the action of natural forces must precede that of the more ambitious one

## THE RISE OF TERRESTRIAL MECHANICS

of *why* they behave as they do. This was so epoch-making a break from traditional modes of thought that it will repay us to hear how it took shape in his own mind: this he shows us in the words of Salviatus, the principal speaker in his dialogues:

I don't think this a seasonable time to enquire into the cause of the acceleration of natural motion; concerning which philosophers have greatly differed in opinion; some reducing it to the approaching to the centre; others to the fewer parts of the *medium* successively remaining to be perforated; and others referring it to a certain extrusion of the ambient *medium* which, upon conjoining upon the back of, or behind the moveable, presses upon, and continually thrusts it more; which conceits, and others of the like nature, 'tis hardly worth while to examine and answer. But as to our author, it is enough at present that we understand, that his business is to search out and demonstrate to us some passions of an accelerated motion (let the cause of the acceleration be what it will) so that the *momenta* of that acceleration may continually increase after the departure from rest, according to that most simple proportion, wherewith the continuation of the time increases; which is as much as to say, that in equal times there are made equal additaments of velocity. And if it shall be found that those accidents which shall hereafter be demonstrated are agreeable to truth, in the motion of heavy bodies naturally descending and accelerated, we may conclude that our assumed definition comprehends or takes in such a motion of heavy bodies. [*Third Dialogue*, Weston's English Trans., 1730, p. 246].

Setting out with a perfectly open mind, Galileo observes that bodies falling freely do so with increasing speed; moreover that on the whole (note that here his mind is not quite so "open": he tacitly introduces the simplifying assumption that they are not to be of such a kind as to be "buoyed up" by the air) their rate of fall depends on *nothing*—neither weight, colour, size, nor shape—but the distance through which they have already fallen.[1] But this was not enough: how, and to what, is the speed related? Only two determinative conditions remained, namely space and time.

In the next stage of the search—namely that in which he had to decide which of these two conditions was *the* condition—Galileo could no longer call upon experience to act as his guide; for since a velocity which is not constant can clearly not be measured (see later Ch. IX), any form of experiment was out of the question.

[1] Tradition has it that he tested this view by the famous Pisa experiment in which two unequal weights dropped from the same height fell side by side and hit the ground simultaneously; strange to say there is no real evidence that this spectacular experiment ever took place!

In the manner in which he attacked this problem may be seen all the essentials of that method of investigation which has yielded us the knowledge known as mathematical physics. This method is not, as has sometimes been suggested, the *only* " scientific method "; nor is mathematical physics the only sort of knowledge we can have of nature; but it has without doubt played the predominant part in modern thought, so it will surely repay us to examine most carefully the nature of its birth and growth.

Having dismissed, by a curiously fallacious line of reasoning, his first assumption, namely that the velocity varies as the *space* traversed, Galileo next assumed that the velocity at any instant varies as the time which has elapsed since the commencement of motion. From this he deduces (in a manner similar to that adopted in modern elementary text books of dynamics) " that the time wherein any space is passed by a movable, with a motion uniformly accelerate from rest, is equal to the time in which the same space would be passed by the same movable, carried with an equable [i.e. uniform] motion, the degree of whose velocity is subduple [i.e. half] to the greatest and ultimate degree of the velocity of the former uniformly accelerate motion."

Having gained this important result Galileo takes nearly two pages to reach his conclusion; by the use of algebraic symbols, which were only just coming into use towards the end of Galileo's life, we may arrive at the same result in as many lines. Thus, if $v$ be the velocity acquired in time $t$, and $a$ be the uniform acceleration, the " equable subduple " velocity is $\frac{0+v}{2}$; but, by hypothesis, $v = at$, and, by definition, $s = $ (equable subduple velocity) $\times t$

$\therefore \frac{0+v}{2} = \frac{1}{2}v = \frac{1}{2}at$; and $s = \frac{1}{2}at^2$, which Galileo expressed in the words, " If a movable descend from rest, with a motion uniformly accelerate, the spaces passed by it in any times whatsoever, are to each other in a *duplicate* proportion of the same times, i.e. as the squares of the same times."

The whole procedure was so far purely an exercise of the intellect; it might have no relevance to nature. But this is where Galileo's genius shines forth so brightly; he was not content to assert this result as a necessary consequence of the most " beauti-

THE RISE OF TERRESTRIAL MECHANICS 65

ful ", " orderly ", or " harmonious " system which alone could reign in a rational world, but set to work to analyse the conditions of the events whereby an experiment might be devised, upon the result of which the truth or falsity of the preliminary assumption would stand or fall. This analysis led to the conclusion that if the fall of a body be retarded by the use of an inclined plane, it will be uniformly so, with the result that the *form* of the relationship between velocity and time will be unaltered. On measuring, " by experiments near an hundred times repeated ", the times taken to traverse $1^2$, $2^2$, $3^2$, $4^2$ . . . units of length (he used a water clock to measure the times), he found that they were of the order of 1, 2, 3, 4 . . . units of time. The original assumption was thus confirmed, and the first law of kinematics established. Moreover in this series of operations—intellectual and practical— is crystallised the principle of *all* scientific method, namely the observation and correlation of particular instances of natural phenomena in an artificially simplified field;[1] the framing of an hypothesis to express briefly the *general* relationship from which all such particulars may henceforth be inferred; the deduction by intellectual processes of observable consequences which must *necessarily* follow if the original hypothesis is correct; and the contriving of experiments to verify or refute those consequences, hence the original hypothesis.

It was thus the merit of this great man, more perhaps than any of his contemporaries, to place in the hands of the human

[1] The simplification of the field was carried by Galileo to an extreme by his rejection of all qualities of objects except figure and motion. In a famous passage in *Il Saggiatore*, speaking about the nature of heat, he says: "As soon as I conceive of a piece of matter or corporeal substance [I conceive] that in its own nature it is bounded and figured in such and such figure, that in relation to others it is large or small, that it is in this or that place, in this or that time, that it is in motion or remains at rest . . . in short, that by no imagination can a body be separated from such conditions: but that it must be white or red, bitter or sweet, sounding or mute, of a pleasant or unpleasant odour, I do not perceive my mind forced to acknowledge it necessarily accompanied by any such conditions; so if the senses were not the escorts, perhaps the reason or the imagination by itself would never have arrived at them." This view was given a more thorough grounding by Boyle in *The Origin of Forms and Qualities*, and Locke in his *Essay*. The wide extension of this mode of thought in the seventeenth century is shown by the following passage from the opening of Thomas Hobbes's *Leviathan* in which he gave the first theory of the origin of the political state: "All which qualities called sensible, are in the object which causeth them but so many several motions of the matter, by which it presseth our organs diversely. Neither in us that are pressed are there anything else but diverse motions (for motion produceth nothing but motion). But their appearance to us is a Fancy . . ."

race the key by which the treasure-houses of nature have been unlocked. But at the risk of being wearisome it must once more be insisted that there is a fundamental difference between the natural science of Galileo and the natural philosophy of the Greeks; and this difference is not only in the method, but extends into the character of the results. This insistence is the more necessary, since to the failure to distinguish the divergences between the two methods is due the fact that many people are thoroughly convinced that modern science (whose results are trustworthy, numerous, and brilliant) successfully answers the *same* question as, with its halting, vague, and unpractical opinions, ancient natural philosophy tried but failed to answer. It has already been urged in these pages that Greek astronomy was " true " within the limited horizon of observation of the Greeks, but, by reason of its needless mathematical complexity, inadequate. In other branches of physical science (i.e. that division of science which deals with matter as such) the views of the Greeks were even less adequate; but when we say that Galileo rid the world of the errors of Aristotle, we are making a statement which is at the best misleading. For Aristotle sought the reason and purpose of the motion of water, fire, stars, and winds; Galileo, on the contrary, recognising the too ambitious nature of this quest, was content to demonstrate how particles of matter move in the absence of all complicating circumstances. In this he was remarkably successful; but we must not make the mistake of supposing that the answer to Aristotle's question can be obtained by merely extending the application of Galileo's method; something essential to the former enquiry has here been left out. In point of fact, in reducing his description of the behaviour of bodies to relationships between space and time, Galileo has reverted to a mode of thought earlier than that of Aristotle, namely that of Pythagoras; for neither space nor time is a *thing*, but a relation expressed by a mere number, though of course the number which refers to the space is to be distinguished from that which refers to the time. The method is beautifully adequate to its own limited field of enquiry; the mistake is to pretend that it can ever yield the *whole* secret of life. To make such an assumption is to place the human spirit on crutches

## THE RISE OF TERRESTRIAL MECHANICS 67

when, as the poets so clearly tell us, it may at certain times soar on outspread wings.

We may now pass on to discover what use Galileo made of this law of nature; for with characteristic sagacity he realised that the *universal* validity of the fact of uniform acceleration was, as it were, the superficial expression of some much deeper relationship. If it be urged, as indeed it should be, that the correspondence of results of experiments " near an hundred times repeated " gives no *proof* of the *universal* validity of the law, it must be admitted at once that the *whole* of the so-called *certainty* of science rests on exactly the same foundation of doubt. Had the experiments been a million times repeated with identical results, it would merely have increased the *probability* of the law, which can never be absolutely certain. But when a law of nature, tested a large number of times, and especially in a wide variety of attendant circumstances, has never been known to fail, its divergence from certainty (whatever that may mean!) becomes merely a matter of academic interest. This must always be borne in mind at every stage of scientific advance; but the " certainty " of to-morrow's sunrise has a probability greater only in degree and not in kind.

The major consequence of Galileo's application of the law of acceleration was his clarification of the concept of *force*. Once more, if we can grasp thoroughly the nature of this clarification, we shall not only have learnt something of the history of science in the seventeenth century, but we shall have made another great stride towards the understanding of the whole nature of science.

The concept of force is as old as man's first *reflection* on the power of his limbs to move objects; but its homely origin was exceedingly unfortunate for its subsequent history. The reason for this is the almost invariable rule that when I cease to push the garden roller on a level lawn, the roller almost immediately stops. The two " almosts " seem to have been overlooked by classical thinkers, otherwise Aristotle would not have felt impelled to introduce a " Prime Mover " into the universe to *start* its motion, nor a heavenly sphere to *keep* the remainder in motion. Even Kepler failed to rid his mind of a belief in the necessity for some force to *sustain* motion. Galileo however clearly saw that reason demanded nothing of the sort; and experiment explained the difficulty of the roller. The result of his consideration of the problem is that so far as reason is concerned we have no warrant for supposing that motion is any less "natural" than rest; from which it follows that a *cessation* of an existing motion must be accounted for just as much as a commencement. The birth of this postulate is the birth of *modern* science; that is, the effort to supply a reason for *change*. The practical difficulty, namely the absence of perpetual motion on earth, was cleared up by Galileo in this way: To verify this new law of motion we must be able to show that a body, once set in motion, would continue to move at the same rate in a straight line for ever, provided that no other force opposed this motion. This we can never hope to do just because it is *impossible* to observe a body acted on by *no* force—it must always be under the influence of its own weight. But what we can and do observe is that if this weight be neutralised by a horizontal plane, the smoother we make this plane the more nearly does the anticipated motion ensue, the best example being a piece of ice on a smooth ice " slide ". The principle is therefore verified, and thus becomes a " law of nature ".

We have not yet done with the matter however: there remains the apparent contradiction between the now accepted notion of force as *reason* for change of motion and Galileo's unambiguous assertion that he seeks not the *cause* of motion. What difference of meaning can there be between " cause " and " reason "? The answer to this question is the climax towards which we have been struggling.

If it be argued that to call " force " the " reason for change of motion " is in effect to call it the " cause ", this can be admitted to be true only if at the same time it is recognised that the conception of " cause " itself has undergone a transformation in Galileo's hands. For there is nothing in his argument to suggest that he is pretending to a knowledge of *why* or even *how* force changes motion; but this is precisely what the Greeks *did* pretend. Galileo's approach is, on the contrary, much less ambitious: we see bodies change their motion; our reason demands that any such change must be accounted for by a cause; we *call* this assumed cause a " force ". What the first law of motion tells us, then, is that it is in the nature of things to move uniformly; and

THE RISE OF TERRESTRIAL MECHANICS 69

in adding that change of motion is brought about by force, it is giving us an unambiguous *definition* of force.

Henceforth it was the business of science to seek not the nature of causes but the correlation of events due to the same—but unexplained—cause. The justification of the change of aim lies in its power of being realised! In the words of one who did

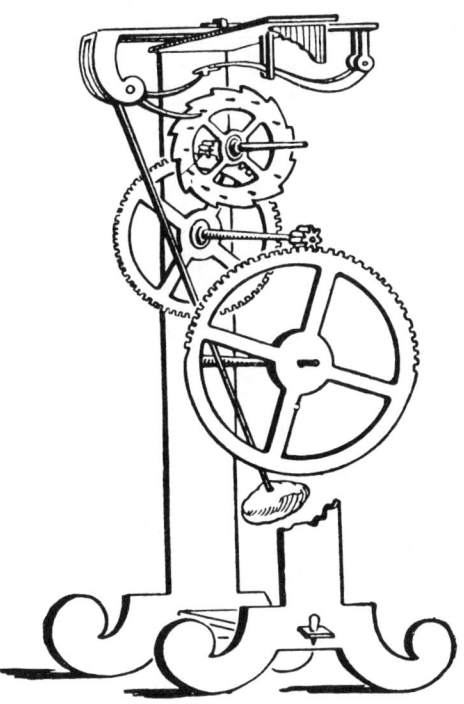

FIG. 12. Galileo's proposal for a pendulum clock (based on a sketch done by his son, Vincenzio, and his pupil, Viviani).

much to clarify the Greek conception of cause, F. M. Cornford, " So long as the fall of the apple to the ground was satisfactorily explained by the impulse of its earthy nature to seek reunion with the kindred mass of earth, the laws of motion were likely to remain a secret " (*The Laws of Motion in Ancient Thought*, p. 39). It must however be borne in mind that this limitation of the field of enquiry carries with it a limitation of the applicability

70 THE GROWTH OF SCIENTIFIC IDEAS

of the results—a corollary wholly disregarded by many of Galileo's successors.

Before closing this chapter we may note another of his discoveries which, though little developed by him, was to play as great a part in modern scientific discovery as the telescope.

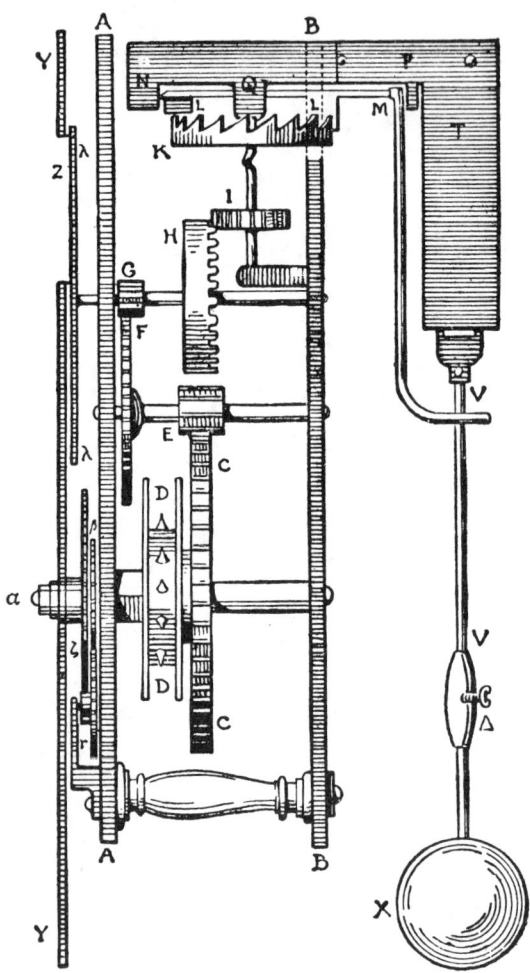

FIG. 13. Huygens' pendulum clock.

It is recorded on good authority that, despite his pious nature, Galileo once allowed his attention to wander during a religious service. But the attention of genius does not wander listlessly; Galileo's gaze was fixed upon one of the great lamps hanging from the roof which was beginning to swing in the up-draft of hot air from a brazier. His accurate sense of time drove him to suspect a uniformity in the time of swing independently of the gradually increasing arc through which the lamp was moving. Aided by the only chronometer available—his own pulse, which he assumed to be regular—he verified his suspicion, and after repeated trials under more rigorous conditions he was able to enunciate the law that, for small swings, the time of oscillation of a freely suspended weight is independent of the amplitude of the swing. It was not till thirty-one years after Galileo's death that Christian Huygens published his *Horologium Oscillatorium*, in which, approaching the matter from the point of view of the conservation of energy, he proved that any suspended rigid body will swing in time with an " equivalent simple pendulum ", whose beats will be isochronous if it follows a *cycloidal* path.

Galileo applied his discovery in reverse by inventing an instrument for measuring the rate of the pulse. He also sketched out a design for an impulse-driven pendulum clock; but it was never constructed. Huygens was therefore able to acquire a patent in 1657—sixteen years before a full account of the theory was published.

## SOURCES FOR CHAPTER VII

F. Bacon's *Advancement of Learning* (The World's Classics ed. contains in one volume this and *The New Atlantis*), a long and leisurely work but full of interesting passages indicating the changing point of view characteristic of the seventeenth century. *New Atlantis* (pp. 264-end, a description of " Salomon's House ")—a remarkable anticipation of the modern form of large-scale research. *Novum Organum*, the exposition of Bacon's method, which has been found less fruitful than he expected. Bk. I, Aphorisms, I-LXVIII, on the *Idols* is well worth reading.

Galileo's *Dialogues*. The definitive English translation is by H. Crew and A. de Salvio, under the title, *Dialogues Concerning Two New Sciences* (New York, 1933); my citations are from Weston's translation of 1730.

E. Mach, *The Science of Mechanics*. The most important historical and critical account of the whole science of mechanics.

CHAPTER VIII

# THE DAWN OF UNIVERSAL MECHANICS

ON Christmas Day of the year 1642 (in which Galileo died) there was born, near Grantham in Lincolnshire, a delicate babe who, during a vigorous life of nearly eighty-five years, extended the laws of motion of Galileo into realms undreamed of by their discoverer; and, triumphing over the immense mathematical difficulties, reduced all the phenomena of motion to manifestations of one simple law. Young Isaac Newton quickly threw off the weakness of infancy, and at an early age took advantage of a stormy night to measure the force of the gale by pacing out the lengths of leaps taken with the aid of the wind and in opposition to it. The story is interesting in that it indicates clearly the germ of the work—the application of calculation to actual experiments with the familiar but little understood forces of nature—by which he gained imperishable fame.

His boyhood was filled with the fashioning of mechanical contrivances such as kites and sundials, and his kindly nature is evidenced by the fact that he often made furniture for the dolls' houses belonging to his little girl playmates. But we must pass over these early years and picture him, as the French philosopher, Voltaire, bids us, sitting in his garden at Woolsthorpe where, driven from Cambridge by threat of the Great Plague, he was passing the summer of 1665 in speculating concerning the motions of the heavenly bodies. Suddenly his attention was arrested by the fall of an apple to the ground. To no one before had such a familiar event seemed in any way connected with the motion of the heavenly bodies; to speak plainly, we cannot be sure that it was the fall of an *apple* which set Newton's mind to work, though his friend Stukeley also said that Newton told him so. In any case we have Newton's own written testimony that it was in 1665 that he began to think of " gravity extending to

## THE DAWN OF UNIVERSAL MECHANICS 73

ye orb of the moon ",[1] and further, that before he set about the calculations he marked the significance of the fact that this force of gravity extended to the deepest mine shafts and the highest mountains—why not then, his bold imagination hinted, to the moon itself?

Before we attempt to trace in outline the use he made of the materials which others had provided, and to learn something of the new intellectual weapons which he himself forged, it is imperative that we clearly discern what it was that Newton was seeking. One thing it certainly was not, though in our childhood we may often have been told so, namely *why* an apple falls to the ground. Nor was it even why the moon moves as it does. His immediate purpose was in fact just that which in another connexion Galileo had proclaimed as the business of science, namely the demonstration that whatever may be the nature of the cause of an apple's fall to the ground, it is a cause of the *same nature* which determines the motion of the moon. Later, when he came to write his Preface to the First Edition of the *Principia*, he was to announce his belief that " the whole burden of philosophy [of nature] seems to consist in this—from the phenomena of motions to investigate the forces of nature, and then from these forces to demonstrate the other phenomena ".

We must further understand that the search itself had been made possible only by Galileo's far-reaching studies; for so long as with the Greeks it was assumed that the motion of the moon in a circle round the earth was the most natural thing imaginable, there was clearly no problem to solve. But unless the moon is governed by influences entirely different from those apparently determining the behaviour of terrestrial bodies, its motion is anything but " natural "; for it should move in a straight line. Now Galileo taught us to distinguish as *forces* those natural influences which cause a body to diverge from rectilinear motion; hence some force must be acting on the moon. It was the genius of Newton to assume that since we know of no influences acting on the moon and the planets other than those acting on terrestrial bodies, it is worth while trying to prove that these will suffice to

---

[1] It is significant for the history of thought that his youthful attempts to master Euclid were prompted by his inability to solve a problem in *astrology*!

account for the behaviour of the former. Later in life he expressed this ideal—the tacitly accepted ideal of all *modern* science—in the following words: " We are to admit no more causes of natural things than such as are both true and sufficient to explain their appearances " (the first of his so-called " Rules of reasoning in Philosophy " placed at the beginning of Book III of the *Principia*). To accomplish this task alone would have enshrined him among the immortals; actually it was but a small part of his ultimate achievement.

To follow Newton step by step on his arduous task would take us far too long even if it were in the compass of our humbler minds. Indeed modern research has made it seem very probable that Newton himself was far from " moving in a straight line " towards a clearly envisaged goal. He worked on the problem at least three times in his life, and on two of these occasions seems to have lost interest in it for years at a time. It will nevertheless be most profitable to appreciate what the central problems were; for in so doing we can gain an insight into the true nature of mathematical physics such as entirely eludes anyone who is content to hear merely the bald statement of the results.

Let us then examine the *facts* as they were known at the time of Newton's first attack on the problem of the moon's orbit. The moon was known to describe an orbit—approximately circular—about the earth as centre. This was not of course a direct fact of observation, but an interpretation of such facts, and of whose truth no competent astronomer doubted. It was also known that the radius of the moon's orbit was about sixty times the radius of the earth itself. How accurately this was known to Newton we cannot tell; indeed at one time it was believed that his information on this point was so defective as to be a bar to his solving the problem, but this now seems unlikely. The only other facts needed by Newton were the period of revolution of the moon—rather more than twenty-seven days—and the distance fallen in one second by a body near the earth's surface when unimpeded by the air—a distance of approximately sixteen feet.

These were the only *facts* used by Newton; to extract from them the relation which he believed to hold between the motion of the moon and that of an object near the earth's surface, he had

## THE DAWN OF UNIVERSAL MECHANICS

to propose a method of *measuring* forces. To do this he assumed the truth of Galileo's law of inertia (p. 68), and explicitly stated what was only *implicit* in Galileo's law of falling bodies. It is a strange thing that it never seems to have occurred to Galileo that if bodies left to themselves remain in a state of rest or uniform rectilinear motion, then it is natural to *measure* forces by the changes of motion which these produce in a given body: in other words, by the accelerations which the body undergoes. It is more than likely, as Professor Andrade has pointed out (*Nature*, 2nd July 1938, p. 19 *et seq.*) that he never quite freed himself from the confused ideas of motion which dominated all thought until his day. In taking this next step Newton completed the work which Galileo had begun.

Newton's statement of this Second Law of Motion was to the effect that forces are proportional to the changes of motion which they produce, whether that change is in the speed with which the body " covers the ground " or of the direction in which it is moving; for it is a simple fact of experience that a weight can be twirled on the end of a string at a steady " speed " only if a constant force be applied. The modern statement of the law is therefore that " forces are proportional to the accelerations they produce ", acceleration being defined as rate of change of *velocity*, which in turn means *speed in a given direction*.

Motion in a circle therefore may be regarded as equivalent to one in a straight line with which is compounded an acceleration towards the centre. Newton easily proved, as had his contemporary Huygens, that this acceleration is in fact proportional to $v^2/r$, where $v$ is the speed with which the body is moving in the circle, and $r$ the radius of the circle. Thus if we accept Galileo's and Newton's conception of force, it follows inevitably that any body moving in a circle must be acted upon by a force which accelerates it towards the centre of the circle. Now it is a fact that the moon is constantly moving round the earth in a path which is approximately a circle centred at the earth, therefore there must be a force acting on the moon accelerating it towards the earth; is it not the most natural, though not of course the only possible, assumption that the earth is itself the origin of the force? The only convincing manner of answering this question in the affirmative is to show that there is a precise *numerical* relationship between the supposed force of the earth on the moon and that on the apple. For numbers, like money, " talk "; that is, there is no disputing their testimony.

The next question therefore is the *form* of this supposed relationship. Here Newton made use of a hypothesis, probably first asserted by Bullialdus a few years before Newton's birth, and certainly familiar to Newton's contemporaries, Halley, Wren and Hooke, that the power of one body, say the sun, to attract another, say a planet, would, if it exists at all, be inversely proportional to the *square* of the distance between them. That this suggestion was put forward by Bullialdus, of whom probably most people—even some men of science—have never even heard, only goes to show how mistaken is the oft-expressed opinion that the great advances in scientific thought are the exclusive products of individual geniuses.

In order to show that his hypothesis was true, Newton had therefore to solve the comparatively easy problem (certain " simplifying assumptions " being made) of showing that if $f_m$, $f_a$, are respectively the forces acting on the moon and on the apple, which are respectively at distances $d_m$, $d_a$ from the earth, then $f_m/f_a = d_a^2/d_m^2$. The first simplifying assumption is that the

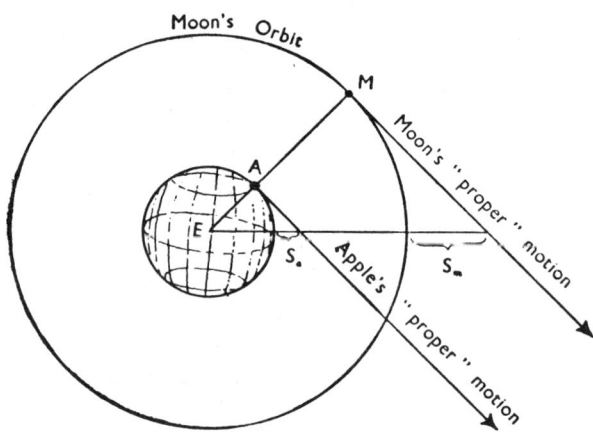

Fig. 14.

## THE DAWN OF UNIVERSAL MECHANICS

moon's orbit is circular,[1] the second is that the distance of each body "from the earth" is to be taken as its distance from the *centre* of the earth; for neither the moon nor the earth can be regarded as a geometrical point, that is as negligibly small compared with the distance between them. And the difficulty naturally becomes far greater when bodies near the earth's surface are being 'considered. This is probably one of the subsidiary problems which Newton later referred to as being "more difficult than I was aware of". He referred to it in a slightly different connexion in III, 8 of the *Principia*; and his inability to solve it at that time, rather than the possible inaccuracy of his information regarding the diameter of the earth already referred to above, probably accounted for his abandoning the formal proof of the whole project. Nevertheless it appears that he *did* make a provisional calculation, probably on such lines as these:

If $a$ be the acceleration of the moon towards the earth, then $a = v^2/r$ and by Galileo's law (p. 64), $s$, the distance fallen $= \frac{1}{2}at^2 = \frac{1}{2}a = \frac{1}{2}v^2/r$ in one second. Now $v = \dfrac{\text{length of orbit}}{\text{time taken to describe it}} = \dfrac{2\pi r}{t}$

$$\therefore s = \frac{1}{2}\frac{v^2}{r} = \frac{1}{2} \cdot \frac{4\pi^2 r^2}{t^2} \cdot \frac{1}{r} = \frac{2\pi^2 r}{t^2} = \frac{2 \times 3 \cdot 14^2 \times 240{,}000 \times 1760 \times 3}{(27 \cdot 3 \times 24 \times 60 \times 60)^2} \text{ feet,}$$

27·3 being the number of *days* taken by the moon to describe its orbit of 240,000 *miles* radius.

The above fraction (in which the quantities used are those of to-day approximated to three significant figures) works out at ·00449 ft. Now if $f_m/f_a = d_a^2/d_m^2$, then, since force is proportional to acceleration, $s_m/s_a = d_a^2/d_m^2$, where $s_m$, $s_a$, are the respective distances fallen in one second by the moon and the apple. But the moon is known to be almost exactly sixty times as far from the centre of the earth as is the apple, so $s_a/s_m = 60^2/1$, that is, the distance fallen by the apple *should* be 3,600 times the distance fallen by the moon, i.e. ·00449 × 3600 = 16·2 ft. The actual value found by experiment is 16·1 ft. Newton probably got 15·5 ft. for the former quantity, which would justify his remark that he "found them answer pretty nearly"

---

[1] In the *Principia III*, 35 scholium, Newton mentions that "Our countryman, Horrox, was the first who advanced the theory of the moon's moving in an ellipse about the earth placed at its lower focus."

While Newton, perhaps dissatisfied with the number of simplifying assumptions he had been compelled to adopt, took no further steps to make his views public, a similar problem was taking shape in the minds of many of his contemporaries both in England and on the Continent. This was whether it could be proved that, on the assumption that the force between the sun and a planet is inversely proportional to the square of the distance between them, the planet would necessarily move according to Kepler's First Law. By 1684 several members of the newly formed Royal Society seem to have proved independently that if it could be assumed, as was unfortunately only roughly true, that the orbits of the planets are circles, then the inverse square hypothesis is implicit in Kepler's *Third* Law. The proof is delightfully simple; for if $r_1$, $r_2$, be the distances from the sun of two planets whose periodic times are $t_1$, $t_2$, then since

$$f(\text{the force}) \propto v^2/r, \frac{f_1}{f_2} = \frac{v_1^2/r_1}{v_2^2/r_2} = \left(\frac{4\pi^2 r_1^2}{t_1^2} \cdot \frac{1}{r_1}\right) \div \left(\frac{4\pi^2 r_2^2}{t_2^2} \cdot \frac{1}{r_2}\right) = \frac{r_1}{t_1^2} \div \frac{r_2}{t_2^2} = \frac{r_1 t_2^2}{r_2 t_1^2}$$

But by hypothesis $\frac{f_1}{f_2} = \frac{r_2^2}{r_1^2}$ $\therefore \frac{r_1 t_2^2}{r_2 t_1^2} = \frac{r_2^2}{r_1^2}$, whence $\frac{r_1^3}{r_2^3} = \frac{t_1^2}{t_2^2}$

But if $r$, the radius of the assumed circular orbit be regarded as an approximation to the "mean distance" in the elliptic orbit, the expression is Kepler's "Harmonic" Law (p. 52). Note that in this case the actual diameters of the bodies concerned are negligible compared with their distances apart, so Newton's great difficulty does not here arise.

Of course no one pretended that this result *proved* the truth of the inverse square hypothesis, but the agreement was too striking to be merely a "fluke". One of these brilliant men, Robert Hooke, who in the intervals of assisting in the rebuilding of London after the fire, found time to invent carriage springs, to make one of the first efficient compound microscopes (see p. 358) and to come very near to discovering oxygen (p. 179), claimed that he could deduce a similar relationship for elliptic orbits, but on being asked to deliver it for inspection, temporised in a manner which suggested that the task was in reality beyond his mathematical powers. The events following on this unworthy display of duplicity were in their way as dramatic as anything in the

## THE DAWN OF UNIVERSAL MECHANICS

history of human endeavour; for had it not been for the resolution of another of this group, Halley, the greatest single discovery ever made might never have been known to any but its discoverer. Halley however determined to consult Newton at Cambridge. So in August 1684 he called on Newton and, without mentioning the tentative views of Hooke and the others, bluntly asked him what would be the path of a planet on the assumption that the force acting on it decreased in the proportion of the square of its distance from the sun. To his astonishment and delight Newton replied that the planet would describe an ellipse. His delight was somewhat tempered when Newton failed to find the papers on which he had, so he said, worked out the proof. However in November the papers arrived, giving two proofs of the theorem that, in his own words, " by a centrifugal force reciprocally as the square of the distance, a planet must revolve in an ellipsis about the centre of the force placed in the lower umbilicus [focus] of the ellipsis and with a radius drawn to that centre describe areas proportional to the times " Newton later claimed to have obtained a proof in 1676, but had apparently lost interest in the subject! Now he set to work again and wrote a small treatise, *De Motu Corporum*, which Halley saw on a second visit to Cambridge. Fortunately for the world Halley realised that here was a " scoop " of the first order and, overcoming Newton's disinclination for publicity, persuaded him to send it to the Royal Society for publication.

The next year, under the stimulus of Halley's encouragement, Newton successfully proved that a spherical body such as the earth exercises its force of gravitation, whether on the moon or on an apple at its surface, in a manner which is mathematically equivalent to the force which would be exerted by the same body regarded as a mathematical point situated at its centre. To effect this wonderful result Newton had been faced by the task of calculating the combined effects of all the particles of matter of which the earth is composed. Such a task appeared impossible of fulfilment; in fact we may say that it *was* impossible until Newton himself had invented a new branch of mathematics— akin to what we now call the " integral calculus "—by which such problems involving the effect of an apparently infinite number of infinitesimal (that is "vanishingly" small) particles could be tackled. This matter will be taken up in the next chapter; meanwhile we may note that he had had recourse to the counterpart of this method when a year or two previously he had made, though did not disclose, the discovery of the theorem from which the planetary orbits could be deduced as special cases. This was to the effect that any body projected with a finite velocity and subject to a force directed from a fixed point and varying inversely as the square of the distance, must describe a path which may be any one of the plane sections of a cone. For a wide range of velocities it will be an ellipse. But just as (see Fig. 15) a circle is geometrically one out of a particular family of ellipses, so, given a definite central force, there will be one initial velocity

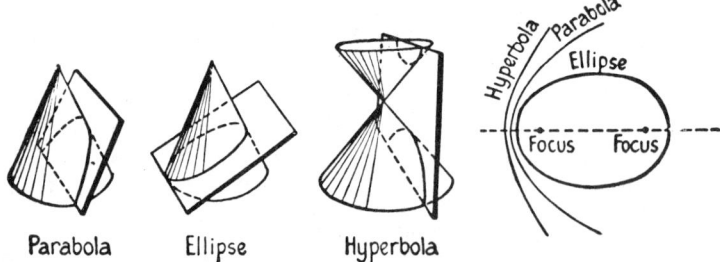

FIG. 15. Curves obtained by plane sections of a right cone. (The circle is obtained as a special case of the ellipse).

for which the body will move always in a circle. Which explains why the planets, moving in accordance with the inverse square law, *might* have circular orbits and still obey Kepler's Harmonic Law, as has been shown already.

The Royal Society was delighted with what it saw of Newton's work, and at once ordered it to be completed and printed at the Society's expense. When the extent of the undertaking was realised however there seems to have been a cooling-off on the part of the well-to-do who largely composed the Society's membership. The exact reason for the delay will perhaps never be known, but the fact remains that Halley, a man of modest fortune and with a young family to bring up, undertook the expense and much of the labour of seeing the work through the press. The Royal Society took the credit for the publication, but posterity

THE DAWN OF UNIVERSAL MECHANICS 81

knows that it was to the vision and self-sacrifice of Halley that it owes the final publication in 1687 of the greatest single monument of human learning—*Philosophiae Naturalis Principia Mathematica.*

SOURCES FOR CHAPTER VIII

The most recent short account of Newton's character and work is J. W. N. Sullivan's *Isaac Newton* (London, 1938).

The *Principia* may be consulted in several English translations, of which the latest is a revision by F. Cajori of Motte's Translation of 1729. This work (Univ. California Press, 2nd printing, 1946) contains a valuable historical commentary by the editor.

## CHAPTER IX
## THE MATHEMATICS OF MOVING BODIES

IN his studies of moving bodies Galileo assumed that the acceleration is *uniform*; by so doing he was able to " pretend " that an accelerated body moves with a uniform velocity equal to the mean of the greatest and least velocities to which it is subject during the time under consideration. Now the whole essence of Newton's law of gravitation is in the hypothesis that the *force* acting on a planet or other body varies according to the latter's distance from another body, hence the acceleration must also vary, so the method adopted by Galileo is wholly inapplicable. It is hardly likely that Newton could have overcome the difficulty of calculating the path of a body moving with a continuously varying acceleration had he not previously invented the " fluxional calculus ". Since in our country it is still possible to be " educated " without ever having been given even a glimpse of this most potent weapon of the mathematical physicist, it may not be amiss to turn aside for a little while to consider in a simple manner the growth of the principles which underlie it.

The method of determining the areas of plane figures bounded by curved lines by inscribing within them regular *polygons* with an " infinite " number of sides is attributed by Archimedes to Demokritos. But it is clear that Demokritos could not " prove " his method, except by assuming that mathematical magnitudes, unlike " things ", are divisible *in infinitum,* thus raising the bogey of the infinitesimal exposed by Zeno. Archimedes however asserted that two of the problems of this kind " solved " by Demokritos were " proved " by Eudoxos by the use of a lemma similar to the one already quoted (p. 26), namely that of " unequal magnitudes the greater exceeds the less by a magnitude such as is capable, if added continually to itself, of exceeding any magnitude of those which are comparable to one another ". The

## THE MATHEMATICS OF MOVING BODIES

method of exhaustion, thus validated by Eudoxos, was used extensively by the Greeks for comparing the areas and volumes of unknown curvilinear figures with those of rectilinear figures to which they could be approximated *to any degree of precision* as in the familiar case of the circle and inscribed polygon.

But Archimedes himself went further. The books of Archimedes were collected from various Eastern sources in the ninth century and were deposited in the Papal Library in 1266. An almost literal translation of them was made by William of Moerbeke in 1269. Finally with the revival of the interest in mechanics in the sixteenth century they were re-translated, made public and widely studied. All the *discoveries* which we attribute to Archimedes derive from this source; but it was a puzzle to the mathematicians of the seventeenth century, for example Wallis, how Archimedes could have arrived at the *results* for which he gave formal proof. It was clear that he must have used some analytical device of greater power than any which had been put on record by the Greeks. In 1906 the historian Heiberg discovered a MS. in Constantinople in which Archimedes' method, only casually hinted at in the previously discovered works, was set out in sufficient detail for a complete reconstruction to be made. With this he had succeeded in determining the areas and volumes of curvilinear figures, such as a segment of a parabola, in terms of other figures—triangles and cylinders—whose area or volume were known.

The method of exhaustion, powerful as it was, was purely geometrical, in the sense that it consisted in showing that by increasing without limit the number of sides of a rectilinear figure it could be made to differ from a curvilinear figure by as small a magnitude as we please; the area or volume of the latter being then taken as the sum of the " infinite " number of equal parts (not necessarily of one kind—they might consist of both equal pyramids and equal prisms, for instance) contained in the former figure. Archimedes' method broke entirely new ground. But superb as it is, its historical interest is limited to the fact that it was born out of due season—it was entirely without influence on the history of science. Since the appreciation of its true nature will be much facilitated by a knowledge of the progress of mathematical thought during the seventeenth and eighteenth centuries, consideration of it is deferred to an appendix of the present chapter.

In the sixteenth century the method of exhaustion was revived by Cavalieri in the form of a device whereby an area was determined by summing an infinite number of " infinitesimal " strips into which the surface was supposed to be divided. The British mathematicians, Harriot and Wallis, translated this, *geometrical* method into *algebraic* symbols, thereby very narrowly missing the invention of one form of the fluxional calculus. Newton is known to have studied Wallis's book, but in a letter to a friend he confesses that " I had the hint of this method [*his own method of fluxions*] from Fermat's way of drawing tangents and by applying it to abstract equations, directly and invertedly, I made it general." Upon this confession we may build a simple account of the whole matter.

Pierre Fermat has rightly been called the " Prince of Amateurs ", for in the comparatively short leisure hours left after the carrying out of the duties of an important public office, he found time to lay the foundations of many branches of modern mathematics. His problem, like that of Archimedes, had nothing to do with moving bodies; it was simply to draw the tangent to any continuous curve at any point on it. The idea underlying it is familiar to anyone having an elementary knowledge of the geometry of the circle: it simply consists in allowing a secant to revolve in such a way that one of the points of intersection slides down the curve until it finally coincides with the other point; the secant is now a tangent. But Fermat went further than this, and actually translated the final state of affairs into algebraical symbols. How can this be possible? Curiously enough *this* part of the problem was the subject of another discovery made by Fermat's contemporary, Réné Descartes. Fermat seems to have made this discovery quite independently of Descartes, but since the latter's is the more systematic exposition, and was certainly the one which influenced Newton, we shall leave Fermat and consider for a moment the geometry of Descartes.

The " analytical " geometry of Descartes, the " key " to which seems to have come to him in a dream, also had its germ

THE MATHEMATICS OF MOVING BODIES 85

in Greek mathematics; but it differs from the geometry of Euclid and the other " orthodox " Greeks in that the former's figures represent always some *numerical* relationship between two quantities, one of which being wholly dependent on the other is, following Leibniz, called a " function " of it. Descartes did this by constructing curves out of a series of points whose positions were determined by measuring their perpendicular distances from a fixed straight line, the distances of the feet of these perpendiculars being measured from one end of the straight line. This use of what we call an " axis " had already been introduced by the Greek mathematician Apollonius to discover the relations between the various sections of a cone; the advance made by Descartes was to link up this " numerical geometry " of the ancients with the newly-developed science of algebra. This he did by noting that a problem may have not one but an infinite number of solutions; and that when this is so, the conditions which each of these solutions must fulfil may be concisely summarised in an equation containing *more than one unknown*. These " unknowns " he was the first to represent by the letters x, y, z, . . ., using a, b, c, . . . for quantities which were supposed constant in value in any particular problem. The method may be illustrated by a very simple case.

Required to find an equation to represent all the possible points equidistant from a given point, this distance having a given value. To solve such a problem Descartes would have taken a fixed point, $O$ (Fig. 16) and a straight line $OA$. A number of

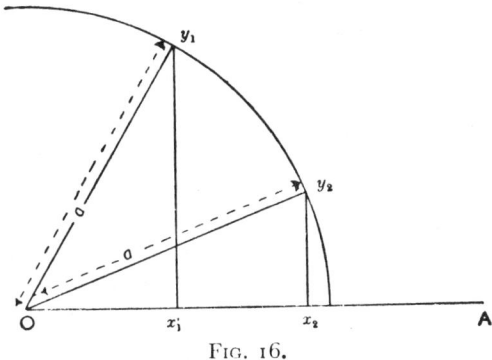

FIG. 16.

points $y_1$, $y_2$, ... equidistant from $O$ would have been chosen, the letters $y_1$, $y_2$, ... representing the actual distances (at a constant angle, usually a right angle) from the points $x_1$, $x_2$, ..., in whose case the letters $x_1$, $x_2$, ... represent the distances from $O$ along $OA$. It will be noticed that there is no Y-axis, such as is always used now; this seems to have been introduced nearly a century later.

Now by Pythagoras' well-known theorem $(Ox_1)^2 + (x_1y_1)^2 = a^2 = (Ox_2)^2 + (x_2y_2)^2$, i.e. $x^2 + y^2 = a^2$. This means that whatever value we give to $x$ (within limits set by the value of $a$) we can find a corresponding value for $y$. This relationship once established, we may dispense with the figure; the geometrical truth is crystallised in the algebraic equation.

The astonishing power of this method was quickly made evident by Descartes' general solution of a famous problem propounded by Pappus (fourth century A.D.) which was solved by the Greeks for one case only. This point deserves more than mere passing notice; it illustrates one of the most fundamental characters of the progress of mathematical science. Progress in the power of mathematical method is largely a matter of increasing the *generality* of its propositions. At an early stage of the development of thought man must have been struck by the similarity of a group of four haddock to one comprising two mackerel and two whiting; each contained the same number of *fish*. Centuries, or even millennia, may have passed before he realised that four haddock had one feature in common with four stones. The recognition of this common attribute—the " number " four— was the birth of mathematics; for the recognition of numbers is mathematics at its lowest stage of generality. The claim of the Greeks to be the founders of mathematics (in so far as it is justified, see p. 21) rests on their recognition of the *form* common to a square field and a square tile—that is the possibility of discovering properties of " squares " irrespective of the material of which they may be made. But as Whitehead points out in his beautiful little book, *An Introduction to Mathematics*, Descartes' geometry was an immensely more powerful weapon than Greek geometry, not only because it applied to the latter algebraic notation, but also because it dealt with more *general* ideas of space.

## THE MATHEMATICS OF MOVING BODIES

For Euclid, as for all the Greeks, a straight line was generally (though not always) conceived as having a definite length. In Descartes' geometry on the other hand, a straight line represented by a particular equation has no such limits. Thus the equation $y = 2x + 3$, which as shown in Fig. 17 is that of a straight line, represents the *whole* straight line having a particular direction with respect to the axes; it is drawn of limited length only because time (and paper!) are limited. Thus we find again and again that the famous " problems " of the Greeks are special cases of much wider relationships.

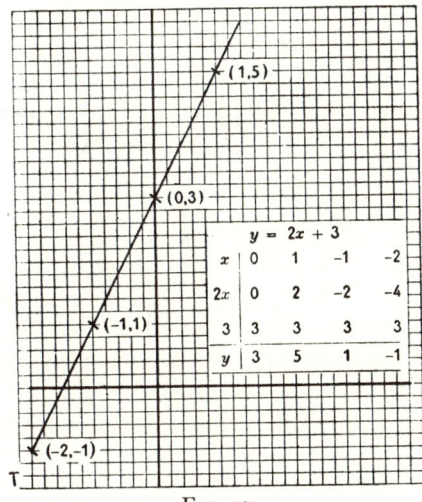

Fig. 17.

We may now observe one other feature of Cartesian geometry which was characteristic of the new outlook on nature, and which we may believe was the one which gave to Newton the clue to his " fluxional " calculus. Descartes, besides being an able mathematician was, with Bacon, largely responsible for the new modes of thought which began to characterise men's views on nature during the seventeenth century. One great difference between the " modern " and the " classical " outlook was the growth of the belief that rest is only a special case of *motion*. Descartes was well aware of Galileo's successful foundation of the science of dynamics on the assumption that not motion itself but *change* of motion, either in magnitude or direction, called for explanation. In this way geometry ceased to be merely the comparison of the innate properties of given shapes, but took on a *dynamical* aspect. A circle was no longer merely a " plane figure bounded by a line, all of whose points are equidistant from a given point called the centre ", but came to be regarded as a " plane figure bounded by a line *traced out by the motion of a point moving in such a way as to be always at a constant distance from a given point* ". In this light geometrical figures are " generated " by points moving according to definite laws; and, as we have seen, it was the business of analytical geometry to frame all such laws in terms of general numerical relationships with respect to two fixed axes. We need not be surprised to learn that Archimedes showed himself to be ahead of his time in defining a special spiral by the motion of a point from a fixed point along a straight line which is revolving at a uniform rate about the fixed point, like the hand of a clock.

We know that Newton had read Descartes' *Geometry* by 1664, and that according to his own statement he hit upon the " method of fluxions " in 1665 (though he made no communication of the discovery to anyone until 1669, and even then declined to publish it), so we may surmise that the reading of the *Geometry* was largely responsible for the manner in which he attacked the problem of calculating the effect of continuously changing influences, such as gravitational forces. This point is of interest, because several other mathematicians, notably Kepler, Fermat, Cavalieri, Descartes, Pascal, and Newton's own teacher Barrow, had each in the solution of specific *geometrical* problems used a method similar to that which Newton devised; but Newton alone (and, as recent research seems to prove, his Scottish contemporary, James Gregory) seems to have realised that the same method may be used with equal success for the drawing of a tangent to a curve as for calculating the varying velocity of a planet: in each case, as will be seen hereafter, we are dealing with a limiting ratio. Thus in his article on the *Quadrature of Curves* he says: " I consider mathematical quantities in this place not as consisting of very small parts, but as described by a continued

## THE MATHEMATICS OF MOVING BODIES 89

motion. Lines are described and thereby generated, not by the opposition of parts, but by the continued motion of points; superficies by the motion of lines; solids by the motion of superficies; angles by the rotation of the sides; portions of time by continual flux: and so on in other quantities. These geneses really take place in the nature of things, and are daily seen in the motion of bodies. Therefore considering that quantities, which increase in equal times and by increasing are generated, become greater or less according to the greater or less velocity with which they increase or are generated, I sought a method of determining quantities from the velocities of the motions or increments with which they are generated; and calling these velocities of the motions or increments *fluxions* and the generated quantities *fluents*."

Before considering a simplified example of Newton's use of the method, let us clarify our ideas as to the exact nature of the difficulties involved. In Fig. 19 are plotted as ordinates (distances on the Y-axis) the distances which a train has travelled from the beginning of its journey at known instants of time. If we can assume (note that the validity of this assumption is really part of our problem) that there have not been any sharp changes in its mode of motion in any given minute, then we may join up the points by a smooth curve (the *possibility* of doing this is an indication, though not a guarantee, of the regularity of the motion). At the point marked on the curve the train entered a certain station; the railway company is anxious to know at what speed it was then travelling. How is this knowledge to be gained? Speed means the rate of travel, that is, the distance the train would have travelled in unit time if its speed had thereafter remained constant. We could get a rough idea of this by dividing the length of the platform by the time taken for the head of the locomotive to pass it. But inspection of the curve shows that the speed was altering rapidly, so this could give us only the *average* speed—a result much lower than the actual speed at the moment of entering the station. For practical purposes the only way of *measurement* would be to determine the distance traversed in as short a time as could be accurately measured; but such a result would still be too low. Now if it had been possible (as is sometimes done in experiments) to remove all cause of *change* in the

Fig. 18.

speed as the engine reached the station, we could then have measured the ratio of the subsequent distance travelled (in any convenient lapse of time) to that lapse of time, for the speed now being uniform, the result would be the same whatever the actual lapse of time chosen. This way of looking at the problem suggests one way of solving it; for to maintain the speed constant is in effect to keep the gradient of the curve constant, that is to make the point, which we may imagine to be tracing out the curve, continue to move beyond $T$, not along the existing curve, but in the same direction, $TS$, as it was moving when it arrived at $T$.

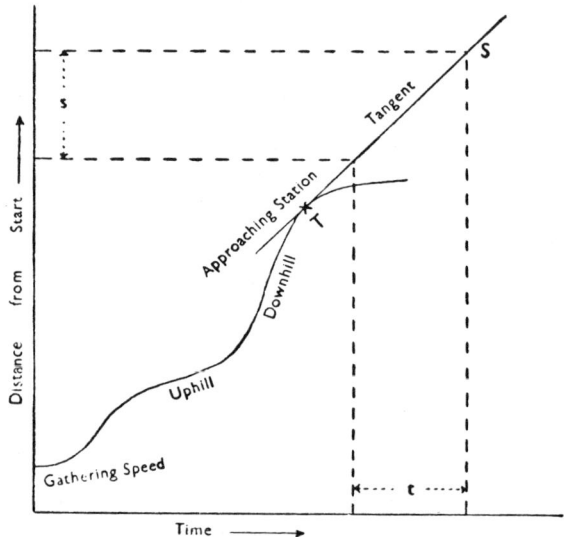

Fig. 19. Space-Time curve corresponding to the motion of the train shown in Fig. 18.

## THE MATHEMATICS OF MOVING BODIES

Such a straight line $TS$ is of course the "tangent" to the curve, since it touches the latter at only one point. Now the gradient of the straight line $TS$ represents a speed of $s/t$, both of which factors can be read off, so we have found a means of obtaining a much more accurate estimate of "speed at an instant" than could be obtained by any direct measurement.

But Newton went further. For suppose our train is replaced by a body such as a freely falling stone which is moving according to some known law—in this case represented by the equation $s = \frac{1}{2}gt^2$; by applying *algebraical* arguments developed by many of his contemporaries Newton was able to show that we can find the velocity at any instant of time, not only without drawing a tangent, but even without plotting a curve! So Fermat's spare time "fun with figures" turns out to be on the high road to the understanding of the motion of the heavenly bodies!

Newton's methods of working out the problem are rather difficult to follow and have for most purposes been superseded by those of his great German contemporary Leibniz, who in the spare time left after cataloguing ducal libraries, writing genealogical histories of shady "noble" families, composing a rounded philosophy of human understanding and the relation of all physical existents to God and the Universe, and conducting a vast diplomatic correspondence in many countries, shared with Newton the honour of the invention of the calculus.

Leibniz argued that to find the gradient of any curve we have in effect to determine the value of the fraction $dy/dx$, where $dy$[1] stands for the "infinitesimal" *increment* in the value of $y$ consequent on an "infinitesimal" increment of $x$—for it must be remembered that this calculus applies only to such cases where $y$ is a "function" of the independent variable $x$. In the case mentioned above—$s = \frac{1}{2}gt^2$—we can find the velocity at an instant by evaluating the fraction $ds/dt$ (here "$y$" $=s$, and "$x$" $=t$).

Now if $s$ increases by an "infinitesimal" amount $ds$ (called by Leibniz a "differential," hence the name "differential calculus" now used) consequent on an "infinitesimal" increase

---

[1] Note that $dy$, $dx$, etc., are convenient symbols for "a vanishingly (but not quite vanished) small increase in the magnitude of $y$, $x$, etc." They do *not* mean $d$ multiplied by $x$.

in $t$ to $t+dt$, wherever we had $t$ in the original function we must write $t+dt$ in the new one; so we get:

$$s+ds = \tfrac{1}{2}g\,(t+dt)^2 = \tfrac{1}{2}g\,[t^2+2t\,.\,dt+(dt)^2]$$
$$= \tfrac{1}{2}gt^2 + gt\,.\,dt + \tfrac{1}{2}g\,.\,(dt)^2$$
$$\therefore ds = gt\,.\,dt + \tfrac{1}{2}g(dt)^2,\text{ since }s=\tfrac{1}{2}gt^2$$

Then dividing through by $dt$,

$$ds/dt = gt + \tfrac{1}{2}g\,(dt).$$

But, argued Leibniz, $dt$ is an infinitesimal increment of time; therefore *in comparison with* $gt$, $\tfrac{1}{2}g\,(dt)$ can be neglected. So the final result is that $ds/dt$ (i.e. the velocity at time $t$) $=gt$—which is the very assumption from which Galileo deduced the expression we started with, namely $s=\tfrac{1}{2}gt^2$!

It is to be hoped that the reader's critical faculties have been sufficiently roused by reading thus far to make him a little more than merely suspicious of the above procedure—as was Bishop Berkeley at the time that the new analysis was launched into the world. In a brilliantly reasoned book—*The Analyst*—he brought many criticisms to bear on the new method. Some being based on his own views as to the nature of time and space, which were themselves open to question, would not even now find universal acceptance. There was one however which no amount of hedging on the part of the supporters of the method was ever able to remove: this was to the effect that in employing the " infinitesimals " $dx$, $dy$, etc. (under the name of " differentials ") Leibniz was running with the hare and hunting with the hounds. A change in the magnitude of the variable, however small, is a definite quantity or it is nothing at all; but it can't be both. Now anyone who has had but a moderate experience of algebra knows that the division of an expression by a term gives a valid result only if the latter is *not equal to zero* (e.g. $ax/x \neq a$ if $x=0$, so $dt$ is certainly *not* equal to zero in the above example when it is used as a divisor. On the other hand we have no right merely to leave out multiples of $dx$ unless it *is* zero, otherwise the result, though perhaps good enough for practical purposes (which being based on measurement are for that reason only approximate) can never be regarded as mathematically rigid. Newton had recognised the trap into which Leibniz undoubtedly fell, and announced

## THE MATHEMATICS OF MOVING BODIES 93

that "in mathematics the minutest errors are not to be neglected"; perhaps for this reason he *defined* his " fluxions " (i.e. rates of change) as continuous changes (see above, p. 89). But it is doubtful whether in making his *calculations* he really avoided the dilemma of the " infinitesimals ".

The reader, if he happens not to have very much experience of mathematics, will naturally feel some impatience at being thus " led up the garden path " to admire the noble blooms brought to fruition by analysis, only to be told that analysis is just as likely to promote the growth of weeds. But strange though it may seem, not only have the new discoveries made possible by analysis been found consistent with observation, but, which is more important, it provides identical solutions of problems already solved by the older, more rigorous, methods. This apparent contradiction was cleared up about the middle of last century when the mathematician, Weierstrass, showed that by a sufficiently thorough examination of the true nature of limits and functions, it is possible to justify the methods of analysis. Without going into technical detail it may be pointed out that in problems concerned with areas, the use of the objectionable (because ambiguous) " infinitesimal " may be avoided by *proper definition*. Leibniz made the mistake of trying to define the infinitesimal (or " differential ", as he called it) and then using it to calculate ratios of infinitesimals (e.g. $ds/dt$), and also areas composed of serried ranks of strips, the area of each of which was the product of the ordinate at the point and the differential, i.e. $y.dt$. Later research showed that by *defining* an area as the limit of the sum of strips of *any size*, provided the greatest tends to zero when the number increases without limit, the " differential " sign $d$ becomes an operator *whose nature is determined by the definition of the area*. An " operator " means an instruction to do something (e.g. the addition operator $+$); in this case it means "form a difference between two values of $x$ such that the number of these differences between two finite values has *no finite limit* ". We are then no longer concerned with the *size* of $dx$; but we can use it to find rates of change, for example, in the sure knowledge that though neither $dx$ nor $dy$ has a determinate magnitude, their *ratio* has. In the case of areas the operation of summing the " strips " was called by Leibniz "integration" and he introduced the sign ($\int$)—really a stylised capital S for "sum"—which we still use. Finally it is necessary to point out that the integration operator is not restricted to the evaluation of areas as such. In general, the integral is the defined equivalent of the limit of a sum of functionally determined units.

All this has a moral to it, namely that it is never wise to deny to men of genius the use of any methods to which their intuition may guide them; they can usually be relied upon to do the right thing, though through the unfamiliarity of the procedure they may give the wrong reason for doing so! It is certain, as Professor E. T. Bell has pointed out, that had the pioneers of the seventeenth century stopped every time that they were unable to give a rigorous proof of the validity of the process, the calculus would never have got itself invented.

That we may get some hint of how this mathematical revolution rendered possible the revolution in man's attitude to nature, we may return to take a glimpse at Newton's attack on the problem of showing that the paths of the planets about the sun are regulated by the same law as that which governs the fall of an apple to the ground.

We have already seen how the velocity of the moon in its path round the earth follows precisely from the assumption that the magnitude of the force which draws all terrestrial bodies together decreases directly as the square of the distance between the gravitating bodies. We have also seen that, *granted* the same form of law to operate between the sun and the planets, Kepler's Third Law would follow automatically—is indeed an inevitable consequence of such a law. By applying the method of "fluxions" Newton was able to show that this result is equally true for an elliptic orbit as for a circular one. We shall now see in a much simplified form how he gave precision to Kepler's pious belief that there *must* be some physical influence subsisting between the sun and the planets.

If we turn back a moment to Fig. 8 on page 51 we shall see that Kepler's Second Law is a concise expression of the fact that each planet moves at its greatest speed when it is nearest to the sun, and at its slowest when furthest away. This *fact* supported Kepler's faith in the attractive power of the sun; but in Propositions I and II of Book II of the *Principia* Newton showed that if bodies are impelled to depart from rectilinear motion by the action of a force, *then they will* move according to this law of Kepler. Once again, then, Kepler's laws are found to be merely concise " regulations " according to which the planets will move if acted on by a force of gravitation exerted between themselves and the sun. We shall give an outline of Newton's proof, as it is an excellent example of how the new laws of motion and the new mathematical method were applied by him to the task of welding the whole solar system into a mechanism governed by one controlling law.

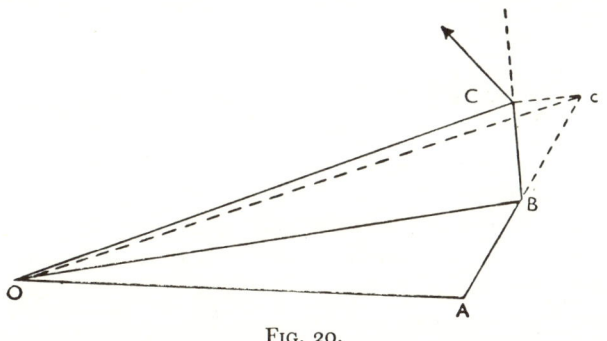

FIG. 20.

Suppose a body $A$ (Fig. 20) to be moving freely (under the action of no force) Galileo's law of inertia (codified as " Newton's First Law of Motion ") tells us that it will move over equal distances in a straight line in equal times; hence if the ends of these equal segments of the straight line are joined to any external point $O$ the triangles so formed, having equal bases and the same altitude, will be equal in area. Now suppose that when the body reaches $B$, having traversed the first segment $AB$, it is suddenly acted on by an impulsive force (that is, a large force acting for a very short time, such as a hammer blow) directed towards $O$, in consequence of which it is driven out of the path $Bc$ into a new path $BC$ such that it reaches $C$ in the same time as it would have taken to reach $c$. Newton had already shown (in a corollary to the laws of motion) that a body moving with a given velocity in a given direction *will be displaced the same distance in that direction,* whether acted on by a second force or not; hence it follows that $Cc$ is parallel to $Ob$, so the triangles $OBC$, $OBc$, on the same base and between the same parallels, are equal in area. If on arrival at $C$ it gets another blow by a force directed to $O$, the whole process will be repeated; from which it follows that lines drawn from $O$ to points $B, C, D$, etc., at which the body is found after equal lapses of time, will form triangles equal in area. This result has a strong resemblance to Kepler's Second Law, except for the fact that planets move in curves and not polygons. Newton however, following the method first hinted at by Archimedes, argued in a " proof ", which is too long to give here, but is not very difficult (nor to a modern mathematician entirely convincing!) that if the times between the successive impulses be reduced without limit, the segments $BC$, $CD$, etc., will be similarly reduced, and ultimately the polygon of motions caused by a succession of impulsive forces will be indistinguishable from a curvilinear motion under the continuous action of a force, continuously altering in direction, but always directed *towards* $O$. From this result the converse follows immediately: that any body (planet) which moves in *any* curve such that the line joining the body to any point on the curve sweeps out equal areas in equal times, must be acted upon by a force directed towards that point (the sun).

We may now fairly ask why, of all the mathematicians who used methods based on " infinitesimals " or their ratios, Newton and Leibniz are alone honoured as the inventors of the " calculus ". To this question Newton himself gave the answer in the passage already quoted (p. 84). Many before him had indeed used an " infinitesimal " method for drawing tangents and another for computing areas. Newton not only used it for both these purposes, but *recognised that there is in fact only one method* which may be applied either " directly or invertedly ".[1] Important as was this step, of far greater importance was that he applied it " to abstract equations " thereby " making it general ". To him alone, and probably to Leibniz, came the full realisation that since areas and curves are but the *geometrical pictures* of the functional relations known as equations, the operations of the calculus can be applied

[1] It is interesting to note that the partial anticipations of the integral calculus are far more numerous than those of the differential. Archimedes alone anticipated Fermat when he showed how to draw tangents to the Archimedean spiral (p. 88).

## THE MATHEMATICS OF MOVING BODIES 97

to the *functions* independently of any geometrical significance they may or may not have. This, combined with Newton's youthful discovery of the possibility of expanding a binomial power function into a series of separate terms, opened up such vistas of mathematical exploration as surpassed anything since the first generalisation of geometrical forms by the Greeks. In the words of Professor Turnbull, " Gregory was right when he stated in the Preface to the *Geometriae Pars Universalis* that the true division of mathematics was not into geometry and arithmetic but into the universal and the particular. Through the genius of Newton and his contemporaries this passage from particular problems to universal methods was once for all effected."

### APPENDIX TO CHAPTER IX[1]

*An example of Archimedes' " Method ".*

$CA$ is any segment of a parabola, with the tangent drawn at $C$. $BD$ is the diameter. $CK$ is produced to $H$ making $KH = KC$. $P$ is any point on the segment. OPNM and AKF are parallel to BD.
Assuming (from properties of parabola) $MO : OP = CA : AO$
$= CK : KN$
$= HK : KN$
and since in any parabola $EB = BD$, $MN = NO$ and $FK = KA$
∴ $MO : KN = PO : HK = TG : HK$.

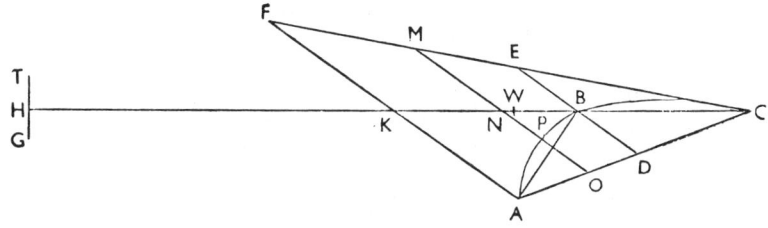

Fig. 21.

Thus if $TG$ ($=PO$) is placed at $H$, the end of a " balance beam ", it will " balance " $MO$ which is placed at $N$. And similarly for all the " lines " of which the segment and the triangle $AFC$ may be con-

[1] This proof and Fig. 21 are adapted from the full account given by Sir Thomas Heath in his *History of Greek Mathematics*, vol. II, p. 29, by permission of the Clarendon Press, Oxford.

sidered to be composed. But the number of these "lines", though "infinite" must be the same in both, ∴ △ placed with its centre of gravity at $W$ balances the segment with its $C$ of $G$ at $H$. Now $CW = 2WK$

∴ $\dfrac{\text{segment } ABC}{\triangle AFC} = \dfrac{WK}{KH} = \dfrac{1}{3}$

∴ segment $ABC = \tfrac{1}{3} \triangle AFC = \tfrac{4}{3} \triangle ABC$

The "mechanical" assumptions made here are (1) that the "weights" of plane figures are proportional to the number of "lines" of which they are composed; (2) the summation of the moments of the "particles" composing the figure about the centre of gravity is zero.

## SOURCES FOR CHAPTER IX

H. W. Turnbull, *The Mathematical Discoveries of Sir Isaac Newton* (London, 1945).

Turnbull, Heath, Bell, op. cit., p. 43.

J. W. N. Sullivan, *Mathematics in Europe*, and *Isaac Newton*.

R. Courant and H. E. Robbins, *What is Mathematics?* (2nd imp., London, 1943).

F. Cajori, *A History of the Conception of Limits and Fluxions* (Chicago, 1919).

## APPENDIX TO CHAPTER IX

Professor Max Born in his *Natural Philosophy of Cause and Chance*—published unfortunately only after these lines were in proof—shows how the "law" of gravitation can be derived by purely mathematical analysis of the facts summarised in Kepler's purely descriptive laws. This is of special interest to us in that it shows how, in an age of greater mathematical sophistication, the "idea" of gravitation might have been, as it were, "spontaneously generated". Professor Born draws another inference of ultimately much greater interest, which is unfortunately too complex to discuss here.

## CHAPTER X
# THE NEWTONIAN REVOLUTION

THE conclusion of the last chapter has brought us to a point where the three laws of Kepler are seen to be necessary consequences of the hypothesis[1] that there exists a force between the sun and the planets which is at all times inversely proportional to the squares of the distances between them. We have reached this vantage point only by following a long and devious route; it will be well to review the position before proceeding further. In the first place it has been shown that whatever force it is which bends the planets from rectilinear paths and makes their *radii vectores* from the sun sweep out equal areas in equal times, that force is always directed towards the sun (Kepler's Second Law). Next, that if it be assumed that the forces acting on each of two different planets be inversely as the squares of the latter's respective mean distances from the sun, then those planets will describe their orbits at rates which are expressed in Kepler's Third Law. And finally that any body, to which has been imparted a definite constant velocity, must, on becoming subject to a force which varies inversely as the square of the distance of the body from the point towards which the force is directed, thenceforth move in a path which is one of the plane sections of a cone (Kepler's First Law being a special case).

We must now take note of an important implication of the law of gravitation. We have seen that the moon's path round the earth can be accounted for on the assumption that the earth draws the moon towards itself with exactly the same force as it does an apple on its own surface, only providing that the magni-

[1] The hypothesis is thus verified, but it does not thereby become a " law ". The term " law " strictly speaking refers only to those relationships which may be *directly observed*. Newton's *hypothesis* became in his hands a fully verified *theory*; but it was not a *law* until 1797 when Cavendish showed with the newly invented torsion balance that two pieces of matter do actually gravitate towards one another with a force inversely proportional to the square of the distance between them.

tude of the force falls off in the proportion of the square of the distance from the centre of the earth. The moon being about sixty times as far from the earth's centre as is the apple, it will in unit time "fall" towards the centre about one three-thousand-six-hundredth part of the distance fallen by the apple. But what enabled Newton to enunciate his Second Law of Motion where Galileo had failed was the fact that the former had grasped the part played by the *mass*—the unchanging degree of resistance to motion, which characterises every material body; what, indeed, constitutes its "materiality". If of two garden rollers standing on the same level path, the one requires exactly twice the effort to give it the same acceleration as the other, we say that the former has twice the *mass* of the latter. Ordinarily we work the other way round: knowing that one *weighs* twice as much as the other, we at once *infer* that it will take twice as much effort to give it the same acceleration. It is so much easier to weigh garden rollers than to set up experiments by which may be measured the forces necessary to give them equal accelerations. Indeed it is one of the most difficult problems in mechanics to contrive the measurement of "naked" forces unassociated with lumps of matter. But it cannot be too strongly emphasised that though in practice we measure masses by weighing them, the two quantities refer to entirely distinct characteristics of matter.

Now in all that we have said so far about the force controlling the motions of the heavenly bodies, we have been able to get on perfectly well without any knowledge of their *masses*. How can we say that the force acting on the moon differs from that acting on the apple only in respect of the squares of their respective distances from the earth which attracts them? It is to be hoped that the watchful reader has been troubled with a feeling of uneasiness about this very point. Bearing in mind that the magnitude of a force causing motion is always proportional to the product of the mass of the body moved and its acceleration (by suitable choice of units it is *equal* to this product), he must have felt that to compare the forces acting on moon and apple by comparing their accelerations (p. 76) is a mere trifling with the problem, seeing that the *mass* of the moon is so enormously greater than that of the apple. Such criticism would be entirely

## THE NEWTONIAN REVOLUTION

justified were it not for the fact that although the moon is real and unique, the apple is a poetic fiction! Newton never mentioned an apple; the argument holds equally well for a grain of sand and an aerial bomb. The whole of terrestrial mechanics rests on the observed fact that all bodies at a given point near the earth's surface are accelerated towards the centre at the same rate. The mass has nothing to do with it. Obviously if we are going to retain our original definition of a unit force (as that which causes unit acceleration in unit mass) we must ascribe to the particular force of gravity the curious property of being always proportional to the masses it is acting upon, providing of course that they are equidistant from the centre of the earth; for only thus can it come about that different masses are equally accelerated at the earth's surface. This is the course which Newton took, his actual words being: " There is a power of gravity tending to all bodies, proportional to the several quantities of matter they contain " (*Principia*, Bk. III, Prop. VII).

It is a curious fact that the men of science of Newton's day and for nearly two centuries thereafter never seem to have been particularly struck by this odd characteristic of the " force " of gravity. How does the earth " know " that it has got to exert a far greater force on the moon a quarter of a million miles away than on the apple on its own surface? And where does this " force " come from? The law of gravitation as applied by Newton is enunciated in our text books in the following words: " Every particle of matter attracts every other particle with a force which varies directly as the product of their masses, and inversely as the square of the distance between them." The implication is clearly that " matter " itself somehow " generates " this " force " and exerts it in some mystical way across millions of miles of empty space. It is significant that Newton himself is much more guarded in his views than modern text books give him credit for. In speaking of this force he avoids the word " attraction ", and speaks instead of " bodies gravitating towards one another ". And in a letter to a friend, Richard Bentley, he writes: " You sometimes speak of gravity as essential and inherent to matter. Pray do not ascribe that notion to me; for the cause of gravity is what I do not pretend to know . . . It is inconceivable that inanimate brute matter should, without the mediation of something else, which is not material, operate upon, and effect other matter without mutual contact." The widely held belief that Newton regarded gravitation as an *innate* characteristic of matter seems to have been due to one or two less cautious sentences in the *Principia* and in Query 31 of the *Opticks* (which however he is careful to correct at the end of the sentence). But more particularly was it due to the remarks made in the "popular" Preface to the Second Edition of the *Principia*, written by Roger Cotes, the editor. It is probable that Newton all along was struggling against the pseudo-Aristotelian tendency to invoke "occult powers" to "explain" everything—the antithesis to what Newton recognised the nature of scientific enquiry to be. His final views on the nature of matter appear in the last pages of the *Opticks*: " It seems probable to me that God in the beginning form'd matter in solid, massy, hard impenetrable, moveable Particles . . . and that these primitive Particles, being Solids, are incomparably harder than any porous Bodies compounded of them; even so very hard as never to wear or break in pieces; no ordinary power being able to divide what God himself made one in the first Creation." The innate qualities enumerated here do not include gravitation.

When following long established custom we speak of the earth's attraction for the apple and for the moon, we have in fact not a shred of evidence that any " force " such as is exerted by a tow-rope on a barge exists between them. If, however, we decide, as is implied by Galileo, that whenever a body suffers an acceleration, there must be a force acting on it, then we must inevitably deduce the existence of a force between earth and apple. And if further we decide, as did Newton, to measure forces by the acceleration they produce in unit masses, then we must infer that since all bodies fall with constant and equal accelerations at the same point on the earth's surface, the " force " which we have inferred to act on them must be proportional to their masses: this conclusion is inescapable *on the assumptions that we have made*.

This aspect of Newton's work is of great importance in assessing its ultimate validity, that is, as " science's last word ".

We shall make some attempt to consider this in a later chapter; meanwhile another point similar to the last claims our attention.

So far our discussion has been simplified by regarding the sun as " attracting " the planets, and the earth as " attracting " the moon and the apple. But Newton made it perfectly clear that the power of gravitation extends to every particle of matter; it is only for convenience that we have singled out one body—the greater—from the pair under consideration. The moon " attracts" the earth with the same force as that which the earth exerts upon it. To an observer on the sun, we may believe, the earth would not appear indifferent to this action: its path in space would not remain the same were the moon suddenly to be blotted out of existence.

The calculation of this mutual effect is made possible by Newton's Third Law of Motion: " To every action there is always opposed an equal reaction: or the mutual actions of two bodies upon each other are always equal, and directed to contrary parts." The operation of this law is a fact of daily experience. It is impossible to step from an unmoored rowing boat on to a landing stage without moving the boat in the opposite direction. We may believe that the same would be equally true of a man disembarking from the " Queen Mary "; but the effect is masked, even apart from the mooring ropes and rigid gangways, by the enormous mass of the ship compared with that of the man. By " action " Newton clearly meant what in the Second Law he had called " quantity of motion "—what we now call " momentum ", that is the product of mass and velocity. Thus if any body is to alter its momentum it can do so only by altering that of some other body or bodies to an equal amount " directed to contrary parts ". The propellent charge which sets the shell in motion can do so only by giving the gun an equal and opposite " motion ". By using the mathematical device of a minus sign for a " contrary " motion—a convention which became familiar in later years—we are able to express Newton's Third Law in concise quantitative terms: " The algebraic sum of the momenta of an isolated system of bodies is a constant quantity." This is the first of the " conservation laws "—really dogmas, owing to their restriction to " isolated systems ", a point to be dealt with later.

Let us now pause for a moment to see what exactly was Newton's contribution to the growth of scientific ideas about matter and motion, and in what way it was in advance of all previous ideas in the same connexion. The simple idea which in Newton's skilled hands wove together the movements of all observable bodies was no more than this: All moving objects, whether planets or cannon balls, are composed of particles of something distinguished by two characteristics, namely an innate resistance to change of motion, the degree of which resistance is called the *mass* of the particle, and a tendency to gravitate together as if impelled by a force which varies directly as the product of the masses and inversely as the square of the distance between them. The " stuff " thus characterised is what we still call " matter ".[1]

Thus stated, the Newtonian revolution sounds prosaic enough. Why then do we hail it as probably the greatest single achievement of the human intellect? It leaves us completely ignorant as to *why* an apple falls to earth; for to suppose this problem solved by saying " because the earth attracts the apple " is grievously to err. Such a " solution " is no better than that of Aristotle to the effect that the " earthy nature of the apple is seeking its natural place ". The law of gravitation, like the laws of motion of Galileo, does not tell us *why* bodies move towards each other, but precisely *how* their movement is determined. And it tells us more; for it discovers to us what is essential in the motion of *all* material bodies, namely the mere fact that they *are* material. Whether we keep strictly to Newton's own words (see p. 101), or whether we simplify the *verbal expression* of the law by speaking of an actual " attraction ", matters but little. Provided we bear in mind what we *mean* by a " force " in the science of physics, we are compelled to say that all material bodies move *as if* impelled by a force operating according to Newton's law. Whether the operation of that force is " caused " by matter itself, or whether with Newton we prefer to await the results of further research, the *consequences* predicted by the law

---

[1] The critical reader may wonder how the unit of mass is to be measured without dragging in force and so making the argument circular. Newton left this point in an unsatisfactory state. It is too difficult to deal with in detail here.

remain completely unchanged. And that is after all what "science" since Galileo's day has stood for. Before Newton brought the light of his genius to bear into the dark places of nature, the sky was regarded as a realm apart, a region of celestial harmonies into which from time to time the shattering discord of a comet was thrown. The paths of terrestrial bodies, such as cannon balls, the contemporary interest in which undoubtedly promoted the study of dynamics, were likewise seen as compounded partly of "natural" motion and partly of unnaturally forced motions; efforts to control these motions of course met with but little success. The law of gravitation swept aside all seeming distinctions; by its simple assumption as to the quantitative nature of the reaction between material bodies, it was able to reveal all "disturbed" motions, whether of moons or cannon balls, as subject to the same laws. Those age-long portents of disaster, the comets, were shown by Newton (on the basis of data assembled by Halley) to return continually to the neighbourhood of the sun; their elliptical paths merely extend much further into space than those of the planets. It is noteworthy that Cotes made much of this discovery in the "popular" Preface referred to above (p. 102). The future was to reveal that nothing but the refinement of the mathematical technique was necessary to enable man to *calculate* the amounts of the *known* disturbances and thus to seek for, as yet unobserved, causes of *residual* disturbances. The law of gravitation not only explains, it also foretells.

Although a century had to pass after the death of Newton before the law of gravitation was used as an instrument of discovery, it must be realised that Newton himself was able to show that many carefully observed but as yet unexplained phenomena were just as inevitable consequences of this universal characteristic of matter as was the nearly circular path of the moon. Thus he began the proof of the extremely complex problem of showing that the oceanic tides are mainly the result of the varying "pulls" of the sun and moon on the mass of water as they change their stations relative to one another. But perhaps the most beautiful interpretation of all was the demonstration that the precession of the equinoxes (discovered by Hipparchos nearly two thousand years earlier—see p. 41); the so-called "inequalities" of the moon's motion, some already known to astronomers, but one of which was foretold by Newton; and the motions of the tides—all are manifestations of the same phenomenon, namely a disturbance of the simple gravitational reaction between two bodies by the action of a further force due to a *third* body in a different plane. Thus the moon instead of moving in an ellipse round the earth, according to Kepler's three laws, really "wobbles" to a small but precisely determinable extent. The plane containing the earth's orbit about the sun is inclined to that of the moon about the earth, otherwise there would be an eclipse of the sun at every "new moon" and of the moon at every "full moon". Now the sun is about four hundred times as far from the moon as is the earth, but it is vastly more massive; so its "pull" on the moon imposes an additional periodic motion on that body. The actual nature of this motion was *proved* by Newton to involve a gradual revolution of the *axis* of the moon's *orbit*, occupying about eighteen years; this is the principal cause of the recurring cycles of eclipses known to the Chaldean astronomers as the "Saros" at least 2,500 years ago. By a superb use of the "principle of continuity" Newton extended this argument to the case of a ring of satellites, then to a ring of fluid surrounding the central body, which in the next case is conceived to swell out until it touched the ring of fluid now supposed to flow in a channel on its surface. The last case of this sequence was that of a *solid ring* such as exists in the equatorial region of the earth. There will be a similar rotation of the axis of this ring which, being rigidly connected to the sphere, will cause the axis of the latter also to revolve—and so the pole of the heavens will describe a circle and the equinoxes will precess!

Those cases we have briefly touched upon are but a few of the many observations of "natural philosophy" whose dependence upon a very few simple properties of matter Newton revealed by the application of "mathematical principles" and of the new scientific method which Galileo had so clearly shown to the world.

In Newton's hands the law of gravitation not only interpreted all the known motions of material bodies but actually foretold the existence of several others which were later observed by astronomers. In this Newton was helped, rather more than he

PLATE II

Greenwich Observatory in Flamsteed's time

gave him credit for, by the Astronomer Royal, Flamsteed. But the greatest triumph subsequent to Newton's death came towards the end of the eighteenth century, when by its application it became increasingly more probable that a hitherto unknown planet existed in the solar system. The whole story is one of the great romances of science; we have space here only to show how it illustrates one of the most valuable scientific methods.

On 13th March 1781, Sir William Herschel, by profession a musician but who became one of the most brilliant observational astronomers, noticed among a group of small stars one whose size increased progressively with the magnification employed in his telescope (the almost " infinitely " distant stars actually become smaller but at the same time *brighter* when viewed through an instrument of higher magnifying power). On succeeding nights its changed position relative to the " fixed " stars confirmed his view that the body might be a comet. Later however, other observations showed its movements to be those not of a comet but of a planet; and in due course it was established that the body had been observed many times during the previous hundred years. This fact enabled mathematicians to determine its " elliptic elements ", that is, the constants by which its orbit about the sun could be determined. But before very long Uranus, for such was the name given to the new planet, began to wander seriously from the strait and narrow path which men had assigned to it. Here was a chance to put Newton's law to the most crucial test by, as it were, " turning it inside out ".

To Newton, as to the great French mathematician Laplace who had applied the law with such thoroughness that there seemed nothing left to discover,[1] the problems of celestial mechanics were always of the form: " Given such and such bodies, to deduce their relative motions." But in the case of Uranus the problem was: "Given such and such irregularities in the motion of one

[1] Laplace's most *original* achievement in celestial mechanics (towards which result Lagrange made several essential contributions) was his proof of the *stability* of the solar system. By 1784 he had proved that however certain characteristics—such as inclination and eccentricity—of the individual orbits might vary, such changes would involve complementary effects elsewhere, thereby rendering the whole system self-correcting. It must however be emphasised that this proof took no account of certain minor factors, such as tidal friction, recently shown to be relevant.

body to deduce the position and magnitude of some unknown body which is the cause thereof ". Such a piece of investigation is an illustration of what logicians call " the method of residues "; for before any attempt could be made to seek the position of the unknown body, all the possible effects of *known* bodies, such as Jupiter and Saturn, had to be calculated and allowed for: the *residual* effect must then be due to some undetermined cause.

To this task the young Cambridge mathematician, J. C. Adams, and the experienced Frenchman, Le Verrier, applied themselves, each in complete ignorance of the fact that the other was seeking the same goal. On 23rd September 1846, the Berlin astronomer, Galle, found the planet which was the cause of the trouble very near the spot where Le Verrier told him to look. Within a few days the Cambridge astronomer, Challis, was able to prove that he had observed the planet in the same region of the sky but in different positions relative to the stars on 30th July, 4th August and 12th August. The new planet was called Neptune. It is perhaps too much to say that but for a purely abstract dynamical law it would never have been recognised by human eye; but it is certain that its discovery would have been indefinitely postponed. The light of Neptune is thus a reflection of the light of Newton's *thought* which shone with such power as actually to bring new material bodies within the range of human ken.

By *assuming* the truth of the law of gravitation it had been possible to deduce the existence of a massive body far beyond the orbit of Saturn but still within the confines of the solar system. In 1862 however an American astronomer, Alvan Clark, observed a new *star* whose existence, unknown to him, had been deduced from the observed irregularities in the motion of another *star*. In all that we have said about the heavens thus far we have assumed the immobility of the stars—Newton, as we know, took the " frame of the fixed stars " as his frame of reference to which all motions could be referred as absolute. But it will be recalled (p. 46) that one of Tycho Brahe's reasons for rejecting the Copernican hypothesis of the revolution of the earth about the sun was his failure to detect any annual parallactic motion of the stars such as should be an inevitable consequence of the

## THE NEWTONIAN REVOLUTION 109

hypothesis. Tycho should have realised that, despite the wonderful precision of his instruments and of his own skill in handling them, any such parallactic motion must, on account of the probably enormous distances of the stars from the solar system, have been quite beyond his powers of observation. By the beginning of the nineteenth century however the minuteness of man's vision had been sufficiently increased by the steadily increased power of the telescope to bring the detection and measurement of parallax within the bounds of possibility.

Among the many beautiful objects in the heavens are those which to the naked eye appear as single points of light, but which by a quite small magnification are revealed to be *pairs* of stars very close together. Galileo had assumed that of any such pair the brighter is most probably the nearer; the double appearance being in fact a "fluke" due to the line of sight from the earth in a given direction passing almost, but not quite, through two bodies separated by maybe countless millions of miles in *depth*. Such double stars should, as Galileo pointed out, provide a delicate test for parallax; since however little they might appear to move, the nearer should at least appear to move *further* than the more distant one. Hence their separation should vary at different times of the year.

Once again William Herschel was prominent in the early efforts to detect this unobserved phenomenon; nor was he disappointed. But, strange to say, the motion he detected was not one such as might be inferred from the fact that his observations were carried out on a moving platform—the earth—but *corresponded to the revolution of the two components of the double star about a common point*. That such movements occur in only *some* of the double stars (e.g. Castor, $\eta$ Coronae) indicates that not all such pairs are "physically connected"; some are in fact merely "perspective" pairs such as Galileo had assumed.

It was not long before astronomers were able to measure the successive positions of the components of a double star with such accuracy as to be able to assert that their motions were in fact governed by the *inverse square law*. Hence the common point about which they revolve is their common centre of gravity.

The most sensational triumph however was still to come.

The German mathematician, Bessel, who in 1838 actually published the first measurements of the long-sought parallax of a single star (61 Cygni) had in 1834 made so many observations on Sirius, the brightest star in the heavens, as to suspect that its proper motion (that is, its own motion among all the other stars which by this time had come to be recognised as moving, as is the sun, at high speeds) was irregular. So just as a new *planet* had been "invented" to account for the irregular behaviour of Uranus which was troubling astronomers at the same time, Bessel, having in ten years failed to detect any cause for the "wandering" of Sirius, "invented" a new and invisible *star*. In a letter to Sir John Herschel (William's son) he wrote: " I adhere to the conviction that Procyon [the 'Little Dog' star, another wanderer, quite separate from Sirius] and Sirius form real binary systems, consisting of a visible and an invisible star. There is no reason to suppose luminosity an essential quality of cosmical bodies. The visibility of countless stars is no argument against the invisibility of countless others."

What a change from the Ionian conception of the stars as being jets of flame bursting through, or fixed to, the crystal sphere which rotates about the earth! Bessel's remarks exemplify in the clearest manner the Newtonian revolution in which " bodies celestial " differ not at all in their effects from " bodies terrestrial "; their presence being deducible from the assumption that, being fashioned from " matter ", they will be subject to the law of gravitation now seen as *universally* valid. Nevertheless, even more than St. Paul imagined, " one star differeth from another in glory "; for as has been hinted above, in 1862 Alvan Clark, knowing nothing of the predictions of Bessel and the more accurate ones of Peters, stumbled by accident upon a faint star in the proximity of Sirius *just where the mathematicians said it ought to be at that time to account for the irregularity in the motion of Sirius itself.* This dark companion, though as massive as our sun, has only one five-hundredth of the intrinsic brightness of the latter.

It is interesting to note that the two greatest advances in the secure establishment of the law of gravitation were the results of " accidents ". William Herschel, examining the smaller stars, discovers a " comet " which turns out to be a planet, whose

## THE NEWTONIAN REVOLUTION

irregular behaviour leads to the discovery of another planet. Once again the same indefatigable " amateur ", striving to detect evidence of parallactic movement, discovers the physical connexion between certain double stars, and so paves the way for the precise verification of the *universality* of the law of gravitation, and of the discovery of dim bodies thousands of millions of miles beyond the reach of our solar system.

These " accidental " discoveries are gravely misunderstood by people not trained in scientific method. William Herschel was an " *amateur* " only in the literal French sense of the word: he *loved* astronomy so passionately as to conquer physical weakness to the extent of sitting at his telescope or polishing his mirrors all night after an ordinary day's work in his profession of music. He was a man of prodigious determination and unusual talent. " He studied science when he could " as George Forbes says in his *History of Astronomy*; but he studied to such good effect that his mind was prepared to make the best use of any " accidental " observation which he might happen to make. " Those that have eyes to see, let them see."

This apparent digression is in fact an essential part of our study, as it throws up into relief a most important characteristic of science, namely as a *connected and reasoned system*. It is only those endowed with genius and prepared by study of the *known pieces of the system* who will recognise the *meaning* of a chance observation. To a savage a diamond is a bright piece of glass; to a jeweller it is an object of surpassing beauty and profit; to an engineer it is the point of a drill which shall cause the first rupture of a resistant rock; to the chemist it is the most astonishing example of the phenomenon of *allotropy*, for diamond and pure charcoal are indistinguishable by any *chemical* test—they are apparently the same chemical element, carbon, taking upon itself marvellously diverse forms by virtue of the different spatial arrangements of its constituent atoms. A " fact " of experience then has a different *meaning* according to the mental context of the observer; and not merely mental, but social also, as we shall see at a later stage.

So ends our study of the greatest sweep of human imagination, from the fall of an apple to the revolution of an invisible star!

Every movement in the universe determined by the application of one simple law, so that " given the distribution of the masses and velocities of all the material particles of the universe at any one instant of time, it is *theoretically* possible to foretell their precise arrangement at any future time ". " Theoretically ", because the mathematical difficulties would far surpass the compass of any human mind however gifted. But even with that qualification this dictum of Laplace has raised philosophical doubts which seem to strike at the root of man's whole conception of the universe; for since every man's body consists of material particles, *their* motion, no less than that of the particles of inanimate matter, must inevitably follow a course already fully predictable to anyone possessed of sufficient knowledge and mathematical skill. And if this be so, then must all human efforts be in vain, all appearance of choice illusory; for the fabric of history which time weaves must be as completely determined as the melody which will emerge from a given gramophone record. Laplace and his contemporaries overlooked the fact that his deduction, like the gramophone record, bears within itself no mark to indicate why it should not be played backwards; but such a possibility makes nonsense of history.

The gradual realisation by the cultured world of this implication of the Newtonian revolution had, and is still having, the profoundest effect on the morals and ideals of Western civilisation. Thus the march of science leads us to positions whence we gain new views of spiritual realms as well as changing catastrophically our physical environment.

In the next chapter we shall briefly consider whether in the light of modern knowledge and criticism the Newtonian world-picture is quite as rigid in its outlines as it appeared to the thinkers of the eighteenth century.

SOURCES FOR CHAPTER X

E. Mach, *op. cit.*, Gen. Bibl.
I. Newton, *Principia, op. cit.*, p. 81 above.
H. Spencer Jones, *Tercentenary of John Flamsteed* (*Nature* vol. 158, p. 648).
W. M. Smart, *John Couch Adams and the Discovery of Neptune* (*Nature*, vol. 158, p. 290).
*The Herschel Chronicle*, ed. C. A. Lubbock (Cambridge, 1933).

## CHAPTER XI
## IS THE LAW OF GRAVITATION TRUE?

IN the autumn of 1919 two topics produced a great deal of discussion in the more serious newspapers: one was the signing of the so-called " Peace Treaties", which left Europe on the brink of war for twenty years, and the other was the announcement that a handful of men of science had returned from a Pacific island with some photographs of a solar eclipse, from which conclusions had been drawn casting grave doubts on the absolute truth of the law of gravitation. Once again it seems the reader has been asked to spend an unconscionable amount of time and trouble being initiated into a scientific discovery which is not true. It is the purpose of this chapter to try to clear up this unsatisfactory situation.

At the present day most people fall into three groups in respect of their attitude towards science. The first group regard science as a messy business, whose study is undesirable for " nice " people; they are dying out fast, though unfortunately not quite fast enough. Of the other two groups, one accepts every statement of " Science ", as our forefathers accepted " Holy Writ ", as absolute truth, to doubt which is a special kind of blasphemy; while the other, bewildered by the constant rewriting of text books which are out of date almost before publication, regard " Science " as their forefathers did the " Evil One ", as the spirit of confusion and deceit.

As we have already seen in connexion with the motions of the heavenly bodies only the study of its history can give us the clue to the unravelling of the apparent confusion in scientific knowledge. A method of interpretation which was perfectly true at the time of its enunciation has to be remodelled—often drastically—in the light of new knowledge. For knowledge is for ever *growing*—not merely accumulating like a heap of pebbles, but changing in outline and perspective like a living thing; the old writer knew what he was talking about when he wrote of the " tree of knowledge ". The kind of knowledge we call science is a system of laws of nature. Laws are *concise* statements of how we believe nature *works*. To achieve this conciseness we have to make use of *general ideas*. The law of gravitation is of universal application simply because it purposely *excludes from consideration* all those properties of natural bodies except the universal property, that is, mass. The sun is so hot that the earth would be completely evaporated were it to approach it; the moon is so cold that its atmosphere is largely liquified and partly solidified. But by regarding these bodies as mere lumps of nondescript " stuff ", each having a certain measurable resistance to change of motion (i.e. mass) and a certain orientation in space with regard to the other two, we find that their subsequent *motions* can be calculated by means of the law of gravitation. The temperature has nothing to do with it.

The law of gravitation therefore boils down to this: Given certain masses at certain known distances from each other, and whose relative velocities are known, we can calculate their subsequent motions.[1] Put in this way, the law of gravitation was absolutely true when it was enunciated, and is very nearly true to-day when the precision of measurement has greatly improved. Nor is it conceivable that it will ever be otherwise.

How can a law be " true " in 1687 and only " very nearly true " in 1948? It is not a question of a possible alteration in the way nature " works "; whether this be so or not is beside the point. What we are here concerned with is the fact that we have " grown out of " the law of gravitation in two ways.

In the first place there was an irregularity in the motion of the planet Mercury which Newton's theory of the solar system failed to clear up. This irregularity was observed only by virtue of an increase in the precision of instruments; latter-day astronomers " knew more " than Newton *could* have known. That was the first growing pain. But there was a more fundamental one.

If we refer to the " boiled down " gravitation law stated above,

[1] The mathematical difficulties become very grave for all possible cases of three bodies and, for four or more, almost beyond the power of man to handle.

## IS THE LAW OF GRAVITATION TRUE?

we shall observe that from this law we can calculate *motions* from *motions*. To do this we assume the truth of " laws of motion " connecting forces and masses; from which assumption it inevitably follows that the " force " of gravitation on any body is *always proportional to the mass of that body*. It is a curious thing that the men of science of Newton's day and for two centuries thereafter never seem to have been struck by this odd property of one piece of matter whereby the " forces " it exerts on other pieces of matter at equal distances from it are precisely regulated so as to make all their accelerations equal. But towards the end of last century it became evident to several men of science that there must be something " fishy " about a theory which ascribes to the same stuff the complete inability to alter its *own* state of rest or motion (First Law of Motion), together with the power to affect, apparently instantaneously, the motion of other matter even millions of miles away. To these thinkers it was clear that Newton, in order to express his theory in words, had had to invent new general ideas, such as " mass ", and adapt old ones such as " force " and " motion "; and that while the new ones were clear and adequate, the old ones passed into his use with all the confused associations of the ages. We have already seen that " force " has a very unscientific vagueness about it, but " motion " is in an even worse position.

After all, what do we mean by " motion " ? We shall at once be told " change of position ". But when, seated in a stationary train, we see a neighbouring train slowly begin to move, how do we know which has " really " moved ? Only by checking our position with respect to the platform. This Newton fully realised: a body " really " moves when it has changed its position with reference to the " fixed " stars. He realised moreover that this matter of a *fixed frame of reference* to which all motions are to be referred is so essential a part of his whole mechanical system that he dealt with it in the famous Introductory Scholium to the First Book of the *Principia*. To avoid any possibility of misunderstanding it is necessary to quote the whole Scholium here:

> Hitherto I have laid down the definitions of such words as are less known, and explained the sense in which I would have them to be understood in the following discourse. I do not define time, space, place and motion, as being

well known to all. Only I must observe that the vulgar conceive those quantities under no other notions but from the relation they bear to sensible objects. And thence arise certain prejudices, for the removing of which it will be convenient to distinguish them into absolute and relative, true and apparent, mathematical and common.

(1) Absolute, true, and mathematical time, of itself, and from its own nature flows equably without regard to anything external, and by another name is called duration: relative, apparent, and common time, is some sensible and external (whether accurate or unequable) measure of duration by the means of motion, which is commonly used instead of true time: such as an hour, a day, a month, a year.

(2) Absolute space, in its own nature, without regard to anything external, remains similar and immovable. Relative space is some movable dimension or measure of the absolute spaces; which our senses determine by its position to bodies; and which is vulgarly taken for immovable space; such is the dimension of a subterraneous, an aerial, or celestial space, determined by its position in respect of the earth. Absolute and relative space, are the same in figure and magnitude; but they do not remain always numerically the same. For if the earth, for instance, moves, a space of our air, which relatively and in respect of the earth remains always the same, will at one time be one part of the absolute space into which the air passes; at another time it will be another part of the same, and so, absolutely understood, it will be perpetually mutable.

Although a full discussion of this difficult matter would take us too long and too deep, yet since the answer to the question set at the head of this chapter depends on a clear understanding of the implications of Newton's assumptions, a brief exposition must be attempted.

The most important point to be noted is Newton's tacit departure from his own ideal of scientific enquiry: he *assumes* the existence of " absolute " time and space " without regard to anything external "—in other words, without any evidence from sensory experience, such as the " vulgar " base their ideas of time on. It is equally important to notice that Newton *had* to assume just this if he was to construct his mechanical System of the World. Finally we must not blame Newton for failing to haul himself up by his braces out of the climate of opinion into which he was born, and so to realise that " when once you tamper with your basic concepts ", as Whitehead puts it, " philosophy is merely the marshalling of one main source of evidence and can not be neglected." The liquidation of the Aristotelian physics involved

## IS THE LAW OF GRAVITATION TRUE?

the liquidation of the Aristotelian metaphysics of the immutable heavens; but Newton was not the man to do it. He was too great a man of science to attempt the impossible. The scientific advance has to proceed like that of an army—by jumps to tactical strong points, sometimes without the clearance of strategic minefields.

As soon as men of science realised that the " fixed " stars, so far from being " fixed " are moving relative to each other (p. 110) they began to look for another " platform ". The most promising candidate was the " ether " whose undulations were supposed to constitute light. Unfortunately Michelson and Morley could find no evidence for the existence of the " ether " under conditions which should have rendered it easily observable. So the search had to start again. A few optimists are still searching, but most have given up the task as meaningless. Now if we can never be sure that a body has " really " moved, we can never be sure on which of two " moving " bodies a force is acting, for which reason it is hardly worth bothering about " force " at all. Following this train of thought, and taking as a fundamental property of nature the fact that weight and mass, though different properties are always equal in the same place, Professor Albert Einstein developed a system of dynamics (i.e. of the nature of motion) from which a new law of gravitation emerged. This law precisely accounted for the irregularity in the motion of Mercury which Newton's law had failed to do. But physicists rightly held that the " truth " of the law could not be finally established until, like Newton's and Galileo's laws, it had been shown that consequences *deduced* from it by pure reasoning were actually to be observed in nature. That is why the British eclipse expedition under the late Sir Arthur Eddington set out in 1919 to photograph some stars whose light was passing close to the eclipsed sun. The data they obtained were within the limits calculated on the basis of Einstein's theory.

Newton's law of gravitation is more " true " than Aristotle's for two reasons, because it enables us to control the movements of earthly bodies and to foretell the positions of heavenly bodies with a marvellous degree of exactitude, which the latter's was powerless to attempt, and also because Newton's concept of the relation between mass and acceleration was more precisely definable (and consequently "usable") than Aristotle's "natures".

Einstein's law of gravitation was derived from a consideration of the metrical properties of the " space " in which bodies move. As Professor Bridgman puts it, he was enquiring "into the meaning of simultaneity, and he was finding the meaning by analysing the physical operations employed in applying the concept in any concrete instance ". Newton had taken such ideas as simultaneity for granted. With the knowledge of the critical importance of the nature of light signals for the timing of physical events, Einstein saw that such an assumption was too simple-minded. His " analysis of the physical operations " yielded first the Special Theory of Relativity, concerned with measurement between bodies moving at *different* but *uniform* velocities. Later came the General Theory which included the case of bodies moving with velocities mutually *accelerated*. The result of this was to render the concept of *force* redundant.

Einstein's law of gravitation is " truer " than Newton's for reasons similar to those which made Newton's " truer " than Aristotle's. It is more than likely that history will repeat itself and Einstein's beautiful theory be found after all to be not quite " true ". But no theory which can reveal to us new secrets of the universe can be " false ", though it may be " out of date ". We do not regard the " Cutty Sark " as falsely contrived merely because it could not equal the performance of the " Queen Mary ". The latter ship is in the main a development of the older; but the " steam wind " which presses upon her turbine plates by being more controllable, is more economically applied. The " Cutty Sark " was the wonder of her age; the " Queen Mary " can do all that she did and more; but thousands of people still derive pleasure in using the graceful methods of the old " sea bird " rather than the more powerful ones of the modern machine. Similarly for all our everyday uses, whether for the design of engines or for the writing of the " time table " of the planets, we use and shall long continue to use Newton's law, which is sufficiently exact to meet all purposes except those in which we are probing more deeply into the heart of the universe.

One question remains on which we may close this rather

## IS THE LAW OF GRAVITATION TRUE? 119

difficult discussion. If Einstein's dynamics is "truer" than Newton's, why do we waste so much time in teaching the latter in our schools? As well ask, why waste a child's time with instruction in the use of the foot rule, when the vernier calipers give us a much more accurate estimate of length? Because over-refinement tends merely to obscure the simple ideas of length and number which every human being must acquire. Einstein's dynamics are Newtonian dynamics subjected to intense refinement. One result of this refinement has been the removal of confusion in the general ideas which Newton, living in an age when Aristotelian views were still accepted by many learned men, had not entirely escaped. The author is one of those who would like to see our school text books subjected to the same refining influence; for there is much in them which reflects Newton's failure in this respect. It will be the chief service of the modern critical study of the history of science to present scientific knowledge as a growing tree which depends for its vigour as much on its roots in the past as upon its fresh young branches; but from which for true health all dead and rotten wood needs to be drastically pruned. In other words it will teach us that new theories do not replace old ones because they are more *true*; but grow out of and embrace the older because they are more adequate.

### SOURCES FOR CHAPTER XI

H. Poincaré, *Science and Hypothesis*, trans. W. J. G. (London, 1905). An outstanding work on the philosophical basis of modern physics.

Magie, *op. cit.*, Gen. Bibl., for Michelson-Morley experiment on velocity of light.

Mach, *op. cit.*, Gen. Bibl.

A. Einstein, *The Theory of Relativity* (Engl. Trans., London 1920).

P. W. Bridgman, *The Nature of Physical Theory* (Princeton, 1936).

CHAPTER XII

# THE CELEBRATED PHENOMENA OF COLOURS

WHOEVER has been fortunate enough to visit the chapel of Trinity College, Cambridge, must have been struck by the compelling beauty of Roubaillac's statue of Newton, which Wordsworth so finely describes as " the marble index of a mind forever voyaging through strange seas of thought alone ".

In his right hand Newton holds a prism; and we might well wonder why the discoverer of the law of gravitation should be thus portrayed. Reference to the records of Newton's life however shows that the study of optics very early claimed his attention, and it is probable that his famous researches on the nature of light were completed before the *Principia* was published. Indeed his first claim to notice by the learned world was the letter to the Royal Society (printed in the *Philosophical Transactions* of February 1672) in which he explained the relation between light and colour.

Newton's work in optics demands our closest attention; for not only is the basis of the modern theory of light here firmly laid, but the methods employed, and described by Newton with a wealth of experimental detail, give us a clearer insight into the true method of physical investigation than almost any other work ever written.

The historian of science is struck by the fact that Newton was first impelled to the study of optics by the purely technical problem of ridding the images formed by the recently invented telescopes of the indistinctness which always blurred their edges. To this end he first tries the method in common use, namely of grinding the lens to a paraboloidal instead of spherical shape. But having, probably in 1669, once observed the effect of a *prism* on a beam of light, he recognised, what every other investigator had failed to do, that the imperfection of the refracting telescope

## THE CELEBRATED PHENOMENA OF COLOURS

lay in a sense in the nature of light itself. This knowledge leads us to the realisation of how intimately the various branches of science are connected, and how theory and application are mutually dependent. Previous to the seventeenth century the several sciences of physics, chemistry, biology and the like were as yet undistinguished as separate disciplines; but during the course of this century we find Kepler writing *Astronomia Nova*, Boyle *The Sceptical Chymist*, and Newton *Opticks*. As years went by the separation became more and more complete; and to-day it is beyond the compass of any human mind to make itself master of more than a small part of one branch of science. Nevertheless this specialisation, necessary up to a point, has its dangers. We have already seen how the acceptance of the Copernican hypothesis was greatly accelerated by Galileo's proficiency in the construction and use of the telescope. The course of our present study will reveal how the attempt to overcome the imperfections of the early telescopes set investigators firmly on the road leading to the invention of an entirely new instrument—the spectroscope —which has turned out to be man's most potent weapon in the investigation of the structure of all material systems from the simplest atom to the most remote island universe of stars in the making.

Before attempting to retrace the steps by which Newton gradually revealed the "mode of formation" of the colours of the rainbow, let us summarise the facts as they were known to him. We have already seen (p. 56) that Euclid had with characteristic genius recognised that the only way to discover what we may call the " laws of motion " of light was to isolate an exceedingly fine beam or pencil of light whose axis we call a " ray ". A ray of light is an imaginary thing like a line; it is in fact the line to which the pencil would approximate were it made to shrink without limit. There are no " rays " in nature, but if we wish to learn about the *directions* in which beams' of light travel in certain circumstances, we must somehow reduce those beams to mere " pointers " without bulk.

Having thus defined the ray, Euclid established the important facts that rays of light are always straight, and that they are reflected from smooth surfaces according to a simple law. Nearly two thousand years later Snell correctly stated the law of refraction by which could be accurately foretold the *change* in direction suffered by a ray in passing from one medium to another. One other property of light was known to Newton at this time, but we shall speak of this later (p. 143).

Granted a knowledge of these experimental laws, these concise statements of how light does actually behave, it is possible by the application of geometry, without further recourse to experiment, to foretell the exact path of a ray of light impinging upon surfaces of various curvatures, or traversing blocks of transparent materials of various shapes. This may be done by practical geometry or by the use of algebraic formulae derived by the application of purely formal reasoning. If to these processes be added the idea of an " image " formed at a point from which rays actually diverge (" real " as in the photographic camera) or from which they only *appear* to diverge (virtual, as in the eyepiece of a telescope), it is possible to construct optical devices of the most various kinds. No further experiments are necessary. But when a beam of light passes through blocks of glass with bevelled edges, the image formed on a screen is found to be bounded by different *colours*. Here was a *physical* problem, early recognised as having some connexion with the colours of the rainbow, whose *shape* had been explained by Descartes by a beautiful piece of mathematical reasoning. It was to this problem that Newton applied the other side of his genius—the power of contriving simple but adequate experiments and of deducing from the results the *essential* conditions of the phenomena.

A full account of these experiments is given in the *Opticks*, first published in 1704; but we shall here quote from the letter to the Royal Society referred to above. In this letter we get the closest possible view of Newton the experimenter probing into nature's secrets; it is less formal than the *Opticks*, and since it can be consulted only with difficulty, we shall quote at length. Imagine then Newton in his rooms at Trinity, and hear how he started on one of the greatest of all experimental researches.

Having darkened my chamber and made a small hole in my window-shuts to let in a convenient quantity of the Sun's light, I placed my Prisme at his entrance, that it might be thereby refracted to the opposite wall. It was at

## THE CELEBRATED PHENOMENA OF COLOURS 123

first a very pleasing divertisement to view the vivid and intense colours produced thereby; but after a while applying myself to consider them more circumspectly, I became surprised to see them in an *oblong* form, which, according to the received laws of Refraction, I expected should have been *circular*.

They were terminated at the sides with straight lines, but at the ends, the decay of light was so gradual, that it was difficult to determine justly what was their figure; yet they seemed semi-circular.

Comparing the length of this coloured *Spectrum* with its breadth, I found it about five times greater; a disproportion so extravagant that it excited me to a more than ordinary curiosity of examining from whence it might proceed. I could scarce think that the various *Thickness* of the glass, or the termination with shadow or darkness could have any influence on light to produce such an effect; yet I thought it not amiss, first to examine those circumstances, and so tryed what would happen by transmitting light through parts of the glass of divers thicknesses, or through holes in the window of various bignesses, or by setting the Prisme without [*i.e. outside*] so that the light might pass through it and be refracted before it was terminated by the hole. But I found none of those circumstances material. The fashion of the colours was in all these cases the same.

Many before had done this experiment—at the beginning of the letter he had referred to it as the " celebrated Phaenomena of colours "—but no one, we believe, had remarked that light from a *circular* object produced, after traversing a triangular prism, an *oblong* image with semi-circular ends; it was as if the circle had been pulled out lengthwise. Contrary to the practice of earlier thinkers Newton did not dispose of the less likely causes merely by argument, but by actual trial. This is the essence of the scientific method: to eliminate one by one the possible causes by varying the conditions of the experiment *one at a time*. First he shifted the prism to and fro to vary the thickness of the wedge of glass traversed by the light, but allowing no other change in the disposition of the apparatus—particularly in the direction in which the prism was inclined to the incoming light. Next he varied the size of the hole through which the light entered the prism. But neither of these conditions was the cause of the lengthening of the image.

Suspecting that the lengthening of the spectrum was due to some irregularity in the prism (strains in glass objects must have been almost universal at that time) Newton tried the effect of two similar prisms placed with their refracting edges pointing in opposite directions so that, by doubling the length of glass traversed, any effect due to unevenness would also be doubled. But the result was a circular image as if no prism at all had been in the path of the rays.

Knowing the accurate law of refraction he calculated what should be the size of the image and found that it was far too great. So he began to suspect that the light failed to travel in straight lines, but moved in a curved path like " a tennis ball struck with an oblique racket ". He could however find no such curvity in the refracted beam.

One by one the possible causes had been removed. What could there be which had not been tried? Thus far any patient and skilled experimenter might have reached; but it is the mark of genius to make the inspired " guess " which is the " morning star of truth ". Newton does not at this stage explicitly state the hypothesis which had dawned on him, namely that sunlight is not pure or homogeneous but consists of rays of various refrangibilities; but it is perfectly clear from the next paragraph that he had realised that the key to the solution was to *isolate* small regions of the spectrum and subject them to further refraction— an experiment which seems to have been carried out earlier by Marée who, however, failed to recognise its full significance.

The gradual removal of these suspitions at length led me to the *Experimentum Crucis*, which was this: I took two boards, and placed one of them close behind the Prisme at the window, so that the light might pass through a small hole, made in it for the purpose, and fall on the other board, which I placed at about 12 feet distance, having first made a small hole in it also, for some of that incident light to pass through. Then I placed another Prisme behind this second board, so that the light trajected through both boards might pass through that also, and be again refracted before it arrived at the wall. This done, I took the first Prisme in my hand and turned it to and fro slowly about its *Axis*, so much as to make the several parts of the Image, cast on the second board, successively pass through the hole in it, that I might observe to what places on the wall the second Prisme would refract them. And I saw by the variation of those places that the light tending to that end of the Image, towards which the refraction of the first Prisme was made, did in the second Prisme suffer a refraction considerably greater than the light tending to the other end. And so the true cause of the length of that Image was detected to be no other than that *Light* consists of *Rays differently refrangible*, which, without any respect to a difference in their incidence, were according to their **degrees** of refrangibility, transmitted towards divers parts of the wall.

Sunlight, and the same is true of candlelight or lamplight, consists of rays of all possible refrangibilities lying between two limiting values; and " as the Rays of light differ in degrees of Refrangibility, so they also differ in their disposition to exhibit this or that particular colour. Colours are not *Qualifications of Light*, derived from Refractions, or Reflections of natural bodies (as 'tis generally believed) but *Original* and *connate properties*, which in divers Rays are divers. . . ."

To the same degree of Refrangibility ever belongs the same colour, and to the same colour ever belongs the same degree of Refrangibility. The *least Refrangible Rays* are all disposed to exhibit a *Red* colour, and contrarily those Rays, which are disposed to exhibit a *Red* colour, are all the least refrangible. So the *most refrangible* rays are all disposed to exhibit a deep *Violet Colour*, and contrarily those which are apt to exhibit such a violet colour, are all the most Refrangible . . .

The species of colour, and degree of Refrangibility proper to any particular sort of Rays, is not mutable by Refraction, nor by Reflection from natural bodies, nor by any other cause, that I could yet observe. When any one sort of Rays hath been well parted from those of other kinds, it hath afterwards obstinately retained its colour, notwithstanding my utmost endeavours to change it. I have refracted it with Prisms, and reflected it with Bodies, which in Daylight were of other colours; I have intercepted it with the coloured film of Air interceeding two compressed plates of glass; transmitted it through coloured Mediums, and through Mediums irradiated with other sorts of Rays, and diversely terminated it; and yet could never produce any new colour out of it. It would by contracting or dilating become more brisk, or faint, and by the loss of many Rays, in some cases very obscure and dark; but I could never see it changed *in specie*.

The quotations given above need little in the way of amplification; indeed one reason for giving such long extracts is to guard the reader against the acceptance of the facile view that the scientific study of nature is necessarily subversive of a good literary style. Allowing for a certain number of archaisms and a leisureliness proper to a more leisurely age, we may grant to Newton the power to wield the English tongue effectively for that purpose which is not the least of the uses of language—the precise communication of significant ideas. We need no figure to enable us to visualise the disposition of the simple apparatus in the *experimentum crucis*; we can clearly discern a beam of each colour in turn picked out by the hole in the first board, sharply delineated by the hole in the distant second board, and refracted as an isolated entity by the second prism on to the wall beyond. And thus the tacit hypothesis is verified: that white light consists of innumerable " lights " each characterised by having the power to be refracted to a definite extent, those " lights " which are " disposed to exhibit a red colour " being refracted to the smallest extent.

From our present standpoint it is difficult to realise how " obstinate an attempt to think clearly " Newton must have made. To anyone a little versed in the history of thought prior to Newton's age, Newton himself gives an unmistakable clue: " Colours are not *Qualifications of Light*." It is impossible to trace here all the implications of this view which in a more general form goes back to Aristotle; but, roughly speaking, light was regarded by Newton's contemporaries as a " stuff " whose " qualities " could be changed by its interactions with other " stuffs ", just as the " rigidity " of silver is " lost " and " fluidity " gained when it is brought in contact with fire. As far as light is concerned, Newton's experiments banished this view for ever. Light which is characterised by one recognisable property (i.e. the power to be refracted to a precisely measurable extent) is also " disposed to exhibit " another property, that is a definite colour. Here for the first time is the realisation of the dream of the Ionian philosophers (p. 16) for *light exhibiting a definite colour is a permanent constituent of the universe*; it may be mixed with other " lights " and so take on the appearance of a change but, in so far as it can be extracted once more and cannot be changed by any known means, it may be regarded as of the nature of an " element ", that is as something retaining its identity through all the vicissitudes of reflection, refraction, and the like. Whitehead uses the term " eternal object " for such ultimate factors, thereby avoiding the ambiguity involved in " element ".

This great discovery at once makes clear the nature of *colour*. Note that Newton nowhere speaks of " red light " or " red rays ", but of " light disposed to exhibit red "; for it is not to be *assumed* that creatures having differently constituted eyes would see the same colour exhibited by rays of a certain refrangibility as we do. Moreover the colours " of " objects are, as Newton found, conditional upon the kind of light falling on them: " The Colours

## THE CELEBRATED PHENOMENA OF COLOURS    127

of all natural Bodies have no other origin than this, that they are variously qualified to reflect one sort of light in greater plenty than another. And this I have experimented in a dark room by illuminating those bodies with uncompounded light [*i.e. light from one part of the spectrum only*] of divers colours. For by that means any body may be made to appear of any colour. They have there no appropriate colour, but ever appear of the colour of the light cast upon them, but yet with this difference, that they are most brisk and vivid in the light of their own daylight colour." Thus in the red light of the spectrum a " red " flower and a " blue " one both appear to be red, but the " red " one, being so constituted as to reflect red light more easily than any other colour, appears brighter. It is now possible to mix dyes in such a way that the resulting " colour " reflects *only* blue (or other) light; such dyes appear perfectly black in pure light of any other colour.

The first account of these discoveries was published in 1672. In the First Edition of the *Opticks* Newton described more than a dozen experiments of the most varied kinds whereby the hypothesis may be verified beyond a shadow of doubt. It is therefore more than a little pathetic that so-called educated people may still be bamboozled by " authentic " stories of " green death rays " and similar marvels. Any light which is disposed to exhibit green is of nearly the same character as that " exhibited " by the lush green pastures of a west country farm; such pastures are not commonly littered with corpses. Newton had taught, though the lesson seems to have been ill-learned, that any " green rays " which might be associated with destructive radiation (whose existence is by no means denied) would have even less to do with its destructiveness than the water in which the murderer administers his arsenic.

The *Opticks* appeared in three editions during Newton's lifetime, and a fourth appeared shortly after his death. It is a telling reply to those people who assert that all scientific works are out of date almost as soon as they are published that the Fourth Edition was reprinted in 1931. Scattered through this work are hints as to the nature of light which have lain dormant for two centuries, but which now once again testify to the fact that, despite certain errors and over-emphasis, Newton had glimpsed just those characteristics of light which modern physics has been able to illuminate and refine. This subject bristles with difficulties, but we may learn from it many valuable lessons without overtaxing our abilities with too great a burden of detail.

## SOURCES FOR CHAPTER XII

An excellent study of Newton's work in optics is M. Roberts and E. R. Thomas's *Newton and the Origin of Colours* (London, 1934).

I. Newton, *Opticks* (4th ed., rep., London, 1931).

A. Wolf, *op. cit.*, Genl. Bibl., gives a very good account of the foundations of optics in the seventeenth century.

## CHAPTER XIII

## ON WAVES AND WAVINESS

THE laws of optics can be ascertained and applied to the construction of instruments without any knowledge of how light actually *works*. One part of science—some people say the whole—is concerned with the complete description of the possible operations of natural phenomena without concern for their essential nature; but the history of science is full of instances to show that speculation cautiously applied leads to the discovery of new laws of operation which would otherwise, perhaps for ever, have remained undiscovered.

Speculation as to the nature of light followed a similar course to that of early astronomy. The Pythagoreans (p. 22) with uncanny insight believed light to be a " something " given off by certain bodies (called "luminous") and to cause vision by entering the eye. With the rapid rise of man's self-awareness (p. 29) begun by the Sophists and critically developed by Socrates and Plato, the opposite view became fashionable, namely that the eye " searches " for objects by sending out rays of light. It is significant that Euclid, the most systematic of Greek writers on optics, held this erroneous view; for the *geometry* of optics, like that of astronomy with respect to the centre of reference (p. 47), is quite indifferent to the *direction* of the rays in its diagram, provided that the convention is consistently applied. The hypothesis came into difficulties however when faced with the problem of explaining the necessary presence of a *third* body, such as the sun or a lamp, before vision is possible. This difficulty, and perhaps the influence of Aristotle and the Atomists, both of whom in rather a vague manner supported the view that light exists independently of vision (Aristotle seems to have thought of it as existing by itself), may account for the more or less tacit assumption of what is called the " emission hypothesis " until the middle of the seventeenth century. In the second half of that century, however, numerous properties of light were discovered, for the explanation of which a simple emission hypothesis was quite inadequate.

The first author (see however p. 143) to propose an alternative hypothesis was probably Robert Hooke who, in his famous book *Micrographia*, published by the Royal Society in 1664, drew attention to the production of colour by the cleavage of *uncoloured* substances into sufficiently thin plates. What particularly struck his acute mind was the fact that the thickness of the plates required to give coloured light lies between two limiting values and seems to be independent of the nature of the substance composing the plate, which may be a thin sheet of mica, a soap bubble, or a film of oil on water. The interest of this observation, as Hooke clearly saw, is that if light can be regarded as something in motion, the colour it is capable of exhibiting must be due to some property depending on the distance it has traversed; in other words the light must have a wave-like character. It is very doubtful if Hooke would have recognised this had not experiments already shown in a clear and decisive manner that another natural phenomenon, sound, is undoubtedly propagated in waves. This illustrates the value of *analogy* cautiously applied, whereby the nature of some obscure agent may be compared with that of a similar one in which the mechanism is more easily observed. To make this point clear we shall consider briefly the history of the investigation of sound, especially in regard to the discovery of its wave-like mode of propagation.

Whereas the speculations of the Greeks about light were vague and of little influence, they made some guesses as to the nature of sound which came remarkably near to the truth. Thus Aristotle clearly states that " sound takes place when bodies strike the air . . . by its being moved in a corresponding manner." When he attempts to explain the *character* of this motion he makes the mistake of supposing that the air is borne along bodily—which harmonises ill with the everyday observation of the great speed with which sounds travel—a speed which would involve a gale of over seven hundred miles an hour. The Roman architect, Vitruvius, is much clearer in his ideas: " Voice is breath, flowing and made sensible to the hearing by striking the air. It moves in infinite circumferences of circles, as when, by throwing a stone into still water, you produce innumerable circles of waves, increasing from the centre and spreading outwards, till the boundary of the space, or some obstacle, prevents the outlines from going further." And he goes on to explain echoes by the reflection of the waves. Nevertheless we must not read too much into this acute but insufficiently developed analogy; Vitruvius does not describe in precise terms the actual movements of water over whose surface a wave is travelling, and if he had done so and applied it by analogy to the motion of sound, he would have been wrong. Analogy, as has been hinted above, must be applied with caution. Vitruvius was right in regarding the propagation of sound as " like " that of a wave on water; such a comparison suggests new lines of investigation of the less evident motion of sound; but when the investigation is made it is found that there is a fundamental difference in the *mode* of motion of air-bearing sounds, and of a water *surface* bearing waves.

An understanding of the various kinds of wave motion is fundamental to the study of science. Both nature and the works of man abound in examples of this particular kind of motion; of their widespread occurrence the ancients had no doubt. We should expect that it would not be long before the school of experimental physicists, of whom Galileo was the grand exemplar, would attempt to determine the exact nature of such motions as may be directly observed, and to attempt to discover to what extent the phenomenon of sound can be similarly explained. It was Mersenne, the most ardent of Galileo's admirers in France, who seems to have undertaken the first systematic investigation of these matters, and to have summarised them in his *Harmonicorum Liber* published in 1636.

Assuming, as was already well known, that sound is the consequence of some body's *vibration*, he sought to demonstrate that the nature of the sound is completely determined by the nature of the original vibration. He repeated the alleged experiment of Pythagoras which shows the connexion between the length of a vibrating string and the pitch of the note it emits; and followed this up by showing that the pitch of the emitted note is inversely proportional to the square root of the thickness[1] of the string and directly proportional to the square root of its tension. The vibrations of the strings emitting the common musical notes are much too rapid to be counted by eye, but using a string giving one of the lowest notes of the organ, Mersenne was able to *count* the number of vibrations, and hence, by applying the above law, to deduce rough values for the notes of the whole scale.

Such experiments and many others with toothed wheels, etc., by Hooke and his contemporaries, established beyond a doubt that sound is caused by vibration, and that the pitch of a note is precisely determined by the number of vibrations in unit time, or " frequency " as it is called. The next problem was to show with equal certitude, what had already been guessed, namely that the *sensation* of sound is merely the impact on the ear of " waves " set up in the air by the vibrating body. The first part of this problem was settled by discoveries concerning the atmosphere and the vacuum recently carried out by Galileo, Torricelli, Pascal, von Guericke and Boyle (the study of which is now a regular feature of elementary science courses). Boyle was able to show that a vibrating body placed under the " receiver " of an air pump continued to vibrate, but quickly ceased to be audible as the surrounding air was removed by the pump—an illustration of how advances in the theory of a subject may be made possible by the invention of new instruments. Thus there is no doubt that air is the necessary " medium " for the propagation of sound, but there remained the much greater problem of determining what *kind* of motion is involved. Rough measurements of the velocity of sound showed that it travels in still air at rather more than eleven hundred feet a second; such a velocity precluded the possibility of complete *translation* of bodies of air to communicate sound. But anyone who has seen a field of corn stirred by a breeze will realise how swiftly a *wave* of motion may pass over a surface which is not at all moved *as a whole*. In face of the great difficulty of making visible the motion of sound-bearing air, Newton applied

---

[1] " Thickness " being a somewhat vague measure of a string, the second law is usually expressed as " pitch varies inversely as the square root of the density of the material composing the string ".

the *deductive* method employed by Galileo when searching for the law of change of velocity in a falling body (p. 63).

In the *Principia* Newton explains how the impact of a vibrating body on a surface of air must drive the particles of air forward, thus, through the inertia of the more distant air particles, forming a " heap " of higher average density than that of the surrounding air. On the rebound of the vibrating body on the other hand a " hole " will be formed. Now experiments carried out a few years previously by Boyle, and described by him in 1662 in his *Defence of the Doctrine touching the Spring and Weight of the Air*, showed that the air is elastic, that is, when squeezed it will tend to expand once again and, if allowed to do so, will, like a pendulum bob let fall, *overshoot* the equilibrium volume and so set up a tendency to contract once again. The process will continue, but the oscillations will rapidly die down, owing to the dissipation of the energy in a continuously increasing space, unless maintained by external force. A vibrating body is just such a source of external force, hence the air in its neighbourhood will continue to oscillate. Returning now to the " heap " produced by the first oscillation Newton showed that the " spring " of the air will tend to release itself in all directions, and since the pressure all round (except in contact with the vibrating body) is less than the pressure of the heap, the particles will fly outwards forming a new " heap " or " *compression* ", and leave behind a " hole ", or *rarefaction*. The next phase will be the expansion of the newly-formed " spring ", which can expand *backwards* into the newly-formed rarefaction. The upshot of the matter is that *pulses* of high pressure are generated and will travel outwards. Using the exact law of Boyle, and making certain simplifying assumptions which were afterwards found to be unjustified, Newton *calculated* that the velocity with which such pulses must travel depends only on the state of the medium (here air) in which they are travelling, and not on the frequency ($\nu$) with which they are being generated. It must follow therefore that the faster they form the shorter must be the distance between them; this distance is now known as the " wave length " ($\lambda$), and the equation $v = \lambda \nu$ is of fundamental importance in all wave motions.

Having thus *deduced* the velocity consequent upon his *assump-*

*tions* (or hypothesis as to the nature of the motion of sound) Newton made rough measurements in the Great Court of Trinity to verify the calculation. The result was a little disappointing; but when nearly a century later Laplace was able to correct one of the assumptions (which has nothing to do with the wave mechanism), Newton's hypothesis was found to be fully justified by the facts.

In modern times experiments have been devised whereby the to-and-fro (or longitudinal) waves by which sound is propagated have been made almost a matter of direct observation. Doubtless the reader is familiar with the beautiful photographs showing a bullet in flight with air pulses spreading out round it; such pulses bear the whip-crack familiar to those who have sat in the butts while rifle bullets have sped singly over their heads.

Before returning to the question of the physical nature of light we must note that whereas sound is invariably *propagated*, whether in air, water, or other medium, by *longitudinal* waves, the *generation* of sound waves is commonly brought about by a vibration of quite a different character. The simplest case is that of a stretched string which, if emitting a note of not too high a frequency, can actually be seen to be vibrating *transversely*. Such a wave is " stationary "; but if a long string made fast at one end is given a " flick ", a transverse wave is seen to *travel along the string*. Here there is no forward movement of the string as a whole, every part of which, if moving at all, is oscillating in a direction at *right angles* to the direction in which the *wave* is travelling. By watching the rise and fall of a cork floating on water, over the surface of which waves are passing, it is possible to see both that these waves are *transverse* in character (see p. 131) and that little water is borne along as a whole.

The partial resemblances (e.g. reflection and echoes) between light and sound were such that we find Galileo early in the seventeenth century trying to measure the velocity of light by instructing an observer to open a lantern carried by him as soon as he saw the light from the lantern held by a second observer some distance away. The latter observer however judged that the lapse of time before the first observer's answering " wink " was visible was no more than that required for any muscular response. The experiment was therefore inconclusive.

In 1676 however the Danish astronomer Roemer, working in Paris, was able to show that an irregularity in the appearance of Jupiter's satellites from behind the disc could be accounted for only on the assumption that light from the sun takes a measurable time to cross the earth's orbit.[1] The general truth of his assumption was verified by several contemporary observers; there was thus little doubt that light does *not* travel instantaneously. Though the actual speed could not at that time be stated with any degree of precision, it was at any rate several hundred thousand miles a second. With the wonderful refinement of modern instruments it has been possible to develop a method not essentially different from Galileo's; in this way the speed of light has been measured by Michelson, between two stations less than a mile apart, as almost exactly 300,000 kilometres ($=186,000$ miles) a second.

The discovery of the *finite* velocity of light was of the utmost importance for the development of the subject; for though the actual velocity is of no immediate consequence, it is clear that only an influence which *takes time* to travel can possibly have a wave-like character. The very idea of a *wave motion* involves the idea of a *periodic repetition*, that is a fixed " pattern of motion " repeating itself in successive intervals of time.

No investigator in the seventeenth century recognised this more clearly than Huygens who, in his beautiful book *Traité de la Lumière* published in 1690 but already communicated to the learned world twelve years previously, first gave a systematic exposition of the view that light is propagated in waves.[2] Pointing out the extreme unlikelihood of any material substance being able to travel at such enormous speeds, he draws attention to our knowledge of the propagation of *sound* as a model by which we may hope to understand the nature of light. Not that *air* could be the medium of propagation of light; for any oscillation

---

[1] A fuller account will be found in most elementary text books on light, which do not however usually mention the interesting fact that the discovery was the result of attempts to make accurate tables of the positions of the satellites for use in fixing longitude at sea.

[2] It is necessary to point out that the concept of " wave " applies to Huygens' theory only in a restricted sense. The property of *periodicity* is nowhere explicitly attributed by him to the *pulses* which constitute the propagation of light; nor does he make any use of it in his interpretations.

of air particles sufficiently rapid to account for the huge velocity of light is inconceivable. But his own experiments and calculations with elastic solids had given Huygens a clue of which he made brilliant use. He asks us to recall what happens when a long column of marbles is sharply struck at one end; it is well known that the marble at the remote end is, apparently instantaneously, made to jump away from the line. Yet that this transmission of motion is instantaneous only in appearance may be surmised from the fact that if polished spheres marked with the dew of breath are made to strike one another, the marks left upon them are large in proportion to the force of impact; there is no doubt that they suffer an actual *flattening* of their surfaces, the restitution of whose shape is the cause of the rebound. Once again we see how skill and experience in one branch of natural knowledge may lead to insight into the mechanism of some process not directly observable.

By geometrical constructions of great beauty Huygens showed how the assumption of a disturbance in an imaginary all-pervading *ether*, composed of particles of perfect hardness and elasticity, would lead to waves spreading out in all directions, suffering

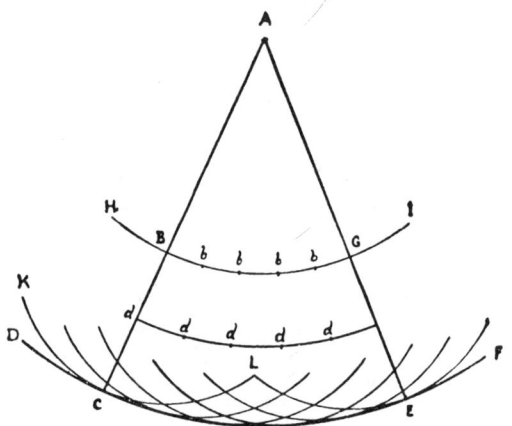

Fig. 22. Huygens' construction to show that light from a source, A, passing through an aperture, BG, will not sensibly bend into the shadow HC or GF. Only along CE will the wavelets sufficiently reinforce each other.

reflection and refraction according to the known laws of optics. Nor did Huygens fail to forestall the obvious objection to such an hypothesis, namely that waves of sound like those on a water surface are known to bend round objects into the " shadow " It is only necessary to suppose that every particle set in motion becomes itself a new centre of disturbance; then although waves *will* spread out into the shadow, they will be too feeble compared with those lying between the straight lines in Fig. 22 to arouse any sensation of sight.

While Huygens was working out this ingenious hypothesis Newton was carrying out experiments with thin plates similar to those performed by Hooke. The observation which led him to take up this study was the appearance of a black spot at the point where two prisms, having very slightly convex faces, were pressed firmly together. On rotating the prisms in such a way that the light was less obliquely incident on the prisms, circles of colour appeared, caused, as Newton quickly realised, by the thin plate of *air* between the glass surfaces.

To avoid the complication of dispersion (p. 124) produced by prisms Newton " took two Object glasses, the one a Planoconvex for a fourteen foot Telescope, and the other a large double Convex for one of about fifty foot; and upon this, laying the other with its plane side downwards, I pressed them slowly together, to make the colours successively emerge in the middle of the circles, and then slowly lifted the upper glass from the lower to make them successively vanish again in the same place." The colours formed a regular series from the centre outwards, and the diameter of the circle of each colour increased as the lenses were pressed together. From this it could be certainly inferred that each colour corresponded to an " air plate " of definite thickness; moreover Newton was able to calculate the actual thickness from the known curvatures of the lenses.

In order to form some hypothesis as to the *cause* of these curious facts, Newton with characteristic insight further simplified the problem by " allowing light of one pure colour from a prism to fall on a sheet of white paper which was then viewed by reflection from the lenses (using the latter as a looking-glass)." Then " appointing an Assistant to move the Prism to and fro about its Axis that all the Colours might successively fall on that Part of the Paper which I saw by Reflexion from that part of the Glasses where the Circles appeared, so that all the Colours might be successively reflected from the Circles to my Eye, whilst I held it immovable, I found the Circles which the red Light made to be manifestly bigger than those which were made by the blue and violet. And it was very pleasant to see them gradually swell or contract accordingly as the colour of the light was changed." But there was a striking difference between these circles and those formed in ordinary daylight (that is light composed of all spectral colours) for, in any one position of the prism, the circles were all of the same colour separated by bands of complete darkness. By placing a paper *behind* the lenses Newton saw circles of transmitted light corresponding to those of darkness formed in the reflection. "And from thence the origin of these Rings is manifest, namely that the Air between the Glasses, according to its various thickness is disposed in some places to reflect, and in others to transmit, the Light of any one Colour."

The last quotation gives Newton's conclusion as to the immediate cause of the systems of rings—" Newton's rings ", as they are still universally called—formed when light traverses thin plates of air in certain circumstances. Although plausible, it must be frankly admitted that it was a little too facile for one of Newton's extraordinary penetration. Newton evidently realised that this was far from being the whole story, for since it was only in the case of light of one colour that the thickness of the plate of air was the *sole* determining factor, it was evident that *light itself* plays some part in the process.

Let us get this matter quite clear. The point of contact between two curved glass surfaces is surrounded by rings of light variously coloured when the incident light is " white ", or by alternate rings of light and dark when the incident light is of one pure colour. Alteration of the thickness of the air film surrounding the point of contact alters the diameters of the rings, as does alteration of the colour of the monochromatic light. So much is *fact*. Newton's statement of the " manifest origin of these Rings " seems to have been regarded by him as an immediate and undoubted inference from these facts. If he had not accepted this

## ON WAVES AND WAVINESS

inference quite so readily, the whole history of the science of optics might have been entirely altered, since, as we shall now see, he realised quite as clearly as did Hooke and Huygens that *some* periodic disturbance was involved; where he differed from these thinkers was in seeking the periodicity elsewhere than in the light itself. It is essential to quote his actual words: " Every ray of light in its passage through any refracting Surface is put into a certain transient Constitution or State, which in the progress of the Ray returns at equal Intervals, and disposes the Ray at every return to be easily transmitted through the next refracting Surface, and between the returns to be easily reflected by it."

There is no ambiguity here: every ray of light acquires a certain state which returns at equal intervals; which seems to be implicit in Huygens' view. But whereas Huygens tacitly assumes that the periodicity is in the motion of light itself, Newton's emphasis is tacitly on the word " put " Moreover Newton goes on to apply this conception to his supposed proof that light is in a perpetual oscillation between the power of breaking through the next surface it meets or of being reflected thereby; whether it will be, as he supposes, transmitted or reflected by it being purely a matter of how far it has travelled since it was first " put " into this state of dither.

FIG. 23. Newton's diagram to illustrate his hypothesis of alternate " fits " of reflection and transmission.

We need not spend much time on considering Newton's views on the nature of the agent which "causes" light to alternate between these "fits" as he called them. "Whether it consists in a circulating or a vibrating motion of the Ray, or of the Medium, or something else, I do not here enquire", upon which, being human, he proceeds, ostensibly for the sake of "those that are averse from assenting to any new Discoveries, but such as they can explain by an Hypothesis", to enquire most ingeniously for a whole page. The substance of his enquiry being whether the rays of light do not stir up the particles of the refracting medium so that the latter propagates vibrations, which move faster than the ray, overtake it, and alternately "push" or "pull" it along and so dispose it to easy transmission or reflection. He returned to the question in the long list of prescient queries appended at the conclusion of the *Opticks*.

Newton, living in an age following close upon the Renaissance when men were renowned more for their enthusiasm than for their discretion in propounding opinions as to the nature of things, had an almost morbid horror of the word "hypothesis". But he used the term in a much more special sense than is the custom to-day. When the possibility of extending "gravity to ye orbe of the moon" first struck him it was no less an hypothesis than that of the wave-like propagation of light; but it was not an *idle* hypothesis. In other words it was capable of verification in the particular instance first considered and of being extended into realms at first undreamed of. On the other hand to speculate upon the *cause* of gravitation (p. 101) was rightly rejected by Newton as idle, in that such speculations as had been put forward led nowhere and were incapable of experimental verification.

It was otherwise with the problem of light. Periodicity was a fact established even more precisely by Newton's experiments than by any cited by Huygens. Surely it was a better hypothesis to suppose the periodicity to exist in light itself than in some hypothetical[1] interaction between light and the hypothetical medium (or "ether") in which it was supposed to travel? That Newton dallied with this hypothesis is proved by Query 8: "Do

[1] In putting forward this suggestion Newton was false to his own *Rules for Philosophising* (p. 74).

not all fix'd Bodies when heated beyond a certain degree emit light and shine; and is not this Emission performed by the vibrating motions of their parts?" and also by Query 13: "Do not several sorts of Rays make Vibrations of several bignesses, which according to their bignesses excite Sensations of several Colours, much after the manner that the Vibrations of the Air, according to their several bignesses, excite Sensations of several Sounds?" But the bias of his thought is shown in Query 28: "Are not all Hypotheses erroneous in which Light is supposed to consist in Pression or Motion, propagated through a fluid Medium?" and in Query 29: "Are not the Rays of Light very small Bodies emitted from shining Substances? For such Bodies will pass through uniform Mediums in right Lines without bending into the Shadow, which is in the nature of the Rays of Light?"

For Newton the value of an hypothesis (in the modern sense) is to be determined solely by its consistency with known facts and those subsequently discovered, and by its power of suggesting new lines of enquiry. There is no doubt that he was over hasty in rejecting the wave hypothesis on the ground that waves, *as we know them in water and air*, spread out round obstacles placed in their paths, whereas light does not appear to bend into the shadow. Huygens with greater perspicacity, was able to show that an *almost* complete absence of "spreading out" is not inconsistent with wave motion as such. Newton here fell into the same trap as had Tycho (p. 109), the instrument (human vision) being ordinarily (but see p. 143) incapable of detecting any "bending into the shadow".

A further test of a good hypothesis is whether it needs a number of subsidiary hypotheses to explain new discoveries. Of reflection and refraction both corpuscular (emission of small particles) and wave hypotheses were able to give plausible explanations; of the spectrum neither gave a very convincing picture, though the supposed unequal bending of corpuscles of unequal weights was perhaps the better of the two. But at the time that Newton and Huygens were fashioning their rival hypotheses three less familiar sets of phenomena had been lately discovered, some understanding of which is essential to the study of the further history of optics.

When light is transmitted through certain kinds of crystals it is invariably refracted, but often in a manner to which the ordinary laws of refraction are inapplicable. The first account of such observations was given by a Dane, Bartholin, in 1670, and was fully discussed by Huygens in his *Traité de la Lumière*. In secreting into rock crevices the particular mineral form of calcium carbonate which occurs principally in Iceland and which was the subject of these observations, nature had provided man with a potent weapon in the investigation of light. Iceland Spar is formed in huge crystals, often almost perfectly transparent, and fashioned in a peculiarly simple geometrical shape. Huygens confirmed Bartholin's observation that when a spot of sunlight is viewed through such a crystal, *two* spots of light are invariably seen, one formed according to the ordinary laws of refraction, the other displaced to one side even when the light falls normally on the face of the crystal. Moreover when the crystal is rotated the " ordinary " image remains fixed, while the " extraordinary " revolves round it. On viewing these images through a second crystal, either similarly orientated or with its principal optical section turned through 180°, the " ordinary " image suffered only " ordinary " refraction, and the " extraordinary " only " extraordinary " refraction; from which Huygens concluded that the two sets of light rays traversing Iceland Spar had by some means or other lost their normal characteristics of sunlight and been separated into two kinds. This hypothesis received a rude contradiction when, on rotating the second crystal into a position at right angles to that of the first, the ordinary image was extraordinarily refracted and vice-versa; moreover for all other positions of the second crystal, each image was split into two, thus giving four altogether, whose several brightnesses varied according to the relative positions of the crystals.

Huygens displayed great geometrical ingenuity in accounting for the splitting of the ray in the first instance by supposing that the distribution of the particles of matter in the crystal gave rise to unequal densities in different directions (it is known that heat travels in the crystal at unequal rates in different directions) and so cause the wave to be of a spherical form in the " ordinary " direction and of a spheroidal (i.e. flattened) form in the " extraordinary" direction. But assuming, as he did, that the periodic motion constituting light is *longitudinal*, and therefore perfectly symmetrical with respect to the axis along which it is moving, he was completely at a loss to explain the variation in its behaviour towards a crystal rotated *about that axis*.

Once again Newton gave evidence of an almost superhuman power of expressing the essential characteristics of a problem in a telling phrase: " Every Ray of Light has therefore two opposite Sides, originally endued with a Property on which the unusual [i.e. " extraordinary "] Refraction depends, and the other two opposite Sides not endued with that Property " (Query 26). The word " side " applied to a ray of light may at first strike us as somewhat inappropriate. But what Newton means is that there is some quality in *light itself* which determines how it will behave when striking a piece of material whose orientation in space can, unlike that of a piece of glass, be fixed by reason of the fact that it has other properties which differ in different directions. Such a quality certainly could not be imagined in an action propagated in pulses such as Huygens had suggested; but it is equally difficult to see how it could be associated with a beam of corpuscles.

One other property of light discussed by Newton in the *Opticks* is that which he calls "inflection", and which is now known as " diffraction ". Newton correctly attributes the first observation of this phenomenon to the monk Grimaldi, who in a book (*Physico—Mathesis de lumine, coloribus, et iride*) published posthumously in 1665, announced that if a thin object (such as a needle) be placed in a fine beam of light the shadow of the needle is apparently broadened by the presence of a number of fine fringes at its edges, while at the same time the existence of fringes within the shadow implies that light can to some extent at least bend round the edges of objects. Grimaldi himself likened these effects to the spreading of ripples beyond an obstacle when a stone is thrown into water.

Here, one would have thought, Newton might have seen some confirmation of Huygens' hypothesis of waves *almost* obliterating each other except in the region outside the straight lines from the source to the object. But his First Query, " Do not Bodies act upon Light at a distance?" shows that he was too prone to extend the principle of gravitating matter into every problem of nature.

During the hundred years preceding the death of Newton almost every fundamental property of light had been discovered. Experiment and interpretation had shown that it is some kind of activity which is propagated in straight lines at a speed of about 200,000 miles a second, is reflected from the surface of bodies, changed in direction when traversing transparent media, is characterised by *some* periodic quality " returning at regular intervals during the progress of the ray ", has *lateral* properties as well as longitudinal, and does appear capable of *slight* deviation round the edges of bodies. The hypothesis which gave expression to its undoubted periodic quality was undoubtedly wrong in so far as it failed to account for the " sides " of the rays. On the other hand the simpler hypothesis of a " jet " of particles missed altogether this most fundamental property of periodicity, which could be explained only by supposing a quite arbitrary interaction between the " particles " of light and those of matter.

The dilemma is most instructive. To Newton, whose major triumph was in the analysis of motion into the interaction of gravitating particles of matter, nothing seemed more natural than to suppose that light is merely a more " subtle " kind of " matter ", that is subject to laws similar to those obeyed by tangible bodies. Seeing the inadequacy of Huygens' hypothesis he failed to recognise that its fundamental feature—periodicity—was in fact inescapable.

To Huygens who, with Wallis and others, had laid the foundations of the theory of elasticity, impact, and periodic motion, the wave-like character of such a high-speed disturbance as that of light seemed sufficiently probable *even in ignorance of the crucial evidence which Newton's " rings " provided.*

Thus temperament and interest largely determine the way in which we visualise the workings of nature. Nor is this the only lesson. For the subsequent history of optics illustrates the consequences of adopting a " wrong " hypothesis. From such ingeniously contrived and honestly interpreted experiments as Newton's " truth will out ". In familiar language we may say that Newton

ON WAVES AND WAVINESS 145

inadvertently let the " periodic " cat out of the bag, but stuffed it in again sufficiently dexterously to deceive his less keen-sighted followers. For though, as modern research has shown, there was hardly anything actually *wrong* in Newton's work on the nature of light, he put into the shade that quality, the study of which was ultimately to be so wonderfully fertile in promoting new discoveries.

## SOURCES FOR CHAPTER XIII

The foundation of our knowledge of the pressure of the atmosphere (and of other fluids) is dealt with in detail in Mach, *op. cit.*, Gen. Bibl.; Wolf, *op. cit.*, Gen. Bibl. A new approach of special interest is contained in J. B. Conant, *op. cit.*, Gen. Bibl.

R. Hooke, *Micrographia* (Alembic Reprint No. 5).

C. Huygens, *Traité de la Lumière* (Leiden, 1690, Eng. trans., S. P. Thompson, London, 1912).

## CHAPTER XIV
## "THE SUBJECT OF THE VERB TO UNDULATE"

FOR a century after the first publication of the *Opticks* hardly any advance was made in the study of light, except in the improvement of telescopes made possible by the invention of the achromatic objective by Chester Moor Hall and later Dollond in defiance of Newton's dictum that it was impossible. On the other hand, with the reading of a paper by Thomas Young to the Royal Society in London on 16th January 1800, in which Newton's discoveries were admired, but his hypothesis denounced as untenable, there commenced a new epoch during which not only was the periodic or wave character of light abundantly confirmed, but its fertilising influence given full scope.

The analogy of sound already mentioned (p. 130) was once again to play a leading part, for Young, who was a qualified doctor of medicine, tells us that his interest in the theory of light was awakened by the similarity he found to exist between the colours in thin plates of various materials to which his attention had been drawn by an early study of the human voice as a machine for the production of sounds. Indeed Young is an admirable example of the wisdom of a wide interest in natural phenomena and human affairs; he was an accomplished classical scholar, master of several ancient and modern languages, and a writer of prose whose elegance and lucidity are a model to those men of science who seem to regard obscurity of diction as a necessary accompaniment to profundity of thought. Like Newton also he gave early evidence of uncommon manipulative skill.

Coming then to a study of light with a knowledge of elasticity in general and of those elastic vibrations of air which we perceive as sound far in advance of that of his British contemporaries, Young drew attention to the complete inadequacy of any purely " corpuscular " hypothesis to explain the well-attested fact that beams of light are commonly refracted in part, and in part

## "THE SUBJECT OF THE VERB TO UNDULATE" 147

reflected, when they impinge upon the surface of a transparent medium like glass. If however light be allowed to possess a wave-like character, then the phenomena of " Newton's rings " and of thin plates can readily be accounted for by drawing upon no further properties than are already known to be possessed by waves on water or waves of sound. The retardation of phase by half a wave length, which Young found himself compelled to assume, and later to explain, forms no exception, since a similar phenomenon occurs in the theory of organ pipes.

The property in question, which both Huygens and Hooke seem to have overlooked, was the tendency of waves to " get out of step " and thus to interfere with one another. Reference to Fig. 24 should make this clear; here it is shown how water at a certain point may remain completely undisturbed if the waves from two sources always have equal and opposite tendencies at that point. Thus if one disturbance is tending to heap the water up to a maximum height, or " crest ", at the same time as an equally strong disturbance is tending to depress it into a "trough", the two tendencies will cancel one another and no motion will result. Moreover if the disturbances be of the same frequency it will evidently follow that if the crest of one coincides with the trough of the other, then as the waves pass on there will *always* be equal and opposite tendencies at the same point. On the other hand if the point of reference is so situated with respect to the sources that the disturbances arrive " in step ", then the effect of the displacement will be double what it would have been as a result of one train of waves only.

This effect of interference is most marked when the waves are emanating from one source but are broken up by an obstacle. They may sometimes be seen beyond a breakwater with two gaps in it. The periodic character of sound however is most commonly revealed by the regular humming or " beating " noise which accompanies the mingling of sounds from two sources whose frequencies differ by a small amount. In this case there is no " stationary " effect as in the water waves, but the difference in frequency results in one wave gradually getting into step with the other, when the sound swells up, and then gradually getting out of step when the sound dies away.

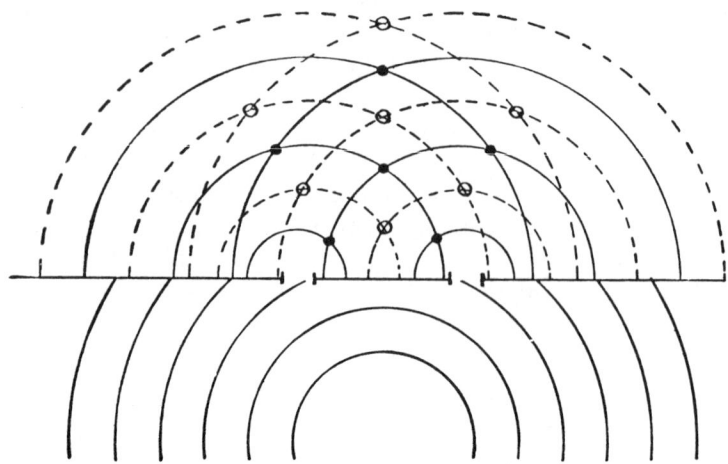

FIG. 24. Interference of waves from common source. Dots and circles show points of reinforcement. At other points of intersection the waves are in opposite phases, so nullify one another.

It was clear to Young (as Fig. 24 shows) that the explanation of Newton's rings needed no subsidiary hypothesis but only the assumption of a periodicity in light itself. And he put this matter beyond any sort of doubt by a simple experiment whose ingenuity would have done credit to Newton himself. By allowing a fine pencil of light from a single source to emerge through two tiny holes very close together he was able to produce alternate rings of different colours. When monochromatic light was used then the rings were of alternate light and darkness. How could two sources of light produce darkness except by interfering with one another? And how could interference occur unless there exists in light an *alternation* of states rapidly following upon each other?

A second experiment even more crucial than the above is described by Young as follows: " I brought into the sunbeam a slip of card, about one-thirtieth of an inch in breadth, and observed its shadow . . . Besides the fringes of colours on each side of the shadow, the shadow itself was divided by similar parallel fringes of smaller dimensions, differing in number according to the distance at which the shadow was observed, but leaving the middle of the shadow always white." This was a most remarkable observation, supplying an answer to Newton's objection that waves, if they exist, must bend into a shadow. Young's experiment shows that, provided the shadow is narrow enough (in conformity with the exceedingly small wave length involved), then they *do*. That the bands of light and darkness within the shadow were produced by two sources of *light* (i.e. light passing *both* sides of the strip of card), Young proved by the ingenious expedient of covering *one* side of the strip, when the bands vanished.[1]

Having satisfied himself of the undoubted wave-like character of light, Young showed how the distances between corresponding points of the periodic disturbance—the wave length—could be readily calculated from the distances between the interference fringes, much in the same way as the wave length or frequency of sounds can be calculated from the speed with which the " beats " follow one another. He found that the wave length varies from twenty-six millionths of an inch in the case of red light to sixteen millionths in the case of violet.

It might be supposed that to the propounder of such an ingenious and economical hypothesis every encouragement would have been given by his contemporaries. But it was not so. It happened that the publication of his views roughly coincided with the foundation of a number of independent journals which aimed at the exposition and criticism of novel ideas whether in science, in letters, or in the arts. The most prominent of these " reviews "—the *Edinburgh Review*—numbered among its contributors a young man who combined with the possession of wide general knowledge and a pungent literary style a very rudimentary sense of common justice (he wrote anonymously) and great confidence in his own immature and ill-founded opinions. He attacked Young with all the cheap weapons of the successful lawyer (as Lord Brougham he afterwards became Lord Chancellor) and though Young replied in a pamphlet written with equal skill and greater politeness, such was the mentality of the

[1] It should be noted that faint bands also occur on the *outer* edges of the shadow. These correspond to Grimaldi's bands (p. 143), and were not fully understood by Young. They were finally explained by Fresnel by means of an extension of Huygens' wave geometry.

" polite " world that the wave theory might have suffered sterility and oblivion had not sounder critics revived it in France. In 1815 Augustin Fresnel, a brilliant young military engineer[1] and mathematician, submitted to the Academy a paper on diffraction, which, as was the custom of that learned body, was reported upon by two members—Arago and Poinsot. The former took up the matter with great enthusiasm and drew Fresnel's attention to the almost identical views of Young published fifteen years previously. Although this was the year of the battle of Waterloo, Fresnel paid a generous tribute to Young and they corresponded frequently until the year 1827 when the death of the former put an end to a career full of great promise.

Meanwhile in 1809 there was published in Paris an account of one of the most remarkable " accidental " discoveries in the history of science. Etienne Malus, a young army officer whose interest had been awakened in the science of optics, happened one evening to be viewing through a crystal of Iceland Spar the image of the setting sun reflected from the distant windows of the Palais Luxembourg. He saw as usual the two images familiar from the time of Bartholin and Huygens; but chancing to rotate the crystal he was surprised to see that they varied in intensity according to the position of the crystal. Thus the light passing through his single crystal was behaving just as had the light which in Huygens' experiments (p. 142) had traversed two crystals whose axes were set obliquely to one another. In other words, as Malus realised, the act of reflection may in certain circumstances impress a one-sidedness on the rays of light. Malus fully confirmed this opinion by causing light to be reflected by a variety of substances; indeed it is possible by careful adjustment to *polarise* completely, as Malus called it, the reflected beam, so that on its traversing a rotating Iceland Spar crystal, one of the images is seen not merely to become dimmer, but to disappear altogether.

The next problem was of course to discover the nature of this

[1] The astonishing output of first-class minds in mathematical physics which occurred in France at the turn of the century was due in no small measure to the Convention's support and encouragement of the school of engineering, founded in 1794 as the *Ecole de Travaux publics* and re-named *Ecole Polytechnique* in 1795. The comparative freedom for research accorded to the teachers at the " Polytechnique " is an example to " socialising " governments of the present day.

"one-sidedness" which, as Newton had correctly pointed out (p. 143), cannot be conceived in a longitudinal wave such as sound can be shown to be. There is however another kind of wave to which that on the *surface* of water approximates; this is the form of periodic motion in which the displacement occurs, *not* to and fro along the axis of the advancing wave, but in a plane at right angles to it. Young recognised and unambiguously announced that some such wave motion must be characteristic of light. The suggestion met with ridicule; and since the medium (or " ether ") whose undulations were to be regarded as producing the sensation of light, as the pulsation of air causes that of sound, would in that event have to possess enormous elasticity (to permit of very rapid vibrations) and very low density (to allow of a very high velocity), we cannot altogether blame his critics.

It is of great interest to the historian that, in 1821, Fresnel admitted that " when M. Arago and I had remarked (in 1816) that two rays polarised at right angles always give the same quantity of light by their union, I thought this might be explained by supposing the vibrations to be transverse, and to be at right angles when the rays are polarised at right angles. But this supposition was so contrary to the received ideas on the nature of the vibrations of elastic fluids " that he hesitated to adopt it.

This problem is so far removed from ordinary modes of thought that it may perhaps be desirable to explain what was the nature of the mechanism which Young and Fresnel had in mind. The conception of a transverse vibration is best visualised by the vibration of a stretched string. Now it is perfectly evident that such a string may be made to vibrate in any plane at will; it is thus to be supposed that any " pencil " of light rays, however small, will ordinarily be composed of individual rays vibrating in planes cutting the axis of the pencil at all possible angles. Now let the pencil traverse some medium, like Iceland Spar, whose properties differ in different directions. It may then be shown, and was so by Fresnel by mathematical arguments of great elegance, that the rays will be, as it were, combed out, those which are vibrating in one plane being transmitted as the " ordinary " ray, those in a plane at right angles to the former will go to make up the " extraordinary " ray. But what of all the other rays vibrating in planes intermediate between these mutually perpendicular planes? It had been recognised at least from the time of Galileo that many of the quantities considered in " mathematical demonstrations " of science can be completely specified only when the *direction* as well as the magnitude is given; such quantities, now known as " vectors ", are exemplified by forces, velocities and, in the simplest case of all, mere displacements in space. One of the interesting features of these vectors is that their joint effect is not deducible by simple addition—you cannot add a walk of two miles due north to one of three miles due east. They may however be " resolved " in two mutually perpendicular directions (or of course any other chosen directions, but as will be seen, this particular pair is the most useful) and then the " pieces" can be added up in the usual way.

It should now be evident that any *vibration* in any plane can be resolved into two vibrations in a pair of mutually perpendicular planes, and the " resolved parts " added up in the usual way. The fate of all the rays entering the Iceland Spar is now clear; the " combing out " is much more complete than might at first have been imagined. And what is so seductive in this hypothesis is that it explains in one scheme both polarisation and double refraction.

Now it is always agreed that the test of an hypothesis is its *generality* and *precision*. Two experimental facts must now be recorded which severally left Young and Fresnel in no doubt that the periodicity of light resembles more closely the transverse vibration of a string than the longitudinal oscillation of a sound pulse.

It is well known that the resolution of vectors in directions at right angles to one another is the most useful, since mutual reaction between such pairs is impossible. A horse walking on a tow-path can urge a barge forward by means of a rope inclined at an angle to the direction in which the barge actually moves; but all the horses in the world pulling in a direction at *right angles* to the bank will not urge the barge forward. When Fresnel and Arago found it *impossible to detect any interference* between rays polarised in mutually perpendicular planes, the former *knew* that a transverse vibration was implied; but Arago allowed his prejudice to overcome the evidence of his senses.

Meanwhile Young had observed that a beam of polarised light produces interference effects by the simple expedient of being made to traverse a plate of certain crystals. Recalling an observation of Chladni that sound travels one and a quarter times as fast *along* the fibres of a fir tree as *across* them, Young correctly drew the conclusion that interference in this case is caused not by any difference in the *lengths* of the paths, such as occurs in thin films, but in the unequal *times* taken to traverse equal paths on account of the differences in the velocities of the ordinary and extraordinary rays.[1] So the rays will, as he had already shown in the case of thin films, " get out of step ", but for a different reason.

By 1818 Fresnel had, independently of Young, proved that light *must* be propagated in a wave-like manner; by 1822 he had proved that the displacements of this wave-like propagation must be transverse in nature. In the few years of life which yet remained to him he arrived at one brilliant verification after another, among which we can mention only the production of interference by an ingenious " bi-prism " of glass which, being free from small orifices, gives rise to no possibility of *diffraction* effects (p. 143), and the *calculation* of a new kind of polarisation which was later produced by experiment. His was as great a triumph as any in the history of science. But to the end his modesty never left him, nor his generous appreciation of his rival's worth. A letter written to Young in 1824 perfectly reveals his character: " For a long time that sensibility or that vanity, which people call love of glory, is much blunted in me. I labour much less to catch the suffrages of the public than to obtain an inward approval which has always been the sweetest reward of my efforts. Without doubt I have often wanted the spur of vanity to excite me to pursue my researches in moments of disgust and discouragement. But all the compliments which I have received from M. Arago, de la Place, or Biot, never gave me so much pleasure as the discovery of a theoretical truth or the confirmation of a calculation by experiment."

---

[1] In this case, since the ordinary and extraordinary rays are polarised in mutually perpendicular planes, interference occurs only if the light emerging from the crystal is viewed through an " analysing " crystal which recombines them.

If his contemporaries could have been satisfied with the beautifully adequate *form* of the propagation of light revealed by the genius of Fresnel, there would have been no opposition. Since beams of light do in fact lose their power of mutual interference when subject to no influence save that of two crystal plates whose axes are mutually perpendicular and whose plane is at right angles to the axis of the light beam, it follows that the vibration which constitutes the light has no component along the axis of propagation. The experimental facts *imply* the theoretical conclusion in a manner similar to that in which the shape of a triangle *implies* the sum of its angles being equal to two right angles. The age of Fresnel's triumph was too close however to the triumphant mastery of " matter " supposed to be implied in the Newtonian system. The " stars in their courses ", if not controllable, were at least predictable; sound was adequately explained in terms of the longitudinal pulses of material particles. Heat, magnetism, and electricity were being marshalled along the same lines; and the temper of the age is admirably revealed by the fact that in the system of physiology completed by Haller in 1765 the action of the *nerves* was " explained " as the conduction by these " tubes " of an " element more gross than fire ".

It was inevitable then that once the scientific world had been convinced that the transmission of light was undulatory in character it would not rest satisfied until it had discovered some " stuff " to undulate; and further, it was inevitable that such a " stuff " would be endowed with properties consistent with the only laws—those of " matter "—which had then been successfully established. The " stuff " was called the " luminiferous ether "; the word " ether " having already a long history behind it, going back to Aristotle who had applied it to a supposed fifth element filling the celestial spaces, and out of which the celestial bodies were fashioned. The next task was to find out how the undulations of this conjectural ether could be made to " work "; or, in technical terms, to discover its dynamical laws.

Now since the ether, if it existed at all, must be sufficiently subtle to penetrate all transparent bodies, even those as hard as diamond; and so tenuous as to suffer the heavenly bodies to perform their ceaseless rounds without measurable retardation, it was naturally conceived as an exceedingly rarefied *fluid*. Unfortunately the conception of " fluidity " is in diametrical opposition to that of " rigidity "; and without rigidity it is impossible to conceive of " deformability ". But the tendency to resist a deformation, to overshoot the mark, and so to set up a further deformation in the opposite direction is just what is meant by a *transverse* wave. In familiar terms, by squeezing or stretching a liquid you can make it oscillate, but only a *solid* tail can wag.

The first response to this dilemma was, as we have seen, to deny the possibility of transverse vibrations on the ground that a medium which could propagate them at the speed of light would have to be endowed with elasticity far greater than that of steel. But when it gradually dawned on men's minds that the " transversality " was an inescapable implication of the *facts* of polarisation and double refraction, a change of view occurred, and on the tacit assumption that the ether must be a bit " queer " an attempt was made to picture it in a form as free from " queerness " as possible. But though many able minds worked at this problem from about 1830 onwards for about half a century, no satisfactory solution was ever found; nor could it be, for the whole quest was founded on a contradiction which has often fogged, and is still fogging, much scientific thought.

The whole trouble arises from the nature of man's thought processes. He discovers that the propagation of light occurs in the *form* of a transverse wave such as he can *see* travelling in a vibrating solid. Quite rightly he seeks further properties of light such as he knows to accompany the *visible* transverse vibrations, one of them being the " solid " itself. But when he fails to discover them he either, like the contemporaries of Young and Fresnel, refuses to face the facts or, like their successors, tries to construct a " picture " of the mechanism whose artificiality and inner contradictions should be only too evident. The unsatisfactory nature of this phase of thought has driven one modern school to reject all " pictures " of the nature of things as superfluous and misleading. But to do so would almost certainly bring discovery to a standstill; after all it was the " picture " of interfering water waves which enabled Young to undermine the prejudice of a century against any form of undulatory theory. What is to be done then? The answer given by Whitehead is that " science is concerned not with the causes but the coherence of nature ". In visible objects transverse vibrations only " go with " rigidity; but it involves only surprised incredulity and not *contradiction* when we find that the *form* of transverse vibration in light cannot " go with " rigidity without making nonsense of the whole thing. *If* there is a stuff having the properties of attenuated *matter* whose " waggings " constitute light, *then* it must be " solid "; but *there is no independent evidence whatever that there is such a stuff*. The ether exists only as an " excuse " for light or, as the late Lord Salisbury brilliantly characterised it, as the "nominative of the verb 'to undulate'." If we can conceive (which is difficult but not *incoherent*) of light being *itself* a disturbance in the *form* of a transverse wave without expecting it to be the result of the undulation of something else, the contradictions disappear. The initial mistake is to suppose that the familiar is also the fundamental; " matter " is familiar, therefore we try to " explain " everything else in terms of the properties of matter. But it may be, as one of the authors of *Genesis* guessed in an inspired moment, that light is nearer to the birth of things, in which case it would be more logical to " explain matter " in terms of light. There is more than a little evidence that this is now being done.

We shall close this long and difficult discussion with a reference to the so-called " crucial experiment " of Foucault. To explain refraction in terms of a stream of gravitating corpuscles (Newton) involves the increase of the velocity in the denser medium, e.g. glass. If a wave system is postulated, the reverse is true. In 1850 Foucault succeeded in comparing the velocities of light in air and in water, the result leaving no doubt that light travels faster in *air*. After the loud and prolonged applause of the protagonists of the wave theory had died away, certain other properties of light were observed which could *not* be explained except on the assumption that light has some sort of corpuscular character; these effects are the effects of droplets rather than those of waves. In hailing the Foucault experiment as a " proof " of the truth of the wave theory and of the falsity of the corpuscular theory, the critics had failed to realise that what it proved was that the wave theory was true so far as it went, and that *some* at least of the postulates of the corpuscular theory were false. It did *not* prove that light is not corpuscular; it *did* prove that it is wave-like— the sort of distinction which has to be made in every scientific dilemma.

## SOURCES FOR CHAPTER XIV

There is an excellent short biography of Thomas Young. F. Oldham, *Thomas Young* (London, 1933).

H. Buckley, *op. cit.*, Gen. Bibl., Ch. IV. and VI.

The detailed account of the corpuscular-wave controversy in Whewell *History of the Inductive Sciences*, Vol. II (London, 1837), is still of great value.

CHAPTER XV

# THE SEARCH FOR THE ELEMENTS OF MATTER

FRESNEL died in 1827. The clearest testimony to the greatness of his achievement lies in the fact that within fifty years of his death the wave conception had so enlarged the scope of the study of the spectrum that man was already able to apply it to the chemical investigation of the sun. But before we continue the story of the expansion of optics, we must spend some time watching with sympathetic eye the growing pains of that branch of the science of matter and motion which deals with those qualities which physics rather superciliously sets on one side as being of no account.

The science of chemistry was of all branches the one least cultivated by the ablest minds in Greece. Histories of chemistry commonly recount the speculations of the pre-Socratic nature-philosophers concerning the " elements " of things; such accounts are justified by the fact that the dicta of these speculators, systematised by Aristotle, continued to exercise a profound influence on European thought right down to the eighteenth century; and even to-day we echo the Aristotelian world-view every time we speak of a person being " in his element ". But it is equally clear that many writers on natural science have failed to realise that the problem of the pre-Socratics (see Chs. I and II) was a philosophical one. It has been one of the great misfortunes of human culture that the same word—element—has been applied to concepts so completely distinct as the " water " of Thales and the " oxygen " of Lavoisier. It has not only befogged much so-called history of science, but its persistence is even at this time only too apparent in the ill-defined belief of many otherwise intelligent people that " qualities " can somehow be " superinduced " upon matter in the same way as a few stage properties can be turned into fairyland. Thus common tap water has only to be suitably

# THE SEARCH FOR THE ELEMENTS OF MATTER 159

" medicated " to acquire " tonic properties " and " new sources of vital energy ". Fifteen hundred years of civilisation seem to have raised us very little above the spurious level of later Alexandrian alchemy.

FIG. 25. Egyptian goldsmiths washing, weighing, and melting gold.

On the other hand the store of *practical* chemical knowledge handed down from generation to generation among eastern peoples has perhaps been underrated. Copper was extracted from its ores in Mesopotamia at least as early as B.C. 3500, and tools of iron, whose extraction is much more difficult, have been found in Egyptian tombs dating from B.C. 2500. Good glass was manufactured at Tell-el-Amarna in the fourteenth century before our era, and the application of natural dyes can be traced back to very early times.

With so much sound chemical knowledge to draw upon it is strange that the Greeks who visited Egypt and who, as seems most probable, *rationalised* the practical geometry they learned there, did not also lay the basis of a rational science of chemistry. One can only surmise that the ingrained dislike and even scorn of manual labour which was one of the weaknesses of the Greek character, precluded them from a science whose very springs lie in the application of reason to the results of *experiments*. Thus it came about when a system of chemistry was at last put into writing, the result was a deformed mongrel combining inharmoniously the sound, if unimaginative, technology of the east with the vague, if daring, speculations of the Greeks. The two points of view had nothing in common, and the records which have come down to us must be regarded, apart from the valuable descriptions of *methods* which they contain, rather as warnings than as guides in the eternal struggle to search out the nature of things.

Our knowledge of the origin and progress of ancient technology is based on the pictures preserved in tombs and temples, and on vestiges of the products and even of the factories themselves. There are no written records except casual allusions in letters and legal " documents ". But concerning the origin of chemistry we have the famous Theban papyrus (written in Greek about A.D. 300, but believed to be a copy of much older Egyptian sources) brought to light in 1828. It was divided into two parts, one of which is preserved in Leyden, and one in Stockholm; both contain detailed recipes for carrying out chemical operations on common objects in such a way as to render them similar in appearance to rare metals or gems. It is perfectly clear that the writer was under no delusion that the change brought about was anything more than superficial.

In the treatises on chemistry emanating from Alexandria from about the beginning of the Christian era there is a complete change of tone. The " goldsmiths " of the Leyden papyrus may have intended to deceive others; the " chemists " of the Alexandrian treatises, men whose minds had developed under the influence of debased Greek philosophy, were in the far more pathetic state of deluding themselves.

With records so scanty and corrupt (many of them written under the " borrowed " names of such famous men as Demokritos) it is difficult to analyse the reasoning underlying this belief in the possibility of " transmuting " the baser into the nobler metals. It is not unlikely however that it is based on the uncritical application of Aristotle's principle[1] that all things partake both of " matter " and " form ", and that the " form " may alter without any fundamental alteration of the " matter ". A crude analogy is that from the same block of marble different sculptors might hew very different figures. It is probable that

[1] We shall not at this stage enter into a full exposition of Aristotle's teaching except to emphasise the fact that modification of the form could come only from " within " the thing itself—an idea doubtless suggested by his deep and varied studies of the development of living organisms (see Ch. XXV).

# THE SEARCH FOR THE ELEMENTS OF MATTER

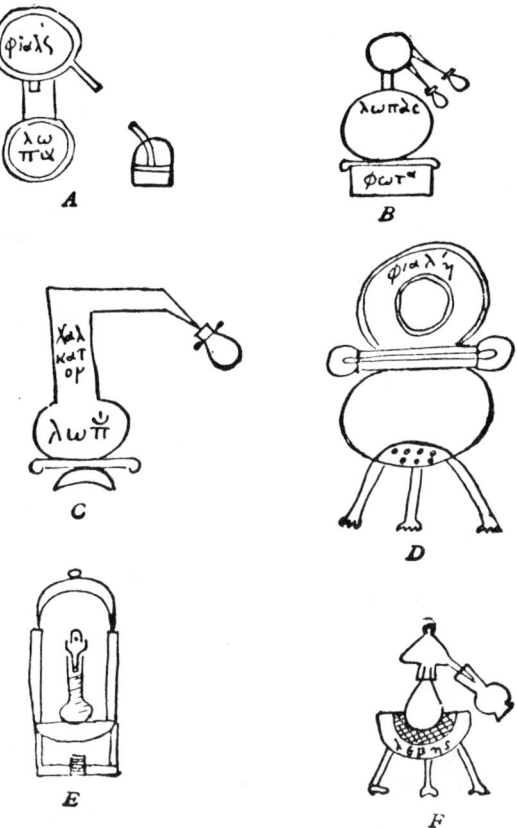

Fig. 26. Illustrations of chemical apparatus copied from Greek MSS of Zosimos and others. A, B, C and F are for distillation and *later* came to be known as *alembics*. D and E show the use of a brazier and sandbath respectively.

the original chemists interpreted Aristotle's teaching in a manner no less crude, and so came to delude themselves into the belief that the whitening of copper by the fumes of arsenic had in truth produced a " kind " of silver. They had failed to realise the principle of individual identity, namely that a thing is what it is, and can in no circumstances be a " kind " of anything else. This misconception is not to be wondered at, since, as has already been noted, many people still find great difficulty in seeing it in a clear light. We shall revert to this matter frequently as it is one of the most valuable lessons to be learnt from the history of chemistry.

Though the theory underlying Alexandrian chemistry was misleading and confused, we should do such pioneers as Zozimos and Olympiodoros scant justice did we not note in passing the ingenious apparatus devised by them for such operations as distillation, crystallisation, and the like. Of the use of this apparatus we have unquestionable evidence in the annotated diagrams contained in early manuscripts believed to have been written by Zozimos and others.

The practice of chemistry in Western Europe dates from the early years of the twelfth century; Albert the Great, Archbishop of Cologne, undoubtedly spent much time in repeating many of the operations reputed to bring about the transmutation of elements. He was a man of an acute and critical mind, so it is not surprising that he finally concluded that any such claim was an empty sham. By this time the word " alchemy " had displaced the earlier word " chemistry ", and this provides an indication of the path by which the records reached Europe from Alexandria. "Alchemy " is the old Egyptian word—*chem* or *chemia*—for " black " in its Greek form with the *Arabic* article " al " prefixed to it. The Greek-Egyptian manuscripts of which we have spoken were quite unknown to the Europe of the Middle Ages; Albert and his followers would have studied *Latin* translations of a Spanish dictation of an Arabic text of the Greek copy of the original Egyptian discovery. It is no wonder that some wit has described these " versions " as " perversions ". But the " versions" of the Greek manuscripts having been preserved and duplicated in Arabic for at least four centuries, it was naturally inferred that the Arabs were the pioneers of the craft. So much so indeed that an influential treatise which became current about the fourteenth century purported to be a Latin translation of a previously unknown work of the famous Arabian alchemist, Jabir (Latinised to the better known form " Geber "); but Jabir had been dead for five hundred years, and no Arabic original of this text has ever been unearthed.

THE SEARCH FOR THE ELEMENTS OF MATTER 163

The Arabs certainly " cultivated " chemistry with great assiduity (many of their terms such as " alembic "—a kind of still—have come down to us) but it is at least questionable whether they made any appreciable progress. The so-called *pseudo*-Geber —the unknown author of the book mentioned above—was the means of introducing a new version of the transmutation theory which played a great part in the shaping of subsequent chemical views. In this connexion it is interesting to note how opinion of the origin of a discovery often oscillates between " earlier " and " later ". By the workers of the sixteenth and seventeenth centuries this theory was assumed to be of Arabic origin; in the late nineteenth century the authorship of the *book* was proved to be not earlier than the fourteenth century, so the *theory* was dated in the same way. But recent research has revealed an almost identical theory in the writings of a secret society of A.D. 950—the Brethren of Purity, at Basra in Mesopotamia. Whatever may have been its origin, its interest for us lies in its connecting the theory of transmutation with a theory of *combustion*—a topic apparently unheeded by the Alexandrians.

It was early recognised that the rarer metals, gold and silver, were characterised by a strong resistance to the action of fire; whereas metals such as lead and tin quickly changed into "earths" when strongly heated. It was also known that certain minerals (sulphides of metals, such as lead sulphide, " galena ") give off a strong smell of " burnt sulphur " when heated in an open vessel; mercury on the other hand distilled almost unchanged as regards its lustre. From such observations probably arose the general hypothesis that all metals are complexes of " mercury " and " sulphur " in varying degrees—the rarer and therefore less combustible the metal the less of " sulphur " it contained. The Graeco-Egyptian view lingered on in the belief that these constituents were not the substances commonly known to ordinary mortals as sulphur and mercury but were the essential " principles ", as it were, of " metallicity " and " combustibility ". While ordinary mercury was held to be but little contaminated with the sulphureous principle, the opposite was true of common sulphur. The hypothesis has all the weakness of the age in which it was born, but its merit is that it is the first step towards *generalisation*: that is, an attempt to explain varied phenomena by one common law.

The reign of alchemy in Western Europe lasted roughly from 1250 to 1500. At the beginning of this period it was exhaustively studied by two great ecclesiastics—Albert the Great in Germany and Roger Bacon in England. Albert quite rightly rejected the pretensions of the alchemists to turn base metals into gold, and perhaps in consequence of his influence the subject seems to have lost interest for the ecclesiastical scholars who at this time were almost the only men of learning. Bacon, perhaps with some insight into the necessity for finding a true basis for chemistry, continued to spend time and money in the prosecution of the subject. The principal fruit of his labour has been said to be the invention of gunpowder; though whether this was an original discovery or had come into his possession from Arabic sources in Spain, cannot now be determined; the Chinese certainly used it prior to Bacon's time (thirteenth century). Its use spread fairly rapidly with the most profound influence on the course of history. Thus it is reported that Warkworth Castle, one of the strongholds of the powerful Percy family, surrendered to Henry IV in 1403 after only a few stone shot had been hurled into the courtyard by the king's primitive artillery. The new weapon thus greatly strengthened the hand of the sovereign against local overlords and thereby hastened the collapse of feudalism.

Not only was gunpowder the fruit of labours directed to quite a different end, but the strong mineral acids, alcohol, and many salts came into use at this time. No one however had the slightest idea as to the relations between these substances, though their properties were made use of by the hordes of charlatans who travelled great distances in the hope of gaining wealth by tricking the half-educated rulers into the belief that by purchasing the secret of the imaginary " philosopher's stone " they would be able to increase their gold reserve at will. Thus there exist in museums to this day coins of a white alloy of silver and gold, which had only to be dipped in nitric acid for the silver to be dissolved away leaving the coin apparently converted into gold.

So the centuries passed, and it was not till the early years of the sixteenth century that a definite change came over the ideals

THE SEARCH FOR THE ELEMENTS OF MATTER 165

of alchemists who thereafter, even if the belief in transmutation lingered on for at least two hundred years, allowed this ill-conceived folly to take second place in their interests. The signal for this change of faith was bellowed forth with great enthusiasm and no little coarse invective against those who did not immediately agree with him, by a German, Philippus Aureolus Paracelsus Theophrastus Bombastus von Hohenheim—usually more conveniently known as Paracelsus. He was born in 1493 and combined the study of mining and mineralogy with that of medicine in which he graduated in Italy. His habit of wandering about the mines, conversing with the workmen and observing nature itself, was symptomatic of the growing intellectual discontent of all sorts of men.

The art of medicine—it could hardly be called a science—was taught from the books of Galen (see p. 330), the greatest systematic physician of antiquity, who died in Rome about A.D. 200, or of the Arab, Ibn-Sina (Avicenna) whose *Canon of Medicine*, written a little later than A.D. 1000, summarised the whole of ancient medicine including that of Galen. Such " physiology " as there was in the medical teaching of the sixteenth century was based on Galen's doctrine of the harmony of the " four humours ". Since these humours were entirely hypothetical, any sort of experimental investigation of their effects was of necessity as sterile as the search for the philosopher's stone.

With an almost uncanny insight Paracelsus recognised that physiology—the science of how living bodies *work*—is to a large extent concerned with the same sort of happenings as the alchemist saw in his alembics; in other words, physiology is largely based on chemistry. Therefore he taught, " the object of chemistry is not to make gold, but to prepare medicines." In the actual *use* of chemical remedies he was anticipated by Arnold Villeneuve in the fourteenth century; but the outcrop of " Herbals " (see Ch. XXVIII) during the sixteenth and seventeenth centuries showed that drugs were still mainly derived from plants.

In his working out of this revolutionary doctrine there is, as history always teaches us to expect, a generous hang-over of superfluous nonsense derived from current modes of thought. With truly modern insight he recognises that the immediate cause of certain digestive ailments is a lack of balance between the various juices which the " chemical laboratories " of the body are able to secrete; but at the same time he must drag in a presiding spirit, *Archeus*, whose " illness " is the " cause " of the imbalance.

Despite his inability to *maintain* a clear train of thought once started, he nevertheless had said enough to set the practice of chemistry on a new road. Doubtless many were the victims of his pupils' chemical enthusiasm—it is a matter of history that the University of Paris had to forbid the indiscriminate use of powerful metallic poisons such as mercury compounds—but great changes of any kind always involve exaggeration.

The immediate influence of Paracelsus on chemistry as such was almost negligible; the addition of another " element "— " salt "—to " mercury " and " sulphur" to account for the infusible, non-volatile, residue of matter being his only contribution to theory. And even this is by some attributed to an earlier worker. But not only had he taken the first step towards the conversion of the mystic art of alchemy into the science of chemistry, but by making this disreputable wench into the humble but honest handmaiden of the " respectable " profession of medicine, he undoubtedly was the means of attracting to its study men of intellect and refinement. It is significant that for two centuries after his death many of the foremost " chemists ", such as van Helmont, Stahl, and Black, were primarily students or even professors of medicine.

From the beginning of the sixteenth century there was a marked progress in such chemical industries as existed in those days: chief of these was metallurgy—the technique of winning metals from the chemical compounds (ores) in which most of them are commonly bound in the rocks—and ceramics—the technique of pottery and glazing. In his compendious and well-arranged book, *De Re Metallica*, Georgius Agricola showed that many advances had by this time been introduced into metallurgical practice: thus gold was being " parted " from silver by dissolving the latter in nitric acid, and the sulphur evolved in the partial roasting of the iron ore, pyrite, was being condensed instead of being allowed to go to waste. Such practices doubtless

## THE SEARCH FOR THE ELEMENTS OF MATTER 167

stimulated the production of the strong mineral acids, which became articles of commerce at this time, and the increasing use of gunpowder would have drawn attention to saltpetre—a substance destined to play a part in the chemical theory advanced in the seventeenth century.

The establishment of ceramics as a distinct branch of technology was almost entirely the work of one man, Palissy, who, living through the greater part of the sixteenth century, has been described as " un des plus grands hommes dont la France peut s'enorgueillir ". His courage and determination in the face of obstacles command our admiration; his clear-headed inferences from carefully verified experiments mark him out as one of those responsible for ushering in the century of scientific practice which was to follow.

Not the least influential of those practical applications which provided so much of the impulse towards the creation of a science of chemistry was the increasing use of " spirit of wine " as a drink. The demand for this commodity led to the improvement in the practice of distillation and rectification (that is the separation of the spirit from the other constituents of the fermented liquor— chiefly water) which quickly spread to the laboratories of chemists. At the time the more plentiful supplies of spirit led to its use by Paracelsus and his followers as a solvent for the extraction of chemical substances from organic material such as flowers and leaves of plants—a prelude to the chemical investigation of natural products which became of increasing importance.

Notwithstanding accumulation of knowledge of distinct chemical processes, there was during the whole of the sixteenth century no progress in general ideas such as could clarify existing knowledge and point the way to new lines of investigation. In the middle of the seventeenth century chemistry was at its best a collection of fairly accurate but unrelated processes for specific technical needs, and at its worst a ceaseless and wordy warfare as to the " true " elements of things. About the year 1661 however, was published a book with the suggestive title of *The Sceptical Chymist*, which holds the same place in the history of chemistry as does Copernicus' *De Revolutionibus* in the history of astronomy. This long, even tedious, book contains little that is new as regards materials or processes; but, what was far more urgently needed, a critical examination of existing beliefs and a programme for future investigation, by which chemistry could follow the realisable objective of searching for the relations between *real* substances instead of the workings of the vain products of confused imagination.

This charter of liberation from the " cobwebs of the mind " was the work of Robert Boyle, of whom the epitaph on his Irish tomb spoke exact truth in describing him as " Father of Chemistry and Uncle of the Earl of Cork ". But it would be a mistake to regard him as a lonely prophet of a new order. We have already (p. 61) had occasion to refer to Francis Bacon's advice to chemists written in 1605; and in the years immediately following on the publication of *The Sceptical Chymist* many of the meetings of the newly formed Royal Society in London were devoted to the investigation of chemical problems in the spirit for which Boyle makes a claim. For these discussions Hooke seems to have been largely responsible. Moreover Boyle himself admits that the works of the Belgian physician, van Helmont, provided him with many of his methods and results. Indeed from the vantage ground of to-day we can see that had van Helmont added to his acute power of observation and resourceful experiment, in which he was probably Boyle's superior, that spirit of detachment and economy of explanation in which Boyle excelled, he might very easily have had chemistry fathered upon himself. It will be evident to the reader that steady progress in chemistry was being held up by the absence of any definite criteria as to the existence of unchangeable residues of matter which could retain their intrinsic nature throughout the vicissitudes of change in the fire and in conjunction with other substances. As long as chemists really believed that iron was actually *converted into copper* by being momentarily dipped into blue vitriol (copper sulphate) solution there was no solid ground upon which the investigator could take his stand. To this end the " medicine " required was not the *magisterium* of the alchemists but a dose of that tonic—accurate *measurement*—which in Galileo's hands was revolutionising the physical study of nature. This " medicine " the physician van Helmont supplied; for in an experiment which has become famous he relates how by supplying a willow shoot with nothing but water he observed it to increase in weight to the extent of 164 pounds, whereas the soil in which it was planted lost only 2 ounces. From which he deduced that *water* is the source of the various materials of which the grown tree is composed. Such an inference would have been perfectly just had not van Helmont himself been the only man of his time to recognise clearly the existence and specific nature of *gases*, and in particular that gas which he called " gas sylvestre " (carbon dioxide) which a century later was found to be the chief agent in the increase of body weight of growing plants. Unfortunately van Helmont came to the erroneous conclusion that gases, though distinct substances, could not be retained in any vessels; so we can understand that it never occurred to him that the absorption of a *gas* could account for an increase in weight. Gases were indeed for him not states *of* matter but a kind of " spirit " *in* matter.

In other actions, such as the supposed " conversion " of iron alluded to above, and the preparation of water glass, from which, on acidification, as he actually proved by appeal to the balance, the same weight of silica could be recovered as had been used in its preparation, he had grasped the very principle which we have seen to be lacking in all the work of his contemporaries. But being unable to throw off the universal obsession with the creation of gold and the prejudice in favour of a few (in his case one) fundamental substances, he just failed to perform that signal service which remained for Boyle.

We must now turn to a brief consideration of the contents of *The Sceptical Chymist* which is written in the form of a dialogue between an Aristotelian (believing in four " elements "—earth, fire, air, and water) a Spagyrist (adopting the three " principles " of Paracelsus—sulphur, mercury, salt) and the " sceptical chymist "; the meeting is supposed to take place under the chairmanship of Boyle himself.

From the beginning it is clear that the problem is to be attacked in a new spirit, for Eleutherius (Boyle) early addresses the others in the words: " I am not a little pleased to find that you are resolved on this occasion to insist rather on experiments than on syllogisms. For I, and no doubt you, have long observed that those dialectical subtleties, that the schoolmen too often employ about physiological mysteries, are wont much more to declare the wit of him that uses them, than increase the knowledge or remove the doubts of sober lovers of truth." He begged them moreover to make it quite clear what they were to argue about; to which the sceptical chemist replied that " they had agreed to use . . . elements and principles as terms equivalent: and to understand both by the one and the other, those primitive and simple bodies of which the mixt ones are said to be composed, and into which they are ultimately resolved." For the first time the protagonists are compelled to define the entities they are defending, and to frame their definitions in such a way as to enable their hypotheses to be tested; for not only are the elements the simple bodies out of which the " mixt " are composed, but also these bodies into which the " mixt " *can be* resolved.

Next it is objected that the alteration of bodies such as wood by fire into earth, fire, water, and air is no proof that the latter pre-existed in the former: " first it must be proved, that the fire does only take the elementary ingredients asunder, without altering them. For else 'tis obvious, that bodies may afford substances which were not pre-existent in them; as flesh too long kept produces maggots, and old cheese mites, which I suppose you will not affirm to be ingredients of those bodies." Boyle has here put his finger on a difficulty which is well known to chemists, namely that mere heat is seldom capable of resolving substances into their elements, but only of rearranging those elements into different compounds.

Having cleared the ground Boyle enters into a minute examination of the evidence adduced by the two sects for their alleged elements; and he rejects the claims of both on three grounds. In the first place there is van Helmont's experiment (independently verified by Boyle himself) that a tree can be grown from nothing but water, and yet its wood is resolvable into oil (" sulphur ") and ash (" salt "); consequently a " chymical principle as well as a peripatetic element may (in some cases) be generated anew, or obtained from such a parcel of matter as was not endowed with the form of such a principle or element before." Second, since the term " salt " covers such a variety of substances, some

## THE SEARCH FOR THE ELEMENTS OF MATTER 173

as still of a " salty " form; but when sulphur appears in the products (identified by a characteristic test) Boyle correctly infers that " oyle of vitrioll . . . may yet be so compounded a body as to contain, besides its saline part, a sulphur like common brimstone." But the problem remains as to how a substance like sulphuric acid, which resembles neither sulphur nor any " sulphureous " (i.e. combustible) form, can be " made of " sulphur, the central problem of chemistry.

Boyle does not tackle the problem in this particular instance, but in another place asks us to take notice of " the sugar of lead, which though made of that insipid metal and soure salt of vinegar, has in it a sweetness surpassing that of common sugar, and divers other qualities which, being not to be found in either of its two ingredients, must be confessed to belong to the concrete itself, upon the account of its texture ". In this passage Boyle establishes the claim of chemistry to be an independent discipline, in so far as it is concerned with what modern philosophers have called " emergent qualities " arising from that peculiarly intimate juxtaposition of elements which we call *chemical combination*. When the chemist tells us that lead is an ingredient of sugar of lead (lead acetate, as we now call it) he means something quite different from the more familiar knowledge that lead is an ingredient of solder. Anyone knowing the qualities of lead and tin could foretell pretty closely what would be the properties of their mixture known as solder. But neither a Newton nor a Boyle could from a knowledge of the properties of lead and acetic acid prophesy that sweet but intensely poisonous crystals are yielded by the mere mixing of these two substances. Nor do these crystals bear the slightest resemblance to lead; they " contain " lead only in the sense that lead is necessary for their production, and may by the exercise of somewhat difficult processes be regained from them. They themselves consitute a new " concrete ".

Nor must it be supposed that by giving a name—chemical action—to the process of formation of such " concretes " we have devised any *explanation* of their appearance. And the same is true of " emergence ", which is merely a concise way of reporting that there exist in nature different *levels* of material complexity, and that although the more complex may be analysed into the simpler in a *chemical* sense, yet in a " psychological sense "—that is in regard to their sensible properties—they are brute facts capable of no analysis.

Nevertheless, as Boyle reminds us, we *can* learn how such emergent levels may be *possible*, even if we cannot explain their individuality. Thus he reminds the Aristotelians that " men will never be able to explain the phenomena of nature, while they endeavour to deduce them only from the presence and proportion of such and such material ingredients, and consider such ingredients or elements as bodies in a state of rest; whereas indeed the greatest part of the affections of matter, and consequently of the phenomena of nature, seems to depend upon the motion and the contrivance of the small parts of bodies ", giving (in another place) as an example that " clouds, rain, hail, snow, frost, and ice, may be but water having its parts varied as to their size and distance in respect of each other, and as to motion and rest ". Here the reader will recognise a sagacious recalling of the speculations of the Greek atomists—speculations which, then but recently revived on the continent by Gassendi, were during the following centuries to be gradually incorporated more and more into the very texture of physical science.

Before taking a farewell of the " sceptical chymist " we should do well to consider what we have learnt from him. His contributions to the progress of actual chemical discovery amounted to very little; but the progress of science depends not so much on the accumulation of facts as on the framing of clear and distinct general ideas which may serve as the framework for new facts. In clarifying (by restricting) the *meaning* of the term element; in revealing the relation of chemical compounds to their constituent elements on the one hand, and to mixtures on the other; and in laying down guiding principles for the recognition by tests of distinct chemical individuals, Boyle made chemistry *possible*.

When allowance has been made for its somewhat wearisome repetitions and rather heavy style, *The Sceptical Chymist* remains as a reminder of the extreme difficulty of the task that Boyle achieved. It emphasises for him who has the patience to read it what our present cut-and-dried text books of elementary chemistry

THE SEARCH FOR THE ELEMENTS OF MATTER    175

quite fail to do, namely that in chemistry there *could* have been no King's Highway as in physics. Once it had been grasped that in *measurement* lay the key to the explanation of motion, then " mathematical principles " at once provided the means for opening up extensive tracts of nature to the enquiring human mind. But chemistry differs fundamentally from physics only in so far as it is concerned with the relations between *qualities* for which there is no " yard stick ". For the investigation of such relations there was needed a mind of quite a different stamp from that which created the *Principia*. It is significant that Newton spent far more time in the study of alchemy than he did in the field of mathematical physics; but so far as we know he failed to make any progress whatever. There lies about the arguments of *The Sceptical Chymist* a fog which is by no means of Boyle's making. In chemistry there are no sharp distinctions; outlines are blurred, apparent contradictions arise at every step; it was by no steady light of reason that Boyle found his way to the fundamental concepts, but by a sort of womanly intuition.

Boyle left behind him several other books on both chemistry and physics but, if the testimony of his contemporaries may be accepted, the monument of his learning was matched only by that of his character. As has been the case with not a few great Irishmen he is typical of what we are pleased to call the best type of Englishmen. " Kindly, courteous, charitable ", Thorpe writes of him in his *History of Chemistry*, " unaffected and temperate in his manner of life, Boyle enjoyed the respect and esteem of all his contemporaries. It was said of him that he was never known to have offended any person in his whole life by any part of his deportment." He was a philosopher in the broadest sense of that term.

## SOURCES FOR CHAPTER XV

J. R. Partington, *op. cit.*, Gen. Bibl.
R. Boyle, *The Sceptical Chymist* (" Everyman " Edition).
J. B. van Helmont, Art by W. Pagel, in *Nature*, vol. 153, p. 675.

## CHAPTER XVI
# THE BURNING QUESTION

SOME day the Learned Societies of the world will raise an international monument to the unknown geniuses whose discoveries have altered the whole course of human progress: the first person to introduce the number *zero* into mathematical notation will stand high in this roll, but higher still will be he who first " played with fire " to good purpose. How fire first came to be applied to human needs we shall probably never know; but that its importance and mystery were realised from very early days is evidenced by the numerous fire cults in primitive religions. It was not, as we know, the first " element " to be named by the Greek nature philosophers as the fountain of all things; in all probability it was the last of the four which came to be preserved in the Aristotelian philosophy for nearly two thousand years. It is not unlikely that its destructive power told against its adoption as a principle of generation; and it is significant in this connexion that it was in the dark sayings of Herakleitos, the first philosopher to recognise the essential status of destruction in the nature of things, that fire achieved its due regard.

With the rise of theoretical medicine the close connexion between heat and life became more and more evident, and in the armoury of the medieval alchemist there was no more potent weapon than the fire over which these " sooty empiricks " bent and laboured night and day in the pursuit of the unattainable.

It was an event of the highest importance when one of these alchemists (p. 163) sought to explain fire as being an essential quality of matter itself. It is easy for the writers of modern text books on chemistry to scoff at the naïve identification of the " fire principle " with sulphur and at the belief that in all save the " noblest " metals there was some admixture of the " sulphureous principle ". The Greek tradition was hostile to the cultivation of experimental studies of any kind, so the most subtle minds were never attracted to the study of alchemy. We cannot wonder then that in the attempt to seek a natural basis for fire or combustibility the Islamic alchemists fell into the error of expressing their beliefs in terms of the most striking combustible solid known to them.

Whether we regard the Arabs of Basra (tenth century) or the *pseudo*-Geber (fourteenth century) as the first to postulate a common basis in all combustible bodies, it seems certain that it did little to clarify ideas about the actual mechanism of burning, for even so acute a mind as Descartes was willing to express the fact of combustion as the release of the " element " of fire from the pores of the body. Modern research seems to make more and more certain the view that it was Hooke who among the great band of seventeenth-century experimentalists first carried out experiments explicitly designed to find out something about the working of combustion.

In his famous book *Micrographia* (published 1665) Hooke foretells that, aided by the lately discovered instruments, " the subtilty of the composition of Bodies, the structure of their parts, the various texture of their matter, the instruments and manner of their inward motions, and all the other possible appearances of things, may come to be more fully discovered ", and to this end what is mostly needed is " a *sincere Hand*, and a *faithful Eye* to examine, and to record, the things themselves as they appear ". In this spirit, he tells us, the members of the Royal Society met, whose discussions led them " to suspect, that those effects of Bodies, which have been commonly attributed to *Qualities*, and those confessed to be *occult*, are performed by the small *Machines* of Nature, which are not to be discern'd without these helps, seeming the mere products of *Motion*, *Figure*, and *Magnitude*; and that the *Natural Textures* . . . may be made in *Looms*, which a greater perfection of Opticks may make discernable by these Glasses; so as now they are no more puzzled by them, than the vulgar are to conceive how *Tapestry* or *flowred Stuffs* are woven."

Hooke and the gentlemen of the Royal Society may have been a little optimistic in believing that any " greater perfection of

Opticks " would ever reveal the dance of atoms whereby is woven the texture of all visible material; but we can imagine with what delight he would have followed the improvements in the quality of steel and other metals which have been attained by examining with the microscope the changes in the grouping of the crystals consequent on an alteration in the composition or heat treatment of the metal. Nor is it such a far cry from the hoped-for revelation by *visible* light to the realised mapping of the atoms in crystals by means of the alterations they cause in the path of a beam of X-rays. Hooke himself made a start in the application of the microscope to chemistry when he proved that the shower of sparks " issuing " from the steel struck by flint is not an emission of heat itself but of particles of molten steel (more probably *oxide* of iron) heated to incandescence by the violence of motion induced by the blow (and of course by the chemical action of whose existence he was hardly aware).

In " Observation XVI " of *Micrographia* under the unrevealing title " Of Charcoal, or *Burnt* Vegetables " Hooke gave an account of the nature of combustion far excelling in clarity and comprehensiveness anything of an earlier date. In this work he merely states his hypothesis without giving any details as to the " Infinite of Observations and Experiments " upon which he said it was founded, and which " will perhaps another time afford matter copious enough for a much larger Discourse ". The " Discourse " never appeared, so we are compelled to guess at the nature of the experiments; it is thought very probable by a modern writer (Dr. D. J. Lysaght, Ambix I. 93) that the experiments recorded in Birch's *History of the Royal Society* may serve as a very fair sample of Hooke's investigations.

In this record of the early meetings of the Royal Society we read that a piece of burning charcoal was shown to stop burning when placed under a glass vessel, but to recommence as soon as air was admitted. From such an experiment he probably inferred " *that the Air* in which we live, move, and breathe . . . is the *menstruum*, or universal dissolvent of all *Sulphureous* bodies ". By " sulphureous " Hooke clearly means " combustible ", according to the usage of the times.

At a later date Hooke made the capital discovery that nitre

## THE BURNING QUESTION

(saltpetre) is as efficacious as air in causing combustion; being led to this by the observation that gunpowder burns as well in a vacuum as in air. From these experiments he seems to have concluded that " *the dissolution of* sulphureous bodies is made by a substance inherent, and mixt with the Air, that is like, if not the very same, with that which is fixt in *Salt-Peter*." This observation was made early in 1667; it is one of the curious tragedies of science that the discovery of oxygen, which Hooke so promisingly began, was not *completed* until over a hundred years later.

In addition to recognising that there is *something* common to both air and saltpetre (about which his contemporary, Mayow, wrote at great length, whether in ignorance of Hooke's views or not is uncertain) Hooke clearly recognised that this " something " was not a vague " principle " but a stuff which could be measured. He effectively silenced one critic who put forward the view that air was necessary for burning merely to " soak up the vapours in its pores ", by showing that burning ceases sooner if air is withdrawn from the combustion vessel and continues longer if the air is compressed. It is interesting to note that both these demonstrations were made possible by the recently invented air pump.

The credit for the discovery that a " burnt " or " calcined " metal is heavier than the metal from which it is made has usually been given to the Frenchman, Rey; but there are several records of experiments having been made by Lord Brouncker, probably at Hooke's instigation.

We may conclude therefore that sufficient facts had been gathered for the following theory of the nature of fire and combustion to be put forward in Sprat's *History of the Royal Society of London* 1667: " That there is no such thing, as an *Elementary Fire* of the *Peripatetics*; nor *Fiery Atoms* of the *Epicureans*: but that *Fire* is only the Act of the Dissolution of heated *Sulphureous Bodies*, by the *Air* as a *Menstruum*, much after the manner as *Aqua Fortis*: or other sharp *Menstruums* do work on dissoluble Bodies, as *Iron, Tin, Copper*: that heat, and light are two inseparable Effects of this dissolution, as heat, and ebullition are of those dissolutions of *Tin* and *Copper*: that *Flame* is a dissolution of *Smoak*, which consists of combustible particles, carry'd upward by the heat of rarify'd *Air*: and that Ashes are a Part of the *Body* not dissoluble by the *Air*."

Making allowance for the form of expression inherited from a thousand years of alchemy, this statement is an outline of the modern theory of combustion. So long as the cause of " fire " was held to lie in the " elementary fire " or " fire atoms ", there could be no progress; such " explanations " merely shift the problem round by restating it in different terms. But as soon as men begin to seek *true analogies* as in the " dissolution by menstruums " (that is, chemical action by strong acids) and *real, verifiable, causes*, as in the " carrying upward by rareify'd air ", they are on the threshold of a scientific, that is a general and verifiable, explanation of the working of things. But, as is well known, the seventeenth and eighteenth century chemists remained on the threshold—when they were not actually driven back—for more than a century. The reactionary hypothesis, which alone seems to have been known on the continent and was ultimately to displace the enlightened views of Hooke and his collaborators in this country, was the famous hypothesis of *phlogiston*. A short examination of this celebrated mare's nest is justifiable on several grounds: first, as an example of how trivial chances may promote error at the expense of wisdom; second, as justifying a belief that a definitely false hypothesis may be better than a less erroneous but also less precise one; and third, that this hypothesis emphasised one aspect of combustion and similar processes which its more accurate successor failed to take account of.

The phlogiston hypothesis, first given systematic exposition by G. E. Stahl in his influential work *Fundamenta Chymiae*, amounts to a not altogether unsuccessful attempt to apply the " common sense " belief in the *giving off of a fiery stuff by a burning body* to actions as widely various, to the " common sense " view, as the burning of a candle, the calcination of a metal (i.e. conversion into an earthy " scale " in the fire), and the respiration of animals. If science can be called, as a modern writer has said, merely " organised common sense ", then Stahl's hypothesis was undoubtedly scientific *to start with*. For though the actual " principle " of phlogiston was purely fictitious, its action was

THE BURNING QUESTION 181

by no means so, as may be seen by consideration of the following schemes in which Ph stands for the principle itself:

Metal = calx+Ph
Charcoal = nothing except Ph
Calx+charcoal = metal

Now since charcoal, as is well known, leaves hardly any residue on burning, it may be considered to be nearly pure phlogiston; hence it might be inferred that the intimate interaction of charcoal with the calx of a metal would, by putting the phlogiston back into the calx, regenerate the metal. But this very fact had been the basis of metallurgical processes for millenia: the hypothesis therefore fits the facts. Moreover Stahl himself provided an even more striking verification when by a series of quite skilful operations he succeeded in regenerating sulphur from a sulphate heated with charcoal, a result which he explained as the putting back of phlogiston into the sulphuric acid obtained in the first instance by the removal of phlogiston (i.e. burning) from the original sulphur.

It must not be supposed that Stahl alone recognised the similarity of burning and respiration. As early as 1660 Boyle had shown that when the air was wholly or even partially removed from a closed vessel by means of his air pump, a candle went out and a mouse died. Nine years later, in his now somewhat neglected book, *Tractatus de Corde*, Richard Lower showed the intimate connexion between the colour of the blood and the time that it has been in contact with air. But it was Mayow who gave the clearest account of the whole mechanism of breathing. He saw how the enlargement of the thorax through the elevation of the ribs by the contraction of the intercostal muscles must reduce the pressure on the lungs and so cause the natural " spring " of the atmosphere to fill them with air. And however confused may have been his ideas about the " nitroaerial particles ", he seems to have had no doubt that it was the *same* particles which caused both combustion and the change of the blood in breathing. Indeed in emphasising that the increased rate of breathing during muscular activity is " to provide for the greater expenditure of the nitroaerial salts in the many effervescences taking place in the contractions of the muscles ", he carried the comparison still further than any of his contemporaries. Although Stahl must be given the credit for saying that if " anything " is added to the blood in breathing, it must be phlogiston, he failed to realise that the principal cause of body heat is akin to combustion, but attributed it in the main to mere friction. The virtue of the phlogiston hypothesis lay in its unambiguous recognition of the *chemical similarity* of many processes which to the common man appear so diverse. Its weakness lay in the fact that it was haunted by one of the old ghosts of " principles " which Hooke and his supporters in the Royal Society had seemingly laid for ever (p. 179). But when we recall the view of Boyle, his close associate, that fire, though not a " principle ", was nevertheless a " swarm of little bodies swiftly moving, which by their smallness and motion are able to permeate the sollidest and compactest of bodies ", and that " I could name to you . . . some particular experiments by which I have been deduced to think that the particles of an open fire working upon some bodies may really associate themselves therewith, and add to the quantity " (*Sceptical Chymist*, p. 119), we can hardly blame Stahl for his failure to realise that he was on the wrong road. Boyle's foolish conjecture, for it was no more, is all the more remarkable, as Professor Partington has recently pointed out, in one who had heard the hiss of the air entering the vessel in which the material had been burnt, and who himself had done more than any man to prove the distinct material nature of air. All of which goes to show how difficult it is for even the most adventurous of minds to shake themselves free from the particular " climate of opinion " in which they are living; and further that the " burning question " was one to which no *complete* solution was likely to be found until the nature of gases on the one hand, and of heat on the other, had been very much more thoroughly understood than was the case when Stahl was teaching *his* view of the matter. It is interesting to note in this connexion that Mayow, some of whose writings suggest an almost complete anticipation of the true nature of combustion as union with the " nitroaerial particles " in the air (p. 181) revealed the confusion of his ideas in stating that " nitroaerial particles agitated with swiftest motion constitute the solar body and fiery chaos ". Thus the same " nitro-aerial particles ", which some modern writers have regarded as Mayow's anticipation of *oxygen*, are here by him identified with " fire atoms ".

The inability of Stahl and his disciples to " produce the body " would not in itself render the hypothesis untenable; the pioneers in the investigation of electrical and magnetic phenomena were in the same case without detracting from the value of their discoveries. But it was not long before the phlogistonists were called upon to explain (1) the necessity for air in combustion, and (2) the fact that the weight of calx was greater than that of the metal from which it was obtained. The replies given, namely that the air is necessary to absorb the evolved phlogiston, and that the latter has a negative weight, are familiar to every candidate for school certificate chemistry.

Even those who, like the author of this book, are anxious to give the phlogiston hypothesis all the credit that is its due, are compelled to admit that the attribution of " negative weight " to " phlogiston " was a retrograde step—a tacit admission that there were cracks in the walls which had therefore to be propped up by artificial means. This judgment our school certificate candidate knows full well; but what he perhaps is not so clear about is just *why* this marks the end of the " scientific " stage in the growth of the hypothesis. It is not, as has sometimes been said, that negative weight is a contradictory idea; weight is force, and repulsive forces are accepted without any show of pious horror in electric and magnetic actions. The sound basis of criticism is rather that so long as, in Stahl's original hypothesis, phlogiston was supposed to be weight*less*, then it was by the same token immaterial and *might* be as real as magnetism and electricity, at that time regarded as "imponderable fluids " (see Ch. XVII). But the increase in weight on calcination pushed itself forward as an awkward fact that had to be accounted for somehow: that is, there must be *some* change in the quantity of " weighable " stuff present. Algebraically it is all the same whether we *add* ordinary matter with a *positive* weight or subtract the hypothetical phlogiston with a still more hypothetical *negative* weight. If the choice had had to be made in the sixteenth century (it is said that Cardan did in fact observe the change in weight) the dilemma would have been real and the supporters of the hypothesis merely unlucky in choosing the wrong alternative; for they would have had no evidence that there was any weighable stuff present besides the metal. In the eighteenth century however anyone not blinded by prejudice or culpably ignorant of the experiments of Torricelli, Pascal, Boyle and von Guericke proving the material nature of air, and of the more than broad hints implicit in Hooke's observations on combustion, must have chosen the hypothesis that calcination involved the formation of a new " concrete" between metal and " air ". To choose the alternative was to introduce two " imaginaries " where none was needed. In other words, phlogiston becomes an obstacle rather than an aid to the elucidation of nature.

The phlogiston hypothesis had gained wide influence long before Stahl's death in 1734; it was more than forty years after this date that any attack was made on it; how are we to account for the long reign of so confused an hypothesis? It was not the entire absence of able minds during this period. For in 1732 Boerhaave published his *Elementa Chemiae* even now regarded as a model treatise on chemistry; and in 1726 and 1732 appeared two volumes of critical essays by a country clergyman, Stephen Hales, on, among other matters, the quantity of air used by growing plants, in which accurate measurement was added to careful observations (see later, p. 350). What was lacking, however, was first the recognition that *gases are distinct chemical individuals*, and not, as even Hales implied, merely " modified " air; and in the second place, a realisation that *heat* is a natural phenomenon whose study demands its own peculiar methods. In other words the investigation of nature was being hampered by too rigid an adherence to the ideal of a single " natural philosophy "; the time was ripe for the specialisation of the means of knowledge and the separation of the special sciences.

The light of a new day dawned, as it has more than once dawned, in Scotland, despite the blanket of cloud which usually envelops that country. There in 1755 was published a paper, written in Latin the previous year, entitled *Experiments on Magnesia Alba, Quick-lime, and other Alcaline Substances*. The author was

PLATE III

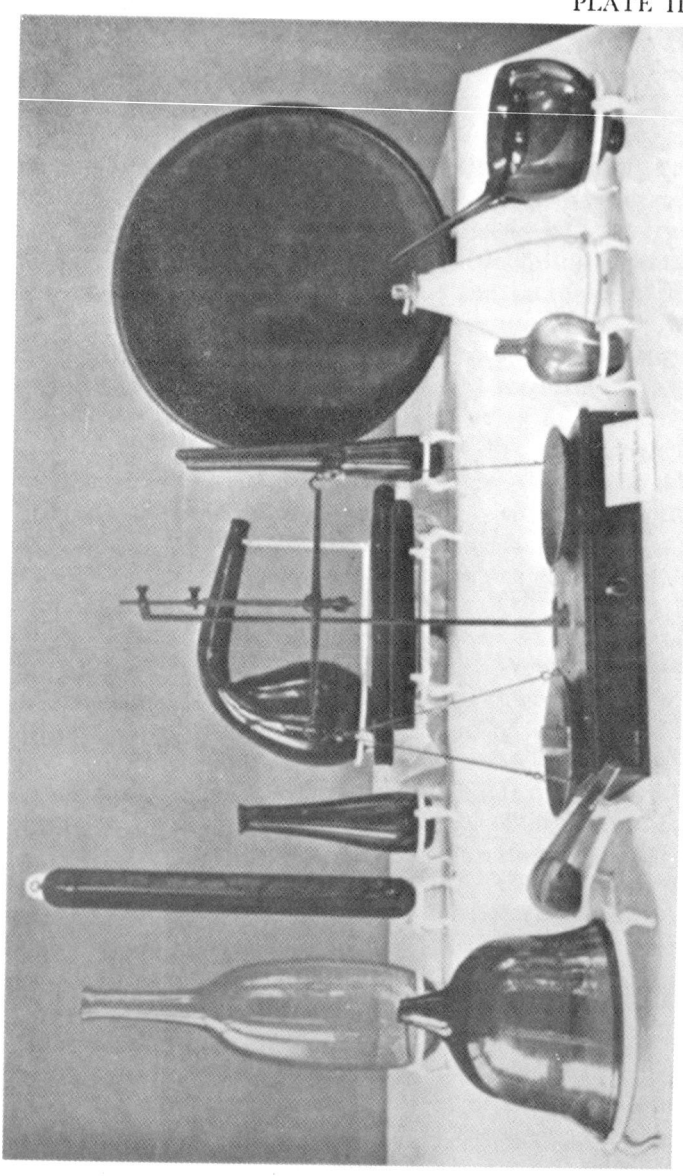

Apparatus used by Joseph Black, from a collection in the Royal Scottish Museum, Edinburgh

Joseph Black who, born in 1728 in Bordeaux of Scottish parentage, became Professor of Chemistry first at Glasgow and later at Edinburgh, in each case in succession to his revered teacher, Cullen. He died in 1799. The city of Edinburgh has its full share of monuments to mediocrity: to Black, one of the greatest of Scottish gentlemen, there is no monument; nor need there be: his name is known in every part of the civilised world.

It has been said, perhaps rather too often, that the verses of the Greek poetess, Sappho, were few but roses; of Black it may be said that his works were small but of crystal clarity. It is a temptation to quote from his *Magnesia Alba*; but since it was fitly chosen by the Alembic Club of the University of Edinburgh to be the first of their famous Reprints, it is (or was) available to anyone for a small sum. Moreover it is a little work which anyone whose knowledge of chemistry does not exceed that required for a school certificate, and whose interest in the science has survived the continuous administration of well-digested pabulum provided by schoolmasters to aspirants to that distinction, would thoroughly enjoy reading. For the sake of those whose appetite has been already sated it will be necessary to summarise the results of Black's simple but elegant experiments carried out in apparatus which may be seen in the Royal Scottish Museum, Edinburgh.

Black resembled Newton in an almost morbid horror of hypotheses, but such hypothesis as he held was that of phlogiston: it was in order to test the view that chalk becomes " caustic " by imbibing phlogiston from the fire that he entered upon this investigation of the similar material *magnesia alba* (magnesium carbonate). With the insight of genius (or perhaps with the unprejudiced eye of the pioneer in this particular study of the effect of fire) he realised at once that the *loss of weight* of magnesia on being heated was the clue to its change of properties, namely that it thereby lost the power of effervescing with acids. Suspecting that the " air " which acids release from unheated magnesia had in fact been driven off by the heat, he heated a weighed quantity of magnesia, dissolved the residue in acid, and precipitated the resulting salt with soda according to the usual practice. On washing, drying, and weighing the precipitate, he obtained a material having all the properties of the *original* and almost exactly the *original weight*. From which he concluded that " the common absorbent earths [e.g. magnesia and chalk] which lose their air by being joined to acids, but show evident signs of having recovered it when separated from them by alkalis [e.g. carbonate of soda or potash], received it from these alkalis which lost it in the instant of their joining with the acid." This is the pronouncement of a modern chemist: the ghostly " principles ", the conveniently vague " fiery " and other " adjectival natures ", have been tried in the (chemical) balance and found superfluous. It is not an *adjective*, but a ponderable *substance* which by its transfer from one " concrete " to another is changing the properties of both. This substance, of whose chemical composition he was entirely ignorant, Black called " fixed air " and identified (by repeating the experiments and testing the product each·time with lime water) with Helmont's " gaz sylvestre ". He recognised not only that it sufficed to explain all the changes so far considered, but what was more important, also the *difference between the " mild " and " caustic " alkalis*. The latter were in fact the former deprived (as for instance by shaking sodium carbonate with lime water and filtering) of their " fixed air "

It was not long before Black discovered (by blowing into lime water) that *expired* air contains considerable quantities of fixed air. When to this fact he had added the demonstration that it is also formed when charcoal is burnt, he might have allowed the wind of faith to blow away the phlogiston theory with the rest of the phantoms. Perhaps his fear of speculation was in this case a regrettable hindrance; but it may equally have been the realisation that his contribution could at the best do no more than explode the existing hypothesis; he had no better one to put in its place.

It was at this time also that Rutherford, a colleague of Black's, discovered that after the whole of the fixed air had been removed by alkali there always remained a large proportion of irrespirable air. This *unambiguous* proof of the existence of what we call nitrogen is not always given its due significance in our text books.

Black had proved the existence of a distinct gas different from ordinary air by showing with the aid of the balance that *some material* was " changing its quarters ". By means of a *specific test*, as advocated by Boyle, he had shown that this substance

## THE BURNING QUESTION

retains its character throughout the vicissitudes of the furnace and the attack of acids. But he never took steps to *collect* this "air" at the time of its formation in a chemical experiment; though strangely enough he does describe a method of " sampling " the air above a vat of fermenting brewer's wort. From about 1770 however his English contemporary, Joseph Priestley—his antithesis in every trait of character save in his devotion to science—was making improvements in the *technique* of " handling " gases which rapidly led to the opening of a new era in the history of chemistry.

Those modern teachers who are contending for the broadening of the science curriculum in schools would do well to remind their sceptical colleagues that it was the study of respiration which set Priestley on the path which led to his discovery of most of the common gases.[1] Priestley's investigations are perhaps the best example of the purely *empirical* method of science that we have. His curiosity was unbounded, his resource and ingenuity at that time almost unparalleled, but his procedure often almost without plan. Thus when he sought to find out how it is that the atmosphere, although being continually " used up " by fires and breathing creatures of all kinds, does not deteriorate, he tried all manner of experiments, even one which his own reason warned him must be foredoomed to failure: this was to grow a plant in air, confined over water. He knew that plants need air to live in, and expected the air to be rendered unfit for combustion; but after several months it " would neither extinguish a candle, nor was it at all inconvenient to a mouse which I put in it ".

Thus thinking " that there was something attending vegetation which restored air that had been injured by respiration, I thought it was possible that the same process might also restore the air which had been injured by the burning of candles.

"Accordingly on the 17th of August, 1771, I put a sprig of mint into a quantity of air, in which a wax candle had burnt out and found that on the 27th of that same month another candle burned perfectly well in it. This experiment I repeated, without

[1] His experiments with gases were first prompted by observations on the " fixed air " issuing from a fermenting vat in a Leeds brew-house—observations which led to the invention of " soda water "

the least variation in the event, not less than eight or ten times in the remainder of the summer."

To Priestley therefore is due the discovery of the simple *fact* that plants, in addition to requiring air for respiration, can also regenerate the latter's respirable quality; he was quite unable to explain this apparent contradiction. A well-known Dutch medical man, Ingenhousz, having learned of Priestley's experiments on a visit to England, published shortly afterwards an account of experiments which revealed *one* of the factors necessary for this regeneration, namely the presence of sunlight. In the absence of light every part of the plant produced air fatal to life; and even in its presence, it was only the *leaves* which could regenerate it. So there is in Ingenhousz's work at least the germ of the discovery of the *second* factor—the pigment chlorophyll. The nature of the *third* factor—carbon dioxide—was not even guessed at until after the *Révolution chimique* which Lavoisier was even then bringing about.

The French chemist Würtz opens his well-known *History of Chemical Theory* with the words " Chemistry is a French science. It was created by Lavoisier, of immortal memory." In the light of what we have seen of the work of Black, such a statement at once stands condemned as a gross exaggeration. But if Lavoisier did not create chemistry, he did at least transform it.

Antoine Laurent Lavoisier was indeed one of the founders of our modern world. Born at a time of intellectual and social ferment, he combined much of the sagacity and innate nobility of the best elements of the *ancien régime* with the sense of practical necessity engendered by the industrial revolution. But the high-handed and ignorant demagogues who, as is usual in such upheavals, ruled the destinies of France for a few years after the political revolution, had neither the wit nor the humanity to recognise the ill they were doing the young republic in repudiating its need of *savants* such as was the organiser of the chemical revolution.

The study of chemistry had become almost a fashionable pastime in the Paris of Lavoisier's youth; a thirst for knowledge of all kinds was everywhere apparent. The " age of reason ", as the first half of the eighteenth century has rather inaptly been called, had left men dissatisfied and restless. The manner of the course of lectures in chemistry, which Lavoisier attended in the enforced leisure too well known to every young lawyer, may not have been typical, but it was certainly symptomatic of the spirit of the age. The precise, dull, and error-ridden recitation by the professor of the " principles " of the science was listened to with bored toleration. On the appearance of the assistant, Rouelle, whose duty was to perform experiments in illustration of the discourse, the house woke up. As a demonstrator Rouelle seems to have been confused and clumsy—Diderot describes how he set himself on fire with phosphorus without having any idea how it had happened—but as a lecturer he was capable of communicating to his audience much of his own enthusiasm, an enthusiasm for the full release of which it was often necessary to rid himself of periwig, cravat, and vest. His criticism of the accepted " principles " was unsparing. If Lavoisier did not create chemistry, Rouelle did create Lavoisier; for the latter abandoned the practice of law and set himself to acquire the technique of accurate observation and measurement. It was in meteorology he served this apprenticeship; he made valuable additions to our knowledge of the distribution of air pressure, and urged the importance of a daily weather forecast based on data from widely separated stations—a task almost impossible of achievement until after the invention of the telegraph.

There was in those days no " endowment of research "; and we may believe that it was partly in the desire to " endow " himself with sufficient means to carry out his investigations that Lavoisier got himself appointed to the Ferme responsible for collecting taxes. The system was undoubtedly corrupt and had been grossly abused; but there is evidence to show that though they became rich beyond their deserts, Lavoisier and his colleagues carried out their work with efficiency and equity. Nevertheless, as is well known, it was his association with this body that brought him to the scaffold.

From meteorology Lavoisier's versatile genius turned in the direction of chemistry and he had already carried out critical researches on solution before he took up the problem of combustion. In 1772 he discovered that when the newly discovered phosphorus was burnt in a *confined* space, air was *absorbed* and the product weighed more than the original phosphorus. " This discovery . . . made me think that what was observable in the combustion of phosphorus and sulphur might equally well take place in respect of all bodies which acquired weight on combustion and calcination. Experiment completely confirmed my conjectures . . ." He rightly recognised this as pointing to a " revolution in physics and chemistry ", and set to work to achieve it. During the next year he made an exhaustive study of the work of his contemporaries. What enabled him to bring about the " revolution " was his application of the quantitative method of Black to the investigation of those " calces " and " airs " qualitatively characterised by Priestley.

Early in 1774 he repeated the experiments of Boyle and Hooke (p. 178) on the calcination of metals. Again, working in the spirit of Black, he discovered that the *weight* of the *whole sealed apparatus and contents remained constant*, though the weight of the calx was greater than that of the tin from which it had been made. He rightly concluded that the calx was a compound body formed by the *gain* of something that the air had lost. But what was this substance? Whether Lavoisier actually discovered, or would in any case have discovered, this for himself we shall probably never know. What is certain however is that he learned from Priestley in October of the same year that mercury calx, when placed in a glass jar which was ingeniously closed by partial immersion in mercury and heated by concentration of the sun's rays with a burning glass, gave off an " air " which allowed things to burn in it with exceptional vigour. Priestley was later to call this " air " " dephlogisticated air ", since it was enabled to absorb " phlogiston " more readily and in greater quantity than ordinary air. Lavoisier ultimately[1] recognised it as the efficient cause of combustion. To verify this hypothesis he performed the

[1] One reason for the delay was the fact that in March 1775 he was appointed to supervise the manufacture of gunpowder. His mind was much taken up with the problems connected with the preparation of nitre. Ultimately this appointment was greatly to the advantage of his purely scientific work, for in the laboratory of the arsenal, in which he and Madame Lavoisier took up their residence, he had at his command skilled artisans who fashioned to his design apparatus which, in the words of his contemporary, Fourcroy, were " of a precision previously unknown in chemical laboratories ".

## THE BURNING QUESTION

most famous experiment in the history of chemistry—an experiment which was virtually a combination of Priestley's two experiments (calcination and decomposition), but carried through in a confined space and checked at each stage by accurate measurement. For twelve days mercury was heated in fifty cubic inches of air, at the end of which period the *appearance* of the red calx was accompanied by the *disappearance* of eight or nine cubic inches of air. With admirable patience and care he collected every grain of calx, and heating it in a separate tube obtained as much of Priestley's " dephlogisticated air " as would suffice to reconstitute the original air both in quality and quantity. The phlogiston hypothesis was rendered untenable, but the new oxygen-addition hypothesis which Lavoisier put forward in its place had still several obstacles to surmount.

The papers describing the crucial experiments were presented to the Academy in 1777. By 1781 he had " explained away " the fact which till then had been the strongest piece of evidence for the existence of " phlogiston "—the regeneration of metals from their calces by heating with charcoal. If we put ourselves in the position of the chemists of the early eighteenth century, having but the vaguest appreciation of the specific material nature of the various gases, we shall recognise how much more plausible must have appeared the view that this regeneration was a " putting back " of a " fiery spirit" from the charcoal. But as soon as Lavoisier had by accurate quantitative analysis identified the " fixed air " of Black with the gaseous product of the " oxidation " of carbon, of which he also showed diamond to be a form, this difficulty faded away: it was the simplest matter to substitute an oxide such as copper oxide for the mercury oxide, which yielded its oxygen by simple heating, to show that the element carbon (of which Lavoisier proved charcoal to be almost entirely composed) could effect what heat alone could not. Incidentally, how different the history of chemistry might have been if *no* metal existed which would take up oxygen at one temperature and release it at a higher! And how important are such " haphazard " investigators as Priestley in discovering just such curiosities by heating, like any school boy, whatever substance happens to take their fancy!

The phlogiston hypothesis had had one head struck off only, like the fabulous hydra, to grow another in its place. In 1767 Cavendish had prepared and collected a new gas which, following the custom of the time, was duly named " inflammable air ". Metals when dissolved in acids released this gas, but their calces (which had " lost " phlogiston) did not; a suspicion arose that here at last was the quintessential phlogiston—a suspicion spectacularly confirmed by the regeneration of metals when their calces were heated in this air. The supporters of the phlogiston hypothesis seized not without justice on this heaven-sent phenomenon. And then two of them, Priestley and Cavendish, proceeded to establish the facts which Lavoisier once again " took down and used in evidence against them ".

Just as the overthrow of the phlogiston hypothesis became possible only as a result of the perfection of the technique of handling gases, so the study of chemical action between gases could have made very little headway but for the rapid strides then being made in the production and use of electricity. Priestley's claim to fame rests as much upon his studies of electricity as upon his chemical discoveries; and it was on firing by an electric spark a mixture of " inflammable air " and ordinary air that he made the important observation that the inside of the vessel became covered with dew. His comment upon this observation was that " common air deposits its dew by phlogistication ". Cavendish however realised that there was far more in it than this, and as a result of a large number of most carefully conducted experiments he was able to announce that:

The bulk of the air remaining after the explosion is then very little more than four-fifths of the common air employed; so that as common air cannot be reduced to a much less bulk than that by any method of phlogistication, we may safely conclude that when they are mixed in this proportion, and exploded, almost all the inflammable air, and about one-fifth part of the common air lose their elasticity and are condensed into dew which lines the glass.

The better to examine the nature of this dew, 500,000 grain measures of inflammable air were burnt with about $2\frac{1}{2}$ times that quantity of common air, and the burnt air made to pass through a glass cylinder eight feet long and three-quarters of an inch in diameter in order to deposit the dew. . . . By this means upwards of 135 grains of water were condensed in the cylinder, which had no taste or smell, and which left no sensible sediment when evapor-

THE BURNING QUESTION 193

ated to dryness; neither did it yield any pungent smell during the evaporation; in short it seemed pure water.

This epoch-making discovery was announced to the Royal Society on 15th January 1784. Cavendish, with perhaps an excess of caution, declined to give up at this stage the phlogiston hypothesis, though in the same paper he refers to " several memoirs of Mr. Lavoisier . . . in which he entirely discards phlogiston . . .; and not only the foregoing experiments, but most other phenomena of nature, seem explicable as well, or nearly as well, upon this as upon the commonly believed principle of phlogiston ". Consequently it was left to " Mr. Lavoisier " to draw the conclusion which the English chemists had not the courage to do. This was simply that " inflammable air " is no less a specific form of matter than any one of the many " airs " by that time recognised as such; that it combines with oxygen to form water; and that it can extract this oxygen from certain oxides as well as from ordinary air. The confirmation of this conclusion needed only the recognition of water as the product of the " reduction " of the oxide by the " inflammable air "; this was observed not only by Lavoisier, but also by Priestley who of course quite failed to realise its significance. Lavoisier however clinched the argument by discovering the now familiar action by which pure water may be made to liberate " inflammable air " by leading it in the form of steam over red-hot iron, a method which was quickly applied to the production of hydrogen, as the gas was henceforth to be called, used to fill the balloons in which Charles and others were then making ascents. Thus did the last of the ancient " elements " cease to exist as such, but came to be regarded as oxide of hydrogen.

There was however one other point which even Lavoisier failed to clear up quite so exactly. If, as he strictly taught, acids are oxides (the name oxygen, which he proposed, means the " acid-generator ") why do metals disengage *hydrogen* from them? The above proof of the constitution of water provides the only answer which could then be supplied, namely that acids are always associated with water, and it is from the latter that the hydrogen comes; it was pertinent to enquire why a metal like zinc should not in that event disengage hydrogen from *water*.

Nevertheless, once this answer has been provisionally accepted, it is clear why *oxides* (unlike the metals from which they were formed) fail to give hydrogen with acids; it is not that they have been " deprived of phlogiston ", but merely that they contain a proportion of oxygen capable of oxidising the hydrogen before it can be expelled.

One of the surest tests for the adequacy of a scientific hypothesis is the ease with which anyone sufficiently familiar with that department of modern science to which it is relevant may read the literature inspired by it. The perusal of one of the memoirs of Cavendish or Priestley is not unlike the reading of a great work in a foreign language with which the reader is so little conversant as to be compelled to translate literally into his own tongue. He is struck by the care and sufficiency of the observations; he is delighted by the acuteness of the author's perception of what is essential; above all perhaps he admires the resolution and ingenuity with which obstacles consequent upon the modest development of laboratory arts are overcome; but to gain an estimate of the *progress* achieved in actual *knowledge* he must be constantly at pains to *translate* the names of substances and the descriptions of processes. In Lavoisier's *Traité Elémentaire de Chimie*, published in 1789, all this is changed. This is a text book of *modern* chemistry; and the Englishman with but a moderate knowledge of French can read it with greater ease than the works of Lavoisier's English contemporaries. It would be gratuitous to suggest that it is entirely free from errors of interpretation (the author expressly warns the reader to distinguish what he states as fact and what he urges as hypothesis) or even of such obscurity as is found in the writings of the phlogistonists; but the foundation stone of modern chemistry is here firmly laid and inscribed in sentences which call for but little modification even at the present day.

Another critical test of an hypothesis, or theory, as Lavoisier's account of combustion may be justly called, is the degree to which it guides posterity in their search for new fields of knowledge. It is no exaggeration to say that within twenty years of the launching of the new theory the additions to sound chemical knowledge far outweighed those of the previous hundred and twenty. This was made possible by the dual achievement of Lavoisier in providing both adequate general ideas and (with the help of his friends Berthollet, Fourcroy and de Morveau) the technical language in which to express them. To a science like chemistry a precise nomenclature is as vital as is an adequate notation in mathematics. Chemists following Lavoisier not only knew the general lines of research likely to be profitably undertaken, but could even name beforehand the unknown but suspected substances they were seeking. An example of this prescience may put the matter in a clearer light.

In the opening of the second part of his *Treatise*, Lavoisier enumerates (in addition to " heat " and " light ", of which more will be said later) thirty-one substances which *at that time* had not been resolved into simpler forms, and had therefore to be regarded as elements. Of these he correctly labels three—the " muriatic ", " fluoric ", and " boracic " radicals—as " unknown "; and five—lime, magnesia, baryta, alumina, and silica—as " simple substances ", with the conjecture that the first four—the " earths "—are perhaps metallic oxides. Within little more than twenty years Davy had isolated calcium, magnesium, and barium; had proved the elementary nature of chlorine (as had also Lavoisier s countrymen, Gay-Lussac and Thenard) and had pointed the way to the isolation of silicon and fluorine—a task achieved before the end of last century. To realise the magnitude of Lavoisier's *révolution* one has only to recall how impotent had been chemists nurtured in the fog of the phlogiston hypothesis to recognise the elementary nature of water even when the facts, so beautifully established by Cavendish, stared them in the face. To do phlogiston justice one may willingly admit that it had anticipated the generalisation of the oxygen theory with everything turned upside down. The rapid flowering of the latter only goes to prove that it is far easier to read he book of nature when not standing on one's head.

This long study of phlogiston and of its supporters has, it is hoped, vindicated the view put forward on p. 180. It was there mentioned however that one of the dual aspects of phlogiston—that of the *heat disengaged* as distinct from the *alteration of material* during combustion—tended to be relegated to a subordinate position by the followers of Lavoisier (though not by him, for his work with Laplace marks the beginning of thermochemistry). To do justice to this aspect of the growth of ideas in the eighteenth century it will be necessary to retrace our steps.

## SOURCES OF CHAPTER XVI

Alembic Reprints: No. 1, J. Black, *Experiments on Magnesia Alba*; No. 7, J. Priestley, *Discovery of Oxygen*; No. 3, H. Cavendish, *Experiments on Air*; No. 5, R. Hooke, *Micrographia*.

[I am indebted to Dr. A. L. F. Smith for drawing my attention to the Memorial Lecture on Lavoisier by Sir Harold Hartley, *Proc. Roy. Soc.*, A vol. 189, which enabled me to correct the above chapter in several particulars.]

An article by Professor J. R. Partington in the Commemoration Volume of the *Philosophical Magazine* 1948 draws attention to Lavoisier's views of the relation of caloric (*vide inf.* p. 203) to combustion.

CHAPTER XVII

## THE " SUBTLE FLUIDS "—I. CALORIC

THE failure of the phlogiston hypothesis was the failure of an hypothesis born out of due season. Of the two aspects of combustion—the material and the thermal—Becher and more particularly Stahl laid the emphasis on the latter. Subtle as is the former, owing to its involving a knowledge of the subtlest form of matter we call *gases*, it is far less subtle than the thermal aspect in which an entirely new concept gradually emerges—the relation between heat and mere motion. It is only too evident that the seventeenth century investigators—with the possible exception of Hooke—never succeeded in separating these two aspects of the problem. Boyle's mistake was to " explain " chemical alteration as an addition of " heat "; Stahl's was in a measure to " explain " both the chemical and the thermal phenomena in terms of a " substance " which at least to his followers was both " matter " and " heat " indiscriminately. The Phlogistonists, like all who try to make the best of both worlds, got the worst of both, and have passed into the category of instructive museum specimens.

It was Lavoisier's luck, and genius, to be born *in* due season, when the climate of opinion in all things was frankly materialistic, the triumph of Newton's " hard, massy, particles " so complete that everything was conceived in terms of matter of varying degrees of subtleness. So it came about that by tacitly ignoring the *thermal* aspect of chemical change, or at least by relegating it to the background, Lavoisier was able to establish the science of *chemistry* on lines which it has ever since followed.

If Lavoisier at first ignored the disengagement of heat in chemical change, he was far from ignoring the problem of heat as such. In his *Treatise* he points out how familiar are the changes —expansion, liquefaction, and evaporation—of materials to which heat has been added; these are to be interpreted as due to the separation of the molecules by the " heat ". In assessing the true nature of this " heat " his genius for framing conceptions adequate to the knowledge, or ignorance, of his times shines most brightly. Wishing to distinguish our sensation of heat from its *cause* he introduces for the latter the name " caloric ", and places it (after " light ") at the head of his table of elements (" *substances simples* "). The hasty critic may see in this a return to the unprofitable ideas which the *Sceptical Chymist* had sought to dissolve; but on careful reading of the text we are on the contrary led to admire the discretion with which he uses the concept. In referring to the states of matter he says: " It is difficult to conceive these phenomena without admitting that they are the effects of a substance real and material, of a very subtle fluid, which insinuates itself between the molecules of all bodies and which separates them: and even supposing the existence of this fluid to be an hypothesis, it will be seen in the sequel that it explains in a very happy manner the phenomena of nature."

For Lavoisier, living in an age when to be " real " was to be " material ", it must have been difficult to conceive it otherwise than he did; difficult, but not impossible, as we shall see. Moreover it *did* " explain in a very happy manner " certain fundamental discoveries made about this time. We shall consider these briefly, not only for their own sake, but because they illustrate even more precisely than in the case of the phlogiston hypothesis how a " false " hypothesis, providing it is adequate to the existing knowledge, may be even more fertile than a " true " one.

The scene is Glasgow; but of the year we cannot be absolutely certain, since nothing was *published* about these discoveries till 1803, when Black's *Lectures on the Elements of Chemistry* appeared (four years after his death) under the editorship of his friend Robison. Though not actually put together by Black they were based on a MS. of lecture notes taken by a student and subsequently corrected by Black—which seems to provide a justification for note-making *in extenso*! In these notes Black is made to say that it was in the year 1760 that he " first began to think on this subject ", namely that " the quantities of heat which different

## "SUBTLE FLUIDS"—I. CALORIC

kinds of matter must receive . . . to raise their temperature by an equal number of degrees, are not in proportion to the quantity of matter in each, but in proportions widely different from this . . ." According to Robison, Black estimated these " capacities for heat " by the method of mixtures of which every school boy is only too painfully aware, deducing that the capacities were inversely proportional to the changes of temperatures of equal weights of the two constituents.

This result clearly accorded well with the transfer of a " weightless fluid " from one to the other, the direction of flow being determined by the relative temperatures, some substances having a greater " capacity "[1] for the fluid than others, that is, requiring more of it to raise their temperatures to an equal degree.

Important as was this result, in 1762 Black read to the Philosophical Club of the University of Glasgow an account of experiments which completely refuted the contemporary view that, once a liquid has been cooled to its freezing *temperature*, a very small and immeasurable further abstraction of heat suffices to solidify it. Realising that melting is easier to study than freezing he carried out quantitative experiments whose simplicity is equalled only by their accuracy; the results conclusively proved that a great deal of heat had " disappeared ". Black soon recognised in this discovery a probable explanation of the familiar fact that the temperature of water ceases to rise once it has started to boil, no matter how fierce the fire under it. He was at a loss to know how to make measurements, until a distiller happened to mention to him that " when his furnace was in good order he could tell to a pint the quantity of liquor he could get in an hour ". So Black had merely to compare the *time* needed to bring a given weight of water to the boil with that required to boil it away. As a mean of several experiments performed by himself " and some friends " he found that " 810 degrees " (the unit quantity of heat had not of course been established) had been absorbed. Regarding the heat " absorbed " both in melting ice and forming steam, he said, " In both cases,

---

[1] The almost exactly parallel concept of the " capacity " of a condenser for the " electric " fluid had arisen about the same time.

considered as the cause of warmth, we do not perceive its presence; it is concealed or latent, and I give it the name of *Latent Heat* ".[1]

Since among his " friends " who are known to have carried out experiments on the same lines was James Watt, it is not too much to say that the whole course of modern civilisation was being determined for better or for worse when Black first put his " cylindrical tin-plate vessels " " upon a red-hot kitchen table " We have not space here to review the whole history of the steam engine, but of the close interplay between " pure " science and its application to the arts, this famous case provides a lesson too valuable to be lightly passed over. There is no indication in Black's writings that his interest in specific and latent heats was prompted by any ulterior motive save that of a deeper understanding of natural processes. On the other hand it is perfectly clear that Watt's interest in these phenomena was the result of his concern to find out the amount of heat " wasted " in warming up the cylinder of the steam engine and the proportion of the heat " stored " in steam which was available for driving the piston.

On the famous " Sabbath afternoon " when Watt first directed his thought to the improvement of the steam engine, the latter had been in regular use for three-quarters of a century. Most of these were of the form designed by Newcomen, a form itself an improvement of the earlier one invented by his master, Thomas Savery. In the Newcomen engine it was the atmosphere which did all the pushing, steam being used merely to reduce the pressure by sweeping out the air after which it was rapidly condensed by means of water jets. Watt, being set by Glasgow University to repair their engine, quickly realised how extremely inefficient such a contrivance was which had to use a great proportion of each charge of steam in re-heating the cylinder chilled by the condensing jets; hence at first his interest in the specific heat of cylinder metal and in the latent heat of steam. Once started, his fertile mind continued its analysis with the result that in 1769 he filed a petition for letters-patent in the following terms:

[1] The term " specific heat " was first used by J. C. Wilcke who, independently of Black, carried out experiments both on this subject and also on the latent heat of fusion of ice. But his ideas never attained the clarity of those of Black.

more abstract point of view became necessary if further progress were not to be hampered. How this came to be we shall enquire later.

Meanwhile we cannot close our account of the *Révolution Chimique* without taking a glance at the *Révolution Physiologique* which was one of its consequences.

One of the strongest claims of the phlogiston hypothesis to the regard of contemporary men of science was its generality. The rapidity with which the " oxygen hypothesis " displaced its well-entrenched rival was due in no small measure to the wide sweep of Lavoisier's speculative genius, as a result of which the new hypothesis was at once brought to bear on every problem to which the older had applied. By means of accurate experiments carried out on devoted assistants, Lavoisier satisfied himself that respiration consists in the slow oxidation of the material of which the organism's body is composed. To this end he developed the method of quantitative analysis still in use for the estimation of the proportions of the elements carbon and hydrogen in organic compounds, that is, by the conversion of these elements into their respective oxides, which are weighed after absorption in suitable chemicals. Moreover, profiting by the discoveries of the " English physicists who were the first to have clear ideas about this matter " he proceeded with the aid of the great mathematical physicist, Laplace, to compare the actual quantities of heat disengaged by the combustion of a candle with the quantity of heat given off by a cock kept quiet for some hours. The heat was measured in these and other experiments by the quantity of ice melted in an insulated vessel devised by Laplace.[1] They were satisfied that to a large extent " animal heat " can be accounted for merely by the oxidation of carbon compounds contained in the blood.

In raising these questions Lavoisier realised that he was venturing into deep waters, which he made no claim to have plumbed; but the methods he introduced—those of exact measurement and comparison with similar events in the non-living world —are those by which physiology has ever since advanced. Deduction from *measurements* presupposes the conservation through-

[1] Black refers to Lavoisier's apparatus, but nowhere mentions " Black's " ice calorimeter, which seems to have been invented by the authors of text books.

out the action of the materials studied; this was fully appreciated by Lavoisier. " Nothing is created ", he wrote, " either in the operation of art or in those of nature." And this led him to ponder on the problem first raised by Lucretius in the immortal words: " Nature never suffers anything to be born until she has recruited herself by the death of some other." " Plants ", in the words of Lavoisier, " take up air and water and mineral matter necessary for their organisation. Animals feed on plants or on other animals themselves nourished by vegetables. Ultimately fermentation, putrefaction, and combustion return to the air and the mineral kingdom the principles borrowed from them by plants and animals.

" By what processes does nature carry out this marvellous circulation between the three kingdoms? How does she form combustible, fermentable, and putrescible substances from materials having none of these properties? These are impenetrable mysteries."

To him, alas, they were impenetrable; for an ignorant and savage bunch of upstarts robbed the world of this fine spirit a few years later. But his work had paved the way so well that by 1804 de Saussure had begun the process of penetration. Using Lavoisier's method of analysis he proved that oxygen appeared in the vessel in which a green plant was growing only when carbon dioxide was present in the first instance.

By the close of the century the phlogiston hypothesis had died out. Priestley alone, writing from his comfortable exile in America (1796) was left to die rather cantankerously in the last ditch; but such was his influence that the leading chemist of the New World, John Maclean, had to warn his students against his ill-considered arguments. Maclean's lectures (1797) are a testimony to the completeness and swiftness of the *Révolution Chimique*, and to the speed of communication between the continents.

Meanwhile, on 8th May 1794, the other revolution had terminated one of the most illustrious of human lives. On the following day the mathematician, Lagrange, sorrowfully remarked " Only a moment was needed to cause that head to fall, and a hundred years perhaps will not suffice to raise another like it."

Posterity has fully confirmed these words.

## SOURCES FOR CHAPTER XVII

D. McKie and N. H. Heathcote, *The Discovery of Specific and Latent Heat* (London, 1935).

A. L. Lavoisier, *Traité élémentaire de Chimie* (Paris, 1789).

H. W. Dickinson, *A Short History of the Steam Engine* (Cambridge, 1939).

CHAPTER XVIII

## THE " SUBTLE FLUIDS "—II. MAGNETIC AND ELECTRIC

THE experimental study of gases by which the "burning question" had been settled and modern chemistry thus established, had been made possible by the manipulation of limited volumes of invisible substances within confined spaces in which chemical action could be induced by the intense heat of electric sparks. The foundation of chemistry, then, had undoubtedly been assisted by the invention of machines capable of producing considerable quantities of electricity; but for the further development of this branch of science, particularly in the isolation of new elements from common materials such as chalk and soda, a further advance in the production of electricity was absolutely indispensable. This advance had, in fact, been made before Lavoisier's death, and its importance quickly realised and put to good use. In pursuance of our purpose to show how the growth of science resembles that of a healthy plant, in which the maturity of one organ can be achieved only by the corresponding development of another, we shall now try to trace the main line of progress in the study of what is perhaps the most wonderful of all natural forces—electricity; and since many of the ways of magnets are easily confused with those of electrified bodies it will be convenient to deal with them together.

Although knowledge of the properties of the lodestone—the natural magnetic ore of iron—and of rubbed amber can be traced back in both Greece and China many centuries before the Christian era, it was not till 1269 that a clearly reasoned account of the properties of the magnet is anywhere to be found; and it is only after the lapse of more than three centuries that the knowledge of electrified bodies was submitted to a critical survey. As is so often the case in the history of science the application of a singular property preceded its thorough understanding. Thus there is no doubt that the art of navigation was receiving its greatest stimulus for several centuries before 1269. Who first distinguished the power of a magnet to set in a given direction from the more familiar property of attracting pieces of iron is a question which the indefatigable labours of many scholars have been unable to answer. There are stories of Chinese land carriages carrying freely mounted figures which always pointed in a direction nearly due south; but though these carriages were said to be in existence from very early times, the earliest written record is dated later than the famous allusion in the book *De Utensilibus* of the English monk, Alexander Neckam, who lived in the twelfth century. The method of using the magnet as described in this book was very different from that of to-day; but it is perfectly clear that it had by that time been introduced as a *subsidiary* aid to navigation when the stars were obscured by cloud.

It is seldom that we can point to a particular event as marking the birth of a branch of science, but in no case can we more justly do so than to the letter written by Pierre de Maricourt (Petrus Peregrinus) on 12th August 1269 to his " nearest of friends ", Sigerius of Foucaucourt. Manuscript copies of this letter were made at the time, but its contents do not appear to have been known outside a narrow monastic circle until after the lapse of three centuries.

De Maricourt's intention was not primarily the improvement of the mariner's compass. Bored, it seems, by the futility of trench warfare before the city of Lucera, he turned his mind to the quite unscientific problem of constructing an engine capable of perpetual motion. The power of the lodestone seemed to him to be the most likely means of encompassing this end. This wild-cat scheme was characteristic of an unscientific age; but the exhaustive study of the properties of the lodestone as a means to this end was such as to give the writer an almost unique position (recognised by his contemporary, Roger Bacon) in the history of science.

When de Maricourt started to make his investigations he probably knew that the lodestone (1) attracts iron; (2) sets in a definite direction; and (3) imparts its power of attraction to previously unmagnetised iron. By carefully contrived experiments he (1) proved the existence of definite " poles ", giving several methods of locating them—one of which virtually implies the idea of lines of force running over a spherical magnet; (2) discovered, and formally stated as a law, the fact that unlike poles attract one another; (3) showed that new poles are created every time that a magnet is broken, and disappear when the fragments are once more joined; (4) that an existing pole may be reversed in polarity by bringing up to it a sufficiently strong opposite pole. Though he naturally failed in his original purpose, yet it is to his credit that he realised that in the use of magnetic fields lay the *possibility* of continuous motion (now effected in the electric motor by continuously altering the position of one of the fields). Moreover by incorporating a fixed scale and a pivoted needle, he much improved the usefulness of the mariner's compass.

All this was achieved in 1269. But little use seems to have been made of it for the next two hundred years. In the first half of the fifteenth century however the Italian city states of Genoa and Venice were bringing to the notice of Western Europe the so far untapped treasures of the East. Great as was this commercial revolution it was soon recognised that the long overland caravan routes and frequent trans-shipments raised the cost of the goods beyond the reach of any but the very wealthy. What was needed was circumnavigation of Africa by the south, or the blazing of a new trail to the Indies by going round the world by the west; for despite the landsmen's belief most mariners were convinced that the world was a sphere. It was largely owing to the improvement of the compass by the Portuguese Prince Henry the Navigator that these two objects were both achieved. Moreover on his third voyage to the West Indies, Columbus discovered the highly important fact that the compass points along a line whose inclination to the true meridian *varies* at different points on the earth's surface. In 1576 Robert Norman found that if a need e is balanced *before* being rubbed by the lodestone it always points downwards at a definite angle *after* magnetisation. Norman was an instrument maker; so once again it was the " practical " man who had made an important advance in our knowledge of the earth's magnetic field.

During the sixteenth century many more discoveries were made concerning the workings of magnets, and the Italian, Cardan, recalled men's minds to the similar but, as he showed, quite distinct attractive force of rubbed amber. In the first year of the new century however appeared one of the most famous books in the history of science; this was *De Magnete, Magneticisque Corporibus, et de Magno Magnete Tellure*, by William Gilbert, a native of Colchester, and physician to Queen Elizabeth. This work is famous not only because it may be said to have founded the science of magnetism, but because it shares with the writings of Galileo and Harvey the claim to be a model of the " new " scientific method of discovery and explanation.

The form and content of Gilbert's book stand out like a watershed in the country of the mind. Reading it from one point of view we cannot help being impatient with the *pseudo*-Aristotelian atmosphere of the whole work. The author is an unstinting admirer of the great philosopher, and comes to his own studies with a determination to see in the *facts* of terrestrial magnetism a further extension of the great Greek's system of nature. But when we turn away from the frame in which Gilbert's observations are set to the observations themselves, we find that the Renaissance preoccupation with signs and wonders has completely disappeared. The compass is no longer magically controlled by the pole star but by the " great magnet the earth ", and this is no mere speculation, but a conclusion arrived at from the careful correlation of observations on the behaviour of tiny pivoted magnets moved over the surface of a spherical lodestone. De Maricourt had made similar observations, but had failed to make clear exactly what the sphere represented.

This complex work of an original genius who, it may be inferred, remains something of an historical curiosity, cannot be profitably studied except with a detailed examination for which we have no space. Nevertheless we must mention one more of his discoveries, since it constitutes the foundation of the science of electricity. Though Cardan had recognised that the attractive force of rubbed amber was different in kind from that of the lodestone, no one before Gilbert suspected that this power, or " soul ", as it had been called from the time of Plato or even earlier, resided anywhere in nature except in this rather unusual material in which not only a " soul " but also the bodies of flies are not infrequently found entrapped. Driven on by one of his Aristotelian fancies Gilbert sought to discover whether the " soul " really *was* peculiar to amber; and in a famous passage he enumerates more than twenty different substances which on vigorous rubbing acquire similar properties to that of amber. The attractive force then is nothing to do with amber, but is an " effluvium " widely dispersed and capable of being generated by friction. This he proved by means of a delicately mounted needle—the first electroscope, though he called it a " versorium ". Despite, as we know now, the unsuitable conditions in which Gilbert worked, he found, apart from metals, comparatively few " non-electrics "—the name he gave to those substances which, after repeated trials, gave a negative effect with the " versorium ".

For nearly two centuries after the appearance of *De Magnete*, progress in the study of magnetism was almost entirely confined to the preparation of maps of the earth's magnetic field, and to the correction of the errors of Gilbert and others, which a careful study of these maps brought to light. But before the end of the seventeenth century scattered experiments by Boyle, Newton, and especially von Guericke, had revealed the far greater variety of the effects of the " electric effluvium ", the study of which consequently attracted a growing number of investigators. An adequate study of the progress of our knowledge of electricity during the eighteenth century would demand a volume to itself; we can as usual do no more than follow what seems to be the main line of advance, bearing in mind that the labours of many men here unnamed contributed much that was of great assistance to that advance.

In 1709, Hauksbee published an account of many varied experiments by which he had established the fact that threads of wool or silk will stand out rigid on the approach of a charged body, and this will occur even if they are placed within an uncharged hollow sphere. A few years later Stephen Gray showed that the electrical influences could be transmitted along hundreds of yards of pack thread when this was suspended by silk, but not at all when *brass* suspensions were used; thus *conductors* and *insulators* were distinguished. In these experiments Gray was aided by his friends Godfrey and Wheler; but the demonstration which had the greatest popular appeal was that in which one of his pupils of the Grey Friars School (Charterhouse) was hung from the rafters by silk threads. After his feet had been touched with a glass tube, charged by rubbing, the boy's face at once attracted feathers in such a manner as only an electrically charged body can do: the human body then is able to conduct electricity. In later experiments with suspended threads Gray showed that a charged thread can *induce* a charge in a neighbouring thread as much as a foot distant from it. By his failure to detect any attraction within a hollow charged body he recognised that the charge lies wholly on the surface.

Gray died in 1736. His own observations, taken with those of Hauksbee and von Guericke, comprise almost the whole of the fundamental properties of stationary ("static") electric charges; but as yet no general hypothesis had been put forward to correlate them. This was done by a brilliant young Frenchman, Charles Dufay who, having in 1733 read the papers Gray had published in the *Philosophical Transactions* of the Royal Society, repeated the experiments and within three years—his last memoir was dated 1737—had established a *science* of electricity out of what had before been disconnected, if ingeniously contrived, observations. His first contribution was to enlarge Gilbert's list of " electrics " to include all rigid substances other than metals. But it was not long before he realised the implication of Gray's silk fibres and resin blocks. Having added glass to the list of insulators he succeeded in showing that metals insulated by glass handles are electrified more readily than any other substances. Thus Gilbert's distinction vanishes; electrification has little to do with the material being rubbed, but only with the nature of its connexion—insulating or conducting—with the earth. This cardinal discovery was expressed by him in the words: " Electricity is a quality universally expanded in all the matter we know, and which influences the mechanism of the universe far more than we think "—a prophecy far more pregnant than he could possibly even have guessed at.

Next Dufay explained the fact first observed by von Guericke, namely that once a body has touched a charged body it is no longer attracted but *repelled* by the latter. Electricity, he thought, repels itself; for if one of the charged bodies loses its charge by contact with the earth it is once again attracted by the charged body.

Then came one of those " accidental discoveries " which have changed the history of science. Up to this time the " electric effluvium " was sharply distinguished from the magnetic in having no essential *duality* such as that which the invariable creation of two poles implies in the latter. But one day, wishing to study the effect of two charged bodies on a gold leaf charged by rubbed glass, he by chance brought near to it a piece of rubbed gum copal. He was " prodigiously disconcerted " to see that whereas the glass, as was usual, repelled the gold leaf, the copal *attracted* it. To a man of Dufay's temper an experiment which didn't work was a challenge. Repeated trials with other pieces of glass and other pieces of gum or amber gave always the same result—the reaction with glass was opposite to that with resins; which led him to " conclude that there are perhaps two kinds of electricities ", thereafter named " vitreous " and " resinous ".

To discover the relationship between the two " electricities " —whether they were, for instance, mutually connected as in the case of magnetic poles—was not possible without more refined means of *measurement*. Such instruments are all based on the mutual repulsion of charged bodies first recorded by Hauksbee. Gray's friend, Wheler, seems to have been the first to have described such an instrument; but the first form in which it attained almost universal use was that devised by John Canton in 1754, in which the intensity of the charge was estimated by the degree of divergence between two threads each carrying a ball of elder pith. From this time onwards improvements of one kind and another were rapidly introduced by Henly, Cavallo, Volta, and Bennet. Of these investigators Volta stands supreme by his recognition of the necessity of introducing an electrical " yard stick " so that investigators in different countries

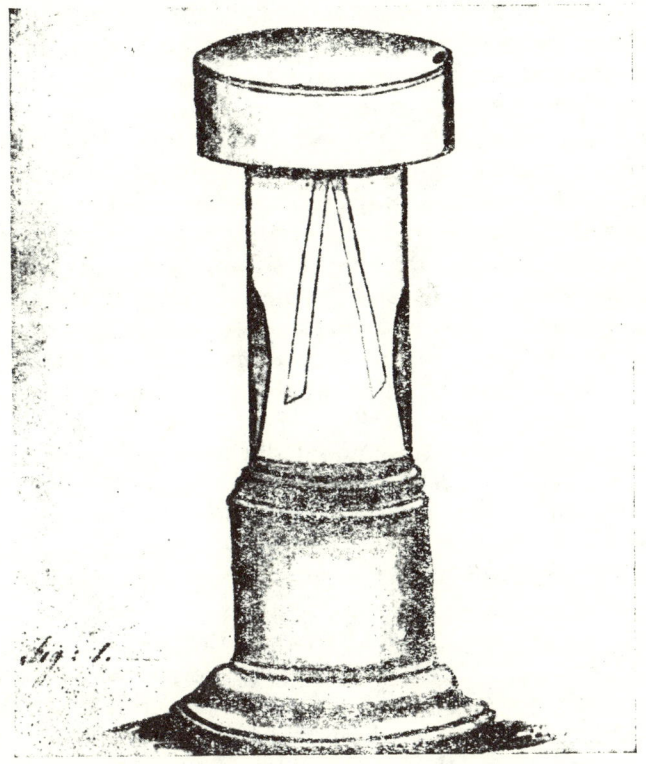

FIG. 27. Bennet's gold-leaf electroscope.

could compare results though using different instruments. He replaced the pith balls by two straws and then calibrated his instrument by actually " weighing " the attraction of two charged plates which gave a convenient angular deflection between the straws. Bennet's contribution was to substitute gold leaves for straws. The first description of this instrument was sent by Bennet from his church at Wirksworth in Derbyshire to Priestley in 1786; this simple contrivance has made possible such modern marvels as the mapping of the atoms in a crystal by the detection of scattered X-rays and of the analysis of radiations from radioactive substances.

The invention of the electroscope had enabled Canton to

prove the true nature of *induction*, namely that the approximation of a charged to an uncharged body causes a charge of the *opposite* kind of electricity to appear on the side of the latter near to the inducing charge, while at the remote side there appears an *equal* charge of the *same* kind as the inducing charge. It is thus quite clear why on the removal of the inducing charge the second body is found to be once again uncharged—the opposite charges have " run together " into electric neutrality.

Perhaps the critical reader is ready to suggest an alternative and, in some ways, simpler hypothesis. Whether this be so or not, Benjamin Franklin from 1747 onwards communicated in a series of letters to his friend and London agent, Peter Collinson, an account of experiments carried out by himself and others in Philadelphia, leading to such an hypothesis which rapidly gained ground throughout the whole of Europe. These contributions are of great interest, showing as they do that the course of scientific discovery is at least to some extent a matter of chance.

It is very probable that Franklin and his little band of associates came to the study of electricity unprejudiced by the confused mass of speculations which had arisen in Europe as a consequence of the widespread activity in this branch of science. All that the colonists had to go upon was the hypothesis of William Watson which had been put forward on the basis of a number of experiments carried out with the " Leyden jar " newly invented by Musschenbroek in 1746. This was at first a glass vessel containing water into which a wire from an electric machine was dipped and which could be charged *continuously* by the operation of the machine. When the operator touched the wire he received a much severer shock than the machine alone could give, and if he at the same time held the jar in his hand the severity was still further increased. Moreover both these effects occurred even after the jar had been disconnected from the machine and carried into another room. This was the first " condenser ", a fuller account of which will be given later.

Watson's hypothesis was that all substances are surrounded and to some extent penetrated by an " elastic electric ether ", which is " disturbed " in a charged body so that a portion of the universal " ether " runs from the nearest conductor (e.g. a man holding a charged Leyden jar) into the " disturbed " region and an equal amount runs from the earth into the body which has lost its own. The hypothesis was short-lived; but it is important historically, because it was in an attempt to render it more adequate to *all* the facts that Franklin put forward his own more famous one.

Not only was Franklin ill-versed (perhaps fortunately) in the current hypotheses, but he seems to have had no acquaintance with the electroscope. In any case his own method of estimating the relative strengths of charges was by comparing the sparks which they could produce. Now a large body may receive a charge without thereafter yielding a spark; but Franklin had rediscovered, and at once applied, in a manner never before attempted, the fact that if a sharp point be held near a charged conductor, the latter produces a spark much more readily. Franklin's innovation lay in his proof that a charged body *can be completely discharged by the mere approach of a pointed conductor held in the hand or otherwise connected to earth.*

While the European enthusiasts were vieing with each other in the production of bigger and better shocks and longer and longer chains of victims joined hand to hand to receive them, Franklin was applying his genius to the elaboration of the simple but adequate hypothesis which has sufficed, at least for the practical electrician, ever since. The *discharge* of a charged body by the mere approach of a pointed, earthed conductor at once suggested to Franklin the analogy of a pipe opening into a cistern of water. Just as all the water in tanks and cisterns in the end finds its way back to the sea, so the earth constitutes the main reservoir of the " common stock " of the " single electric fire ". The charging of a body with " vitreous " electricity is, he suggests, the raising of the body's content of the fire *above* its normal share; while the so-called imparting of a charge of resinous electricity is to be regarded as the removal of some of the body's " stock " of electric fire whereby its content falls *below* the normal. The application of this hypothesis to an actual experiment may be best understood by considering the following quotation of Franklin's own words:

Imagine three persons, each having his normal equal share of electric fire. A, who stands on wax and rubs the tube, collects the electrical fire from himself into the glass; and his communication with the common stock being cut off by the wax, his body is not again immediately supplied. B (who stands on wax likewise) passing his knuckle along near the tube, receives the fire which was collected from the glass by A; and his communication with the common stock being likewise cut off, he retains the additional quantity received. To C standing on the floor, both appear to be electrified; for he having only the middle quantity of electrical fire, receives a spark upon approaching B, who has an over-quantity; but gives to A, who has an under quantity. If A and B approach to touch each other, the spark is stronger, because the difference between them is greater. After such touch there is no spark between either of them and C, because the electrical fire is reduced to the original quantity. If they touch while electrizing the equality is never destroyed, the fire only circulating.

Franklin's hypothesis was one among dozens which went the rounds of the learned societies of the eighteenth century; but it was as superior to all the others as was Newton's theory of gravitation to all those ingenious but useless fabrications of his day; and for this reason, that it merely expressed in unambiguous terms the *form* of the working of electric phenomena without any attempt to decide what electricity " is ". It is true that Franklin himself begged the question by calling it the " electric fire"; but this was force of habit engendered by his use of the *spark* as his test of its presence. As his hypothesis became established as a sound and, for all the phenomena then known, adequate *theory*, it was quickly realised that in describing its behaviour as he did, Franklin was postulating for the electric " fire " the well-known properties of a *fluid*. Thus if water is ladled from one vessel into another of equal size and shape (both being half full to start with) and the two are subsequently joined by a pipe, the water will rush back until the " common stock " is once more shared between them. So Franklin's came to be known as the *one-fluid theory*.

The value of this theory is that it enables us to visualise how a very great number of diverse electrical phenomena *work*. The speculations of his contemporaries on the other hand in ascribing to the " effluvium " a " sulphureous ", " igneous ", " aeriform ", " etherial ", and heaven knows what other perverted form of *known* property, pretended to reveal much more of the nature of electricity, but in fact achieved hardly anything at all. For the most part such speculations are neither true nor false; they are merely empty phrases. Franklin's on the other hand, in so far as it *does* accurately reveal the form of electrical phenomena, is " true "; but a moment's consideration shows that it is but a beginning; it is far from being *adequate*. For in the analogy of the two vessels of water just alluded to we had to premise that they should not only be equal in capacity but of the same shape—the two conditions which must hold if equal quantities of liquid are to stand at the same *level*. Before Franklin's hypothesis could be at all *adequate* with respect to the facts of the Leyden jar then exciting so much attention, it was necessary to discover some characteristic of the electric " fluid " which strictly corresponds to that of " head " or " level " in that of *material* fluids, and at the same time to define the corresponding " capacity " of a conductor for that " fluid ". Two years before Franklin's first letter to Collinson there was born in Italy the man who more than any single investigator was destined to clear up this difficult matter.

Before passing on to consider this fundamental step in the progress of electrical theory it is interesting to note how little was the early study of electricity related to man's practical needs. Whereas the progress in magnetism was clearly the result of Europe's need to expand markets—an expansion which could take place only by an improvement in the speed and safety of navigation—a hundred and fifty years after Gilbert's recognition of the universal distribution of electricity, Watson was compelled to admit that " we are not as yet so far advanced in these discoveries as to render them conducive to the service of mankind ". Except in so far as the study of electricity began as a sequel to that of magnetism this seems a sufficient answer to those " whole hoggers " who to-day protest so loudly that the direction of scientific progress is determined *solely* by economic needs and social conditions. Probably the first technological product of electrical study was Franklin's lightning conductor—a direct *result* of his disinterested experiments on point discharges. His " defying " and " bottling " of the lightning is one of the great epic stories of science. In its bringing of the " special instrument of Divine wrath " within the realm of ordinary natural phenomena it helped profoundly to bring about that change of attitude towards religion which is characteristic of the modern world.

## SOURCES FOR CHAPTERS XVIII

Park Benjamin, *The Intellectual Rise in Electricity* (London, 1895).

W. Gilbert, *De Magnete* (London, 1600); Eng. trans., *W. Gilbert on the Lodestone*, etc., P. F. Motteley (London, 1893).

PLATE IV

Galvani's Experiments on "Animal Electricity"

CHAPTER XIX

# HOW THE STRANGE BEHAVIOUR OF A DEAD FROG LED TO THE DISCOVERY OF NEW ELEMENTS

THE history of science, not to be outdone by the history of kings and peoples, is punctuated with a number of Good Stories which, though probably not true, are sufficiently plausible and even instructive to be worth serving up in a sober opuscule such as this—always with the proverbial grain of salt. One of the best of these is the tradition that when the Italian physiologist, Galvani, was skinning some frogs to make a tasty soup for his invalid wife, he noticed a sudden convulsive movement of one of them every time he touched a nerve with a scalpel. As there are at least two other versions of the story, we may take it that they are probably imaginative reconstructions of Galvani's formal account published in 1791 under the title *De Viribus Electricitatis* in *Motu Musculari Commentarius*. Since these original observations have been frequently misrepresented they will be set down here in some detail.

He had dissected a frog so that the legs were joined to the stump of the vertebral column by the sciatic nerve alone. When an assistant (probably his nephew, Aldini) touched the nerve with a scalpel the muscle of the leg contracted violently. The contraction occurred only when a neighbouring electric machine was working and when the assistant's fingers were in contact with the *metal* part of the scalpel, that is when a conducting path joined the nerve to earth. Since the contractions occurred when the body was completely *insulated* from the machine, Galvani repeated the experiment one day when " thunder was about " and found that atmospheric electricity produced the same results. It is difficult to understand why Galvani persisted with these experiments, as the contraction of muscles under the influence of electric " shocks " had been only too familiar since the invention of the Leyden jar, and experiments had been made on isolated animal preparations as early as 1754. Galvani's experiments however were probably carried out five years before the paper was published, so there may have been other circumstances which he had by then forgotten. Be that as it may, his " second experiment ", as it is called, broke entirely new ground in the history of *electricity*. Here is his own description of the simple observation which acted as so great a stimulus on investigators all over Europe:

> Therefore having noticed that frog preparations which hung by copper hooks from the iron railing surrounding a balcony of our house contracted not only during thunder storms but also in fine weather, I decided to determine whether or not these contractions were due to the action of atmospheric electricity. . . Finally . . . I began to scrape and press the hook fastened to the back bone against the iron railing to see whether by such a procedure contractions might be excited, and whether instead of an alteration in the condition of the atmospheric electricity some other changes might be effective. I then noticed frequent contractions, none of which depended on the variations of the weather.

He repeated these experiments indoors using a variety of metal supports for the body and hooks for the spinal cord. Not only did contractions generally occur, but *their intensity varied according to the nature of the metals constituting the " pair "*. Moreover no contractions occurred when non-metallic substances were used.

Galvani's procedure had thus far been in the best traditions of scientific enquiry: the isolation of the *essential* conditions by means of experiments carried out in freely varied circumstances. But Galvani was primarily a physiologist, and worse still, he was specially interested in the nervous system, the physiology of which was still haunted by " subtle fluids ", " aethers ", and " acid spirits " (p. 453); so it is not surprising that in his mind " the idea grew that in the animal itself there was an indwelling electricity. We were strengthened in such a supposition by the assumption of a very fine nervous fluid that during the phenomena flowed into the muscles from the nerves, similar to the electric current in the Leyden flask."

While admiring Galvani for his determination to find a

## THE DISCOVERY OF NEW ELEMENTS

relation between muscular irritability and the electric discharge, it is perhaps kinder to draw a veil over his subsequent conclusions. When men of science allow " suppositions " to be strengthened by *assumptions*—especially of thin nervous fluids (nerves are not tubes)—they must be left to work out their own salvations, if they can.

An off-print of this paper however came into the hands of Volta, then Professor of Physics at Pavia, who seized upon all those aspects of the experiments whose significance Galvani had missed.

It is almost a platitude to emphasise the degree to which the progress of science is determined by the production of instruments of precision; but it has not been generally realised that the epoch-making invention which Volta was able to make as a result of his grasp of the new ideas latent in Galvani's observations was a consequence no less of the remarkable delicacy of response to which the electrometer had been brought by the labours of himself and others during the previous twenty years. The principal cause of this increase in sensitivity was not so much the introduction of lighter conductors, such as gold leaf, but of a new method of using the instrument based on the elucidation—mainly by Volta—of the nature of the electrical *condenser*.

We have already seen (p. 217) that Franklin's " one fluid " theory was inadequate in so far as it failed to make explicit the characteristic of the " electric fluid " corresponding to " level " in mechanical fluids: how should a small body which must have acquired only a small *absolute* amount of " fluid " above its " normal " share of the common stock discharge into a large body which may quite easily have a much greater *absolute* surplus than has the small one? Such questions were cleared up by Cavendish in 1776, and by Volta in 1778. The former charged up a jar by a machine and measured the largest distance at which a spark would just pass—the discharging distance; he also, through his body, discharged the jar. Next he charged up two such jars and made them share their charge with two others. Whereas the shock of the four was rather greater than of the one, the discharging distance was only half; he concluded that " the strength of the shock depends rather more on the quantity of fluid which passes through our body, than on the force with which it is impelled."

Volta summarises these ideas in an even clearer manner. Recognising that for the electric " fluid ", just as much as for milk or water, there must be a measure of what a body can " hold ", he introduced the term " capacity ".[1] If a given quantity of electricity be given to a conductor, the latter produces a certain deflection of an electrometer; if the *surface* of the conductor be then increased, the deflection is reduced; hence the given quantity may be considered to be spread over a larger capacity, whereby its intensity is reduced. In the same way a pint of milk will stand at a lower level when transferred to a quart vessel having the same *height* as the pint vessel. At first sight the " fluid " theory of electricity seems to be much strengthened by this extension of its concepts, but Volta also drew attention to the remarkable fact that the intensity of electrification of a charged conductor is much reduced by the mere approach of an " earthed " conductor. For convenience we say that the capacity of the former has been increased, and the whole contrivance is called a " condenser."; but it is surely a strange pint of milk which presses less strongly on the bottom of the jug as soon as an empty jug with the bottom knocked out is placed beside it on the table! So what the new views had given to the theory with one hand, they took away with the other. Moreover the recognition that the Leyden jar is, though singularly complicated by the accident of its birth, nothing but a condenser proved that the interposition of a layer of some material other than air between the two plates of a condenser in general increases still further the capacity of the condenser. Let him who can devise an analogy for the pint of milk.

It was difficulties such as these which led to the gradual replacement of the crude but serviceable " fluid " theory by the modern dynamical theory of *potential*. But these difficulties in theory fortunately did not prevent Volta from combining with the gold leaf electroscope the principle of the condenser,

[1] The mutual influence of various branches of science is well illustrated by Cavallo's suspicion that different bodies have different capacities for holding the electric fluid as they have for holding elementary heat. It should be recalled that at this time (1795) heat was very generally regarded as another " imponderable fluid "

whereby he was able to obtain the results similar to those of Galvani,[1] but *without any frogs*. In other words he proved that the contractions " are the effects of a very weak artificial electricity which is excited in a way of which no one had any suspicion, namely by the application of two coatings of different species of metals ". In modern terms the hook and the plate were the battery, the animal only the electrometer.

When Galvani heard of Volta's denial that the animal had anything to do with the *production* of the electricity he replied in an anonymous pamphlet (1794) to the effect that a contraction of the muscle had since been *observed in the absence of any metal contact* merely by allowing the sciatic nerve of one leg to touch the muscle of the other; so there is " animal electricity " after all!

The story of these events is as instructive as it is entertaining; for it shows how important it is when proving the existence of one kind of action not to exclude the possibility of others. Galvani had unconsciously " discovered " *three* distinct kinds of electric phenomena, namely the propagation of electric waves from a discharging machine (not fully explained till very much later), the contact difference of electric potential between dissimilar metals, and the difference of potential established within an animal body, part of whose muscle has been injured—an effect directly demonstrated by Nobili in 1827.

Interesting as Volta's demonstration of contact electrification is, it was only the prelude to a much more important discovery. By touching his condensing electrometer with pieces of different metals in turn, Volta was able to arrange the latter in the following order: silver, copper, lead, tin, zinc; the metals being thus in descending order of charge given to the electrometer considered *algebraically*, that is, silver gave a larger *positive* charge than copper, zinc a larger *negative* charge than tin, lead being almost without effect. By using frogs as indicators he was able to show that the muscular contraction could be best brought about by using a pair of metals, as far apart as possible on the list, e.g. silver and zinc; the source of the electric " stream " being, it

---

[1] Actually Bennet seems to have proved the existence of residual charges on metals *before* Galvani's results were published, so Volta's famous discovery might have been made independently of the latter's.

seemed, nothing but the difference of " intensity of charge " in the two metals. On 26th June 1800 the Fellows of the Royal Society heard read Volta's famous letter to Sir Joseph Banks, their President, describing a new electric battery which would " act continuously, or whose charge after each explosion is re-established of its own accord ". Volta had sufficiently grasped the idea of " intensity " to realise that by arranging suitable pairs of metallic plates in order it should be possible to build up, step by step, a battery twenty, thirty, or more times as strong as the feeble source constituted by only one such pair. It was only necessary that these pairs of plates should be put in conducting communication by means of pads of suitable material thoroughly soaked in water, or better, salt water, since the latter is a better conductor. This contrivance was the famous " Voltaic pile ", which Volta compared with the electric organ of the fish known as the " Torpedo " (*Gymnotus*), whose structure had recently been revealed by dissection to consist of just such a number of separate units. He describes its use for giving shocks, making sparks, and for producing on the tongue an *acid* taste at the positive pole and an alkaline one at the negative. Thus was born the electric *current* whose application has transformed our civilisation. Volta had no doubt in his mind that not only does the battery differ from a body charged by friction or induction in that it regenerates the charge of its own accord, but also that its *discharge* is a continuous process; the evidence that he adduces for this view was the *persistent* taste on the tongue for so long as the battery was connected to it.

Within a few months of the reading of this letter, great " piles " had been constructed in England, and before the end of the year Nicholson and Carlisle had shown that by dipping the poles of such a pile into slightly acidulated water the latter was continuously decomposed into oxygen and hydrogen—an experiment which confirmed by *analysis* the composition of water revealed by Priestley, Cavendish, and Lavoisier by *synthesis*.

Here evidently was a most potent weapon forged by the labours of physiologists and physicists now made available for the use of the chemist in his, so far, rather uncertain attack on the problem of the ultimate elements of matter. We have already

## THE DISCOVERY OF NEW ELEMENTS 225

seen (p. 195) how the prophetic insight of Lavoisier had by analogy suspected that the common strong alkalis—potash, soda, lime, etc.—were in fact oxides of unknown metals whose constituents were bound together by chemical forces so strong as to be impossible to overcome by any methods then known to chemical science. But the same was *almost* true for water (Lavoisier had himself caused red-hot iron to extract the oxygen from steam); yet here were the elements of water torn apart by the mere dipping of two wires into the liquid.

The investigators who most quickly realised the immense potentialities of this new weapon and who at once applied it to the study of chemistry, were Davy in England, and Gay-Lussac and his friend and collaborator, Thenard, in France. Within twenty years of the introduction of the pile these three men had rid Lavoisier's theory of its few serious errors and thus, save for one important aspect, completed the foundation of modern chemistry.

If any man deserved the title of genius it was Humphry Davy. He was born in Penzance in 1778; at twenty years of age, with typically light-hearted courage, he discovered the anaesthetic properties of nitrous oxide by inhaling the various gases which Priestley and others had recently discovered. At twenty-three he was assistant at the newly-founded Royal Institution in London. Before he was thirty-five he had discovered five new elements, cleared up the confusion in regard to the nature of chlorine and the nature of acidity, and discovered many curious and some highly dangerous new compounds. At forty-two he was President of the Royal Society, though continued ill-health had much curtailed his activity after 1813. Though one of the most brilliant experimenters who ever lived, he was also one of those who saw the growing influence of science on industry and social conditions: he lectured to large and fascinated audiences on the application of chemistry to agriculture and, as is well known, invented the miners' lamp. He was handsome, witty, imbued with unusual social grace, a poet of distinction and, as his diaries reveal, a philosopher in the true sense of the word. He died in 1829.

In 1807, using the great pile which had been constructed for the Royal Institution, Davy attempted to split potash into its elements. A vigorous action ensued, but the recognisable products, hydrogen and oxygen, were exactly the same as if water alone had been used. So Davy next tried solid potash free from water; in this case however no action at all occurred. Most investigators would quite reasonably have concluded that potash *was* after all an element; not so Davy, who had sufficient knowledge of electricity to realise that it was necessary to be sure that a *circuit* had been established. He therefore allowed the potash to absorb sufficient water vapour from the air to become a conductor. " Under these circumstances ", he tells us, " a vivid action was soon observed to take place. The potash began to fuse at both its points of electrization. There was a violent effervescence at its upper surface; at the lower or negative surface there was no liberation of elastic fluid; but small globules having a high metallic lustre, and being precisely similar in visible characters to quicksilver, appeared, some of which burnt with explosion and bright flame as soon as they were formed, and others remained and were mainly tarnished and finally covered with a white film which formed on their surfaces."

Here was a remarkable fulfilment of the prophecy of Lavoisier. Who would have thought that the colourless crystalline mass, so greedy for moisture, was in fact a compound of a soft brilliant metal? With quick realisation of the implication of this discovery, Davy repeated the attack on all those substances which were strongly alkaline, using sometimes the alkali itself, or in the cases such as lime, baryta, etc., where the alkalis are infusible, salts of the alkalis. In every case a lustrous, highly active metal was thrust into the world by the new electric magic—a natural magic far more potent than that dreamed of by the alchemists. Thus sodium, barium, strontium, and magnesium quickly followed potassium; and to-day the wonderful virtuosity of Davy is reflected in the great hydro-electric units whose huge annual output of aluminium, wrenched from its earthy source of bauxite, passes into the homes of the poorest, lightening their labour and, alloyed with Davy's magnesium, into the chassis of myriads of aeroplanes likely, through man's inability to control his own folly, to destroy all the wonders he has achieved.

By putting the highly reactive metals, sodium and potassium, into the hands of chemists, the invention of the pile had also an indirect influence on the history of chemistry. One of the first effects was the correction of Lavoisier's statement that all acids are *oxides* and that the reason why certain metals disengage *hydrogen* from most acids is sufficiently clear when we recall that this action takes place only when the acid is mixed with water. The difficulty which Lavoisier could *not* explain was that from the strongest acid—muriatic, or the " marine acid " as it was called—no one had ever succeeded in extracting oxygen by any means; he could only conclude that *chlorine*, being obtained from the acid by heating it with substances like manganese dioxide, rich in oxygen, must itself be an oxide of a higher order— " oxymuriatic acid ". To his successors he left the task of extracting the unknown " element " from this " oxide ".

Before we consider what conclusion his successors did in fact arrive at, let us note how the discoverer of chlorine, Scheele, viewed these matters in the uncertain light of the phlogiston hypothesis. Scheele was himself the earliest discoverer of *oxygen*, which he first obtained some months before Priestley, but by heating manganese dioxide by itself, the very stuff with which, by heating with the " marine acid ", he shortly after discovered chlorine. Since he regarded oxygen as *dephlogisticated* air (actually he called it " fire air ") he rightly regarded chlorine as " dephlogisticated marine acid air ". Now since in the last form of the phlogiston hypothesis " phlogiston " came to be identified with *hydrogen*, we have to admit, what has not perhaps been sufficiently emphasised by historians, that in this particular the phlogistonists gave an account which was, with due allowance for altered nomenclature, essentially the present view; while Lavoisier, with his obsession with oxygen as the " principle of acidity ", was not only wrong, but a false guide to his successors.

Fortunately Davy was sceptical from the first; and having failed to obtain any evidence for the existence of oxygen either in " muriatic acid " or " oxymuriatic acid " he proceeded to a crucial experiment, namely to act on the product of *dry* hydrogen and " oxymuriatic acid " with his newly-discovered sodium, whereby he obtained common salt and hydrogen equal in volume to *half* the " muriatic acid " used. By the action of sodium on "*oxy*-muriatic acid" he obtained common salt alone. These experiments do not prove that oxygen is absent from all these substances; they do however prove that the assumption of its presence is as gratuitous as it would be to assume the presence of phosphorus or any other element. Davy consequently restated the facts in their modern form, and first gave to oxymuriatic acid its present name of chlorine. Gay-Lussac and Thenard repeated Davy's experiments, but at first shrank from spoiling the supposed beautiful harmony of Lavoisier's system. By rare good chance however Gay-Lussac was investigating a new element—iodine—and having found that a precisely similar interpretation applied to it and to its hydracid, he earned the right to share with Davy the credit for completing—and correcting—*La Révolution Chimique.*

## SOURCES FOR CHAPTER XIX

Magie, *op. cit.*, Gen. Bibl.

H. Davy, *The Decomposition of Fixed Alkalies and Alkaline Earths* (Alembic Reprint, No. 6).

## CHAPTER XX

# WHEN IS AN ATOM NOT AN ATOM?

ONE of the facts mentioned in connexion with Davy's demonstration of the elementary nature of chlorine was that the hydrogen replaced by sodium occupied exactly half the volume of the hydrogen chloride originally used. The use of *quantitative* evidence of this kind, which is entirely lacking in the *Sceptical Chymist*, became more and more frequent after *La Révolution Chimique*. Lavoisier's teaching is everywhere inspired by the realisation of the decisive character of data based on measurement; it was however left for others to establish those quantitative laws of chemical combination which set chemistry beside physics as an exact science.

Though Richter had displayed remarkable industry and care in collecting data concerning the relative weights of different acids and bases which are capable of mutual neutralisation, he failed to set forth his results with sufficient clarity to attract the attention of chemists. As has happened more than once in the history of science, an ingenious pioneer has missed the mark by failing to act on the truism " first things first "; Richter's work could hardly impress itself upon a generation, some of whose most brilliant minds were still in doubt as to the *constancy* of composition of all chemical compounds. To Lavoisier this assumption seemed so obvious that he never bothered to establish it by any experiments devised expressly for the purpose; but his fat, good-natured, and ingenious friend, Berthollet, thought that he had sufficient evidence to cast doubt upon this assumption. In his *Essai de Statique Chimique* (1803) he put forward the view that by varying either the conditions of chemical reaction or the proportions of the reacting substances, it was possible to cause a variation in the *composition of the product*. His argument is a model of resource and insight; he was forging a weapon which a century later was to play a big part in industrial civilisation; he was laying the foundations of an independent branch of science; what he was *not* doing was to cast any permanent doubt upon the constancy of composition. Fortunately for the history of chemistry, his compatriot, Proust, was content to walk before he could run; so having obtained from a fixed weight of silver the same weight of silver chloride irrespective of the method used for the conversion, he tried to convert Berthollet to a belief in the *law of definite proportions*, which as everyone knows (or ought to know) states that " every chemical compound no matter how prepared always contains the same elements in the same proportions by weight." Berthollet took no exception to the first part of this enunciation, but he stoutly maintained that whereas in ordinary laboratory operations sodium carbonate forms calcium carbonate from soluble calcium salts, yet on the shores of the Dead Sea sodium carbonate is found mixed with sodium chloride as a result of the action of the huge excess of the latter on the solid calcium carbonate rock. Without knowing it Berthollet was bringing forward evidence for the now well-established " law of mass action "; Proust knew nothing about this law, but he had enough horse sense to see that Berthollet's argument concerned *not the proportion of elements in chemical individuals* but the *proportion of certain chemical individuals formed in complex mixtures*. In further arguments which Berthollet put forward Proust detected a failure on the former's part to recognise that his products were not chemical individuals at all. By sticking to the point Proust brought the controversy to a timely conclusion; but as always happens when a great man has erred, Berthollet's confused recognition of a vital element in chemical reaction—a dynamic equilibrium easily upset by altering the conditions—was thrust into the background for more than half a century.

While this was going on more than one man of science was trying to puzzle out " the law behind the law ". By what mechanism does it come about that pure chemical substances, so different from plum cakes, are formed only of definite proportions of elements, an excess of any of the latter being left over? The man who had the greatest influence on the thought of his contemporaries was undoubtedly the Manchester schoolmaster, John

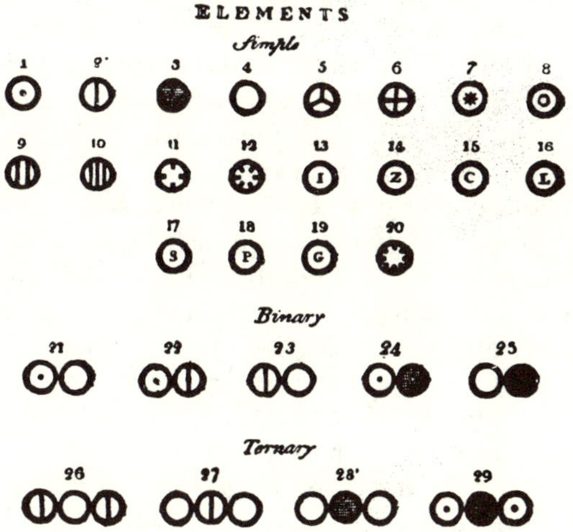

## WHEN IS AN ATOM NOT AN ATOM? 233

**ELEMENTS**

*Simple*

*Binary*

*Ternary*

| Fig. | | | Fig. | | |
|---|---|---|---|---|---|
| 1 | Hydrog. its rel. weight | 1 | 11 | Strontites, | 46 |
| 2 | Azote, | 5 | 12 | Barytes, | 68 |
| 3 | Carbone or charcoal, | 5 | 13 | Iron, | 38 |
| 4 | Oxygen, | 7 | 14 | Zinc, | 56 |
| 5 | Phosphorus, | 9 | 15 | Copper, | 56 |
| 6 | Sulphur, | 13 | 16 | Lead, | 95 |
| 7 | Magnesia, | 20 | 17 | Silver, | 100 |
| 8 | Lime, | 23 | 18 | Platina, | 100 |
| 9 | Soda, | 28 | 19 | Gold, | 140 |
| 10 | Potash, | 42 | 20 | Mercury, | 167 |

21. An atom of water or steam, composed of 1 of oxygen and 1 of hydrogen, retained in physical contact by a strong affinity, and supposed to be surrounded by a common atmosphere of heat; its relative weight = · · · 8
22. An atom of ammonia, composed of 1 of azote and 1 of hydrogen · · · · · · · · 6
23. An atom of nitrous gas, composed of 1 of azote and 1 of oxygen · · · · · · · · 12
24. An atom of olefiant gas, composed of 1 of carbone and 1 of hydrogen · · · · · · · · 6
25. An atom of carbonic oxide composed of 1 of carbone and 1 of oxygen · · · · · · · · 12
26. An atom of nitrous oxide, 2 azote + 1 oxygen · · · 17.
27. An atom of nitric acid, 1 azote + 2 oxygen · · · 19
28. An atom of carbonic acid, 1 carbone + 2 oxygen · · · 19
29. An atom of carburetted hydrogen, 1 carbone + 2 hydrogen 7

FIG. 28. Dalton's illustration of his atomic theory.

it must be presumed to be a binary[1] one, unless some cause appear to the contrary" . . . " From the application of these rules, to the chemical facts already well ascertained, we deduce the following conclusions: first, that water is a binary compound of hydrogen and oxygen, and the relative weights of the two elementary atoms are as 1:7, nearly . . ."

Finally Dalton, realising the " novelty as well as the importance of the ideas ", appended a plate of diagrams to illustrate in a formal way the mode of combination of the atoms.

The above quotations provide a condensed, but I believe fair, account of Dalton's hypothesis. To what extent was it an advance on contemporary views and how far has it stood the test of time? Though Dalton did not invent either the concept or the word " atom ", it is clear that he extended its range. Previous to Dalton the word " atom " stood for the supposed *ultimate* particles into which lumps of matter could be sub-divided. There was no direct evidence for the existence of any such particles; in fact so astute a mind as Descartes' had created a picture of the physical world on the assumption of continuity, that is the *denial* of the existence of atoms. The notion of atoms was born in Greece from purely metaphysical needs and was revived by Boyle, Gassendi, and Newton mainly for mathematical convenience. Dalton's hypothesis is to the effect that these ultimate *physical* particles are also the ultimate units of *chemical* synthesis. We now know (p. 237) that this is not necessarily true; so if it were all, it would be a poor foundation for Dalton's fame. But it is not all.

The objection to the " classical " atoms was the absence of any means by which they could be brought even indirectly within the realm of the senses. By pointing out the constancy of the *weights* of his " chemical " atoms relative to that of the hydrogen atom taken as unity, Dalton gave them a place in the " solid ground of nature " previously denied them. It is of course true that the actual values of these relative weights depended on his assumption, made with the foolhardiness of the pioneer, that the simplest *combining* ratios were necessarily the *atomic* ratios, but the inaccuracy of Dalton's weights does not

[1] That is one atom of each element combined with one atom of the other.

alter the fact that it was he who first showed that such weights *can* be arrived at.

What gives Dalton's hypothesis its place of honour in the history of chemistry is undoubtedly the law of multiple proportions. The experimental fact that with a given quantity of carbon either *one* or *two* parts by weight of oxygen can unite to form two distinct substances, but that any intermediate quantity of oxygen will produce a mixture of these two, constitutes *in itself*, if rightly considered, a guarantee that this action is atomic in character. For, without making any assumptions as to whether oxygen and carbon atoms " exist ", it does unambiguously imply that oxygen can enter into chemical association with carbon only in " sealed packets " which cannot be split.

The law of multiple proportions, established by Dalton for several instances of gaseous compounds and extended by Wollaston to a number of solid salts, is an experimental fact which anyone may verify for himself, not only in the case of those substances investigated by the great pioneers but in numerous other cases rendered more tractable by modern technique. The *implication* of the law, that chemical action is atomic in character, is therefore as certain as the law itself. If Kepler's laws are held to imply a " law " of gravitation, the law of multiple proportions just as certainly implies an " atomic law of chemical action ". This is so clear to us to-day that many writers of elementary text books have assumed that Dalton's atomic hypothesis arose in his mind as an explanation (or implication) of his previously discovered law of multiple proportions. We have insufficient evidence to warrant an explicit denial of this view; but if it were true, then the subsequent development of Dalton's thought betokens a much more muddle-headed nature than we believe Dalton to have been. It is much more likely that the concept of atoms arose in Dalton's mind in the form of Newton's " hard, massy, particles ", but now endowed with Newtonian forces. Having then by a bold flight of genius glimpsed chemical action as an interplay between such corpuscles, he was delighted to find in the law of multiple proportions a striking verification of his speculation. If this be a correct reading of history, we are perfectly right in speaking of "Dalton's atomic *hypothesis*" (or in the light of its continued verification, *theory*) for though the experimental law of multiple proportions necessarily implies that the process of chemical union between elements takes place only between " packets " of a limited degree of smallness, it does *not* imply that these ultimate quantities exist as *free* particles. An illustration from a humbler sphere may perhaps clarify the point. An observer from another planet might, in happier days, have established the empirical law that the transfer of sausages from grocer to consumer always occurred in packets of a limited degree of smallness; but he would find out later that such an atomicity of sausage-transfer was not incompatible with the subdivision of a sausage-atom for purposes of consumption—it is not necessary to cook a whole pound packet at one go.

In this homely illustration (conceived, it is true, when the grocer's sausage-atom *was* a pound!) the atom of transfer is seen to be divisible into smaller units for other purposes; thus it might have happened that the atom of chemical action might be divisible into smaller units when acting in a non-chemical capacity. While Dalton was excogitating his theory that great French experimentalist, Gay-Lussac, had discovered an experimental law which was ultimately shown to imply the reverse, namely that each of the " hard, massy, particles ", an unbelievably numerous swarm of which constitutes a bubble of oxygen, is in fact composed of at least two such units of chemical action.

Gay-Lussac's memoir on the volumetric composition of gases was read in 1808 and published in 1809. Partly from experiments in which the volumes were directly measured, and partly by converting the ratios by weight into ratios by volume by the simple expedient of dividing the weights by the respective densities, he was able to announce his well-known law that " the compounds of gaseous substances with each other are always formed in very simple ratios, so that representing one of the terms by unity, the other is 1, or 2, or at most 3 ".

It will be recalled (p. 234) that in order to fix the relative weights of the atoms Dalton had to *assume* that when only one compound was known (e.g. water, ammonia), it was composed of only one atom of each element, " unless some cause appear to the contrary ". This was patently a weak spot in his theory—

how weak, was to be realised only after half a century of confusion. No blame is to be attached to Dalton for making the assumption that *in the absence of evidence* the simplest composition is the most likely; but unfortunately he seems later to have forgotten that it *was* an assumption, and to have treated it as an integral part of the theory. In no other way can we explain his attitude to Gay-Lussac's law. Gay-Lussac contented himself with stating the facts; Dalton, one might have thought, would have given these facts the most careful consideration in the hope of being able to find a connexion between this law of simple integral volume relations and the similar integral relations which he believed to subsist between the numbers of atoms in familiar compounds. But for no apparent reason other than the fact that the acceptance of the actual volume relations found by Gay-Lussac would have compelled him to modify his own *assumptions* concerning the " binary " character of such compounds as water and ammonia, he so far departed from the ideal implied in all scientific investigation as to *reject the facts* rather than his own hypothesis, until there should, in his own naïve words, be some " reason " discovered for them.

Thus a regrettable prejudice prevented Dalton from recognising this heaven-sent means of clearing up the obscurities in his own system. Meanwhile in 1811 there had appeared a monograph for whose misunderstanding and neglect chemists had in the subsequent half century to pay dearly indeed. This was *An Essay on the Manner of Determining the Relative Masses of the Elementary Molecules of Bodies, and the Proportions in which they enter into these Compounds*. It was written by the Italian physicist, Amedeo Avogadro, and within the space of twenty odd pages so completely corrected and amplified Dalton's atomic theory that it should be rightly regarded as forming with the latter the basis of all subsequent theory of chemical structure.

The basic *assumption* of Avogadro's hypothesis, which he believed to be inevitable in view of Gay-Lussac's law, was " the supposition that the number of integral molecules in any gases is always the same for equal volumes, or always proportional to the volumes ". From this assumption follows very simply the inference that the relative weights of the molecules of any two gases are to one another as the densities of the gases under similar conditions. Avogadro at once recognised the difficulty which arose when the hypothesis was applied to compound substances, " for instance, the volume of water in the gaseous state is, as M. Gay-Lussac has shown, twice as great as the volume of oxygen which enters into it." How can this be so unless the molecule of oxygen be divided between two molecules of hydrogen? Dalton with characteristic insight had sensed this very difficulty *before* Avogadro had approached it, and had in consequence rejected the very hypothesis which the latter was now putting forward. To Dalton the ultimate particles of elementary bodies (what Avogadro calls " molecules ") were unbreakable; any hypothesis which involved their disruption must be *ipso facto* false. Avogadro however took a larger view: " We suppose, namely, that the constituent molecules of any simple gas whatever (i.e. the molecules which are at such a distance from each other that they cannot exercise their mutual action) are not formed of a solitary elementary molecule, but are made up of a certain number of these molecules united by attraction to form a single one." Avogadro went on to explain how this modification may be applied to the case of chemical combination. He has been accused of obscurity of diction and ambiguity in the use of terms. His occasional guilt in respect of the former accusation is such as to preclude lengthy quotation here; but although he uses the term " molecule " to mean several different things, he qualifies it in such a way as to make it quite clear to what it refers in each particular context. He gives several examples of the application of his hypothesis, the most important being that " the integral molecule of water will be composed of a half molecule of oxygen with one molecule or, what is the same thing, two half molecules of hydrogen."

In his exposition Avogadro made the mistake of not coining a new term for his " solitary elementary molecules "—what we now call " atoms "; if he had done so the subsequent history of chemistry might have been far different from what it actually was. But at the same time it is remarkable that not one of the ingenious chemists then living had wit enough to unravel the slight tangle of terms with which he presented them. It must be supposed that the authority of Dalton, who had got in first, blinded them to any contribution which even *seemed* to contradict his views. Avogadro himself suffered from no such diffidence: " Dalton ", he wrote, " on arbitrary suppositions as to the most likely relative number of molecules (*atoms*, in Dalton's phrase) in compounds, has endeavoured to fix ratios between the masses of the molecules of simple substances." What follows is Avogadro's crowning achievement. By means of carefully worked-out examples he shows that not only are the *molecular* weights of gaseous elements and compounds in proportion to their densities, but also that the relative weights of the *atoms* of these elements which form gaseous compounds may be deduced from the volumetric relationships involved in those compounds. Thus, since two volumes of hydrogen and one volume of oxygen form two volumes of water, the molecule of water must contain a half molecule of oxygen united with two half molecules of hydrogen; if therefore one part by *weight* of hydrogen unites with eight parts by *weight* of oxygen, it must follow that the " elementary molecule " of oxygen weighs sixteen times as much as that of hydrogen. But these particles are what Dalton meant, or ought to have meant, by " atoms "; so we have a means of determining certain atomic weights free from those *arbitrary* assumptions which ogadro rightly complains of and Dalton came to regard as sacred.

Avogadro himself at the end of his memoir recognised that his " hypothesis . . . is at bottom merely Dalton's system furnished with a new means of precision from the connexion we have found between it and the general fact established by M. Gay-Lussac." But Avogadro stood alone. As to his contemporaries, Dalton was at least consistent in rejecting both Gay-Lussac's Law (p. 237) and Avogadro's mediating hypothesis; other chemists, with a peculiar lack of logic, ignored the hypothesis while admitting that unless Gay-Lussac was a liar, his law must stand. The results of this extraordinary situation may be read with great patience and no little boredom in the struggles of the great chemists of the first half of the nineteenth century to find a consistent system of atomic weights. The history of this branch of chemistry is of interest chiefly as a commentary on the human mind; it is a succession of ingenious and even heroic attempts to find a way out of a situation which, but for prejudice and purblindness, need never have arisen. Into this jungle of controversy we do not intend to lead the reader; for after half a century of nightmare in which a chemist might, and often did, have to alter the formula of the same compound several times during the course of the study of its derivatives, a fellow countryman of Avogadro, Cannizzaro,[1] took the chemical world gently by the hand, and led it back to an understanding of the words addressed to it, but disregarded nearly fifty years before. All that Cannizzaro did was to make clearer the relation between the atom and the molecule which, as we have seen, was insufficiently explicit in Avogadro's memoir, and by the application of technique invented during the intervening years, to extend the latter's notion of molecular weights to elements like sulphur, phosphorus, and mercury, which do not normally exist as gases. Baldly stated it seems a small title to fame; but one has only to apply a test similar to that referred to in the case of Lavoisier (p. 194), namely to compare original memoirs written before and after Cannizzaro's famous lecture of 1858. In this case it is *formulae* to which one pays special attention. In the earlier papers we have the same symbols (modified by Berzelius from Dalton's original suggestions) as we use to-day, but combined in a staggering confusion. Thus water was represented by $HO, H_2O, \text{H\!O}, H_2O_2, \frac{H}{H}O_2$ according to the school of thought of the author. Within a few years the mists of confusion vanished, and the familiar inorganic compounds appeared in their present-day attire.

The whole development of the theory of chemical structure is based on these words of Cannizzaro: " The different amounts of one and the same element contained in different molecules are all of them whole multiples of a certain quantity which, since it is always found undivided in compounds, is rightly denoted an atom." Here at last is the true definition of the atom; true, because it *works*; and here also is the relation between the atom and the only other unit of structure, the molecule. Whereas the latter is the ultimate unit of matter in bulk, *division of which is impossible without altering the properties of that matter*, the atom is

[1] It should be noted that two Frenchmen, Gerhardt and Laurent, nearly arrived at Cannizzaro's system by independent reasoning, that is without reference to Avogadro.

## WHEN IS AN ATOM NOT AN ATOM? 241

the ultimate unit of chemical action, division of which was at that time unimaginable, because meaningless. For chemically compounded matter the atom and the molecule are in a sense the same—to divide a molecule of water is to destroy it *as water*. For elementary matter the atom and the molecule may be, and in certain cases, such as mercury and the inert gases, actually are, the same; but in the vast majority of cases they are not. Divide a quantity of oxygen as often as you like, and you will ultimately arrive at a *molecule*; further division into particles of oxygen is impossible even in imagination. This point has not always been made sufficiently clear, even in modern text books; for if a molecule of oxygen be further sub-divided the *atoms* which result would not be " oxygen ", as the molecules are, but " oxygen atoms ". This is not a mere verbal quibble; for if it be suggested that a collection of " oxygen atoms " is identical with " oxygen ", how could it ever come about that oxygen is always found in diatomic molecules? The difference is here stressed from the point of view of logic, because it is always difficult to be sure when, as experimentalists, we are dealing with simple atoms; but experiment and theory seem to reinforce one another in the view here emphasised, namely that oxygen atoms do behave differently from oxygen; and it is its *behaviour* which determines the nature of a scientific object.

We are now in a position to answer the question put at the head of this chapter: " When is an atom not an atom? " The answer which Avogadro urged in rather misleading terms was, " When it's a molecule." But from our modern standpoint a better answer would be, " It all depends."

The word " atom " means literally " uncut-able ". In so far as the atoms of Lucretius applied only to the drying up of puddles, the deposition of dew, the smell of the flowers in a neighbouring field, and the gradual wearing away of the stone steps of an ancient castle, then the atomic (better " molecular ") theory of the Greeks is even truer to-day (apart from certain inessential details) than when it was first put forward; truer, because it has been applied to an ever-widening field of phenomena, and found adequate for their explanation.

The " hard, massy, particles " of Newton are the Greek

atoms endowed with the quantitative attributes of mass and weight. The concept of " centre of gravity " by which the " force " apparently inherent in every particle of matter is treated as if it were concentrated at a geometrical point, is atomism *par excellence*; for is not a " point " the essence of indivisibility?

Dalton's atomic hypothesis, in so far as it expresses the *facts* summarised in the law of multiple proportions, maintains its essential truth despite all the sensational atom-splitting which has become the commonplace of modern physics. Thousands of experiments performed by an army of chemists throughout the world depend for their usefulness on the unquestioned acceptance of the belief that chemical reaction does actually occur between units whose weights are characteristic of the several elements, but which show no deviation with the passage of time or the variation of the conditions of the experiment. The atomic theory of Dalton then, with the necessary qualifications revealed by Avogadro, remains as true to-day as when it first took its final shape. But it is no longer as *adequate*. The atoms still perform their ceaseless rounds, weaving new patterns and dissolving old ones; chemistry is still the science of these atomic patterns. Within the context of chemical events the atoms are ultimate; but during the last half century forces have been unleashed which were never dreamed of by Dalton nor even by Cannizzaro. In the stress of hyper-electric fields, in the shattering impact of the α-particle or the neutron, compared with which an aerial bomb on the same scale would be a mere ping-pong ball, the units of *chemical* change break asunder into smaller units. To Dalton, as to Newton, the atom was an ultimate simplicity; Faraday evidently knew better, for in 1853 with prophetic genius he wrote: "As to the little solid particles which are by some supposed to exist independent of the forces of matter . . . as I cannot form any idea of them apart from the forces, so I neither admit nor deny them." For Faraday, matter and its atoms *are* the " forces of matter "—there is no " dead " residue which " exerts " forces on other " dead " residues. And it is thus that we are beginning to regard those enormously stable but *complex* entities, the chemical atoms.

With these words the reader may feel that he is being pulled

# WHEN IS AN ATOM NOT AN ATOM?

into deep water, in which there is no firm ground of observation to rest on: how can matter *be* force? This is indeed a profound mystery; but it is no more of a mystery than the " commonsense " view of a featureless matter " exerting " forces. Moreover it is a mystery whose steady contemplation gives place, as do some of the mysteries of religious experience, to a light in which our perplexities appear though more profound, yet at the same time less confused.

The history of this mode of thought is the history of the electromagnetic field; so we shall now retrace our steps to the University of Copenhagen, where in 1819 the Danish physicist, Oersted, first saw the much sought-for relation between the magnetic and electric " fluids ".

## SOURCES FOR CHAPTER XX

Alembic Reprint, No. 4, *The Foundations of the Molecular Theory*, gives adequate extracts from the works of Dalton, Gay-Lussac, and Avogadro. No. 2, *The Foundations of the Atomic Theory*, performs a similar service; the papers of Thomas Thomson, Dalton, and Wollaston are cited at length.

A. N. Meldrum's *Avogadro and Dalton* (Edinburgh, 1906) seems to me to explore the subject most thoroughly and with freedom from bias. I doubt whether it has been as widely read as it deserves to be.

## CHAPTER XXI

# THE NEWTON OF ELECTRICITY

ON account of the many resemblances—creation by friction, induction, polarity, and many more—between the electric and magnetic " fluids ", it is not surprising that, as Oersted himself reminds us, " certain very celebrated physicists " had attempted to discover a reaction between electricity and the magnetic needle. But in vain; for the excellent reason that between static electric *charge* and magnetism there is no connexion at all. On the other hand it is difficult to understand why up to the year 1819 no one had been more successful by using the " voltaic current ", which, as we have seen, was employed by Davy, for instance, to release new elements. In his first memoir on the subject (July 1820) Oersted remarks that in experiments set on foot at the classes for " electricity, galvanism, and magnetism " held during the previous winter, " it seemed to be shown that the magnetic needle was moved from its position by the help of a galvanic apparatus ". Realising the importance of the subject, and thus wishing to remove all doubt as to the suspected effect, he gathered together a body of " most learned men " (whose distinctions, decorations, and offices are recorded by him with the same precision as the apparatus) to witness a repetition of the experiment.

In his description of these experiments Oersted shows himself to be imbued with the true scientific spirit. Wires of different metals were used and their place and direction altered in every conceivable way. The upshot was—which must at first have surprised him more than he admits—that the effect is at all marked only when the conductor is held parallel to the axis of the needle, whether above, below, or at the side; and the direction of the motion can be summarised in the words " the pole over which negative electricity enters is turned towards the west, that under which it enters towards the east." From all of which he recognised quite correctly that the effect cannot be one of simple attraction such as occurs between magnetic poles or electric charges. As to the cause he is less clear; for he nowhere speaks of an electric *current*, but only of an electric " conflict " which must " perform gyrations " in the space surrounding the conductor. The idea of a conflict is the inevitable consequence of prejudice in favour of a two-fluid hypothesis. In attributing to it the power of gyration he seems not to distinguish between the motion of the " conflict " and the direction of the reaction between it and the magnetic fluid. All things considered however, Oersted claims our respect for having provided his contemporaries with a sound basis of observation on which to construct their own more elaborate enquiries.

The effect on the European Academies was literally electric, but nowhere so great as in France. *Within four months* of the publication of the above paper, three further contributions to the science of electro-magnetism had been read to the *Académie des Sciences*; they were published in the same volume (XV) of the *Annales de Chimie et de Physique*.

The first of these contributions was made jointly by Biot and Savart and was concerned with the development of a quantitative expression of Oersted's qualitative observation. To put this in proper historical perspective it will be necessary first to make good an omission in the account we have so far given (Ch. XIX) of the early development of magnetism and electricity. Just as Volta's elucidation of the mechanism of the " galvanic " circuit was made possible by the refinement of an instrument of precision —the electrometer—so the establishment of a complete theory of electro-magnetism would have been far slower but for the invention of the torsion balance by Coulomb. Coulomb was one of the first of the long line of military engineers who played so great a part in the advance of physics in France during the transition from the eighteenth to the nineteenth century. In 1784 he described to the Academy a piece of apparatus consisting of a cylindrical wire stretched vertically by a series of narrow cylindrical weights whose oscillations about the axis of suspension under various conditions could be timed. He was thus able to deduce a law connecting the degree of torsion with the dimensions of the wire, the material of the latter determining a numerical constant. By using wire whose torsion when submitted to a given force could be calculated from its dimensions, he showed how extremely small forces could be measured. The production of such an instrument illustrates incidentally the advance in metallurgy which permitted of the drawing of a sufficiently uniform wire. In the following year (publication was delayed until 1788) he described the application of this balance to the determination of the laws of reaction between both magnetic poles and electric charges. As every student of elementary electricity and magnetism knows, these laws are of the same form as one another, and as the law of gravitation, namely that the reaction is inversely proportional to the square of the distance between the reacting centres. It is interesting to note that Coulomb had first to demonstrate the existence of such " centres "—the analogues of the centre of gravity in Newton's law. Once this law had been established it was possible to treat the magnetic reaction like any other mechanical force, and so compare the strengths of magnets by measuring the force they could exert on, for instance, another magnet oscillating at a given distance. It was therefore a natural step to take to enquire whether the law of *electromagnetism* was of the same form. This Biot and Savart did by measuring the rate of oscillation of a suspended magnet when placed at various distances from a conductor joining the poles of a voltaic pile. The result was to show that, in their own words, if " from a point where the molecule of austral or boreal magnetism is, a perpendicular be drawn to the axis of the wire, the force which acts on the molecule is perpendicular to this line and to the axis of the wire. Its intensity is inversely as the distance ".[1] The force is therefore of the familiar form except for the startling fact that the force on the " molecule " of magnetism acts not along the line joining it perpendicularly to the conductor, but at right angles to this line. The reference to " molecules " of magnetism reveals the dominance of the idea of atomicity in regard to all the constituent factors of nature—an idea derived from the

[1] The force due to an *element* of circuit is inversely as the *square* of the distance, but Biot and Savart were concerned only with the force due to the whole wire.

astounding success of the Newtonian law of reaction between " particles ", and receiving a new access of strength from its extension by Dalton to the case of chemical action. But in its demonstration of the *tangential* action of the " electromagnetic force " the new law was sounding the death knell of this domination.

The second contribution came from Arago. In repeating the experiments of Oersted he noticed that " the same current produces a strong development of the magnetic condition in bars of iron or steel which were before entirely devoid of it ". The " electromagnetic " force therefore resembled that of a simple magnet, not only in its power of reaction with an existing magnet, but also in the ability to induce magnetism in unmagnetised materials.

The last of the contributions referred to was to mark the beginning of one of the most superb flights of the human genius. It was entitled *Concerning the Action exerted on an electric current by another current, the terrestrial globe, or a magnet*. The author was André Marie Ampère who at an early age showed such a deep insight into the mathematical theory of games of chance as to attract the attention of Delambre. Later he independently recognised the principle generally known by Avogadro's name, and became one of that great company of the Ecole Polytechnique to which reference has already been made.

What distinguishes Ampère's paper from all contemporary contributions was his immediate recognition of the possibility of mutual mechanical reaction between two *currents*—no magnet, nor even magnetic material being necessary. Before describing the beautifully contrived apparatus by which this result was demonstrated he was at great pains to clarify the existing confusion of thought concerning the nature of a " current of electricity ". The term " current " doubtless derived from the conception of electricity as a "subtle fluid" which, as it were, flowed in the form of a spark from a body highly charged with it into one less charged. The large voltaic piles in use at this time would of course have provided the means of passing a spark between the terminals. But there was as yet no clear recognition of anything continuously flowing along a conducting path between these terminals—a state of affairs quite different from that in which the terminals were made to operate an electrometer. At the outset of his paper Ampère defines the former state as *electric current* and the latter as *electric tension*. In the former case " there is no longer any electric tension, light bodies are not sensibly attracted, and the ordinary electrometer can no longer be of service to indicate what is going on in the body; nevertheless the electromotive action continues; for if, for example, water or an acid or an alkali or a saline solution forms part of the circuit, these bodies are decomposed . . . and furthermore as M. Oersted has recently discovered when the electromotive action is made by the contact of metals, the magnetic needle is turned from its direction when it is placed near any portion of the circuit; but these effects cease, water is no longer decomposed, and the needle comes back to its ordinary position as soon as the circuit is broken, when the tension is re-established and light bodies are again attracted ". Here in a nutshell is the clue to the *modern* interpretation of electricity. Gone is the duality between " frictional " and " galvanic " (or " voltaic ") electricity; the latter is merely the *continuous* flow of the former whose behaviour towards an electrometer represents the " tension " which is the cause of the flow. From this " tension " came the term " electromotive force ", a term still in use though much to be deplored, since it is not a " force " but a state of being, better described as a " potential ". And it is clear, as Ampère himself emphasised, that it is only the *current*, that is electricity in motion, which generates magnetic effects. Conversely the magnetic needle, suitably mounted, provides the counterpart to the electrometer, namely an instrument to measure the current, an instrument for which Ampère himself proposed the name " galvanometer ".

The first use to which Ampère put this new instrument was to demonstrate that the current flows in a *circuit*—that is not only from the terminal of the pile outwards through the conducting wire or solution, but also *through the pile itself*, only in the opposite sense as shown by the opposite deflection of the needle. It is only here that a modern electrician finds Ampère's account difficult to follow—though perfectly consistent with itself—for he uses the " two fluid " theory of electricity and so speaks of a "double current, one of positive electricity, the other of negative electricity, starting out in opposite senses from the points where the electromotive action arises, and going out to reunite in the part of the circuit remote from these points". One cannot help feeling dissatisfied with these opposing streams; where do they meet, for instance? If at a point midway between the poles, why does the *whole* conductor acquire new properties? Such questions do not seem to have troubled the peace of mind of Ampère and his contemporaries; he was looking for the *form* of the law of reaction between two neighbouring currents and this, like the quantitative expression of the phenomena of heat, is independent of the *nature* of the electric current.

Thus far there is nothing fundamentally *new* in Ampère's contribution, though by his clear distinction between "current" and "electromotive tension" he has clarified the whole situation. But the next step is revolutionary, namely his demonstration that two delicately balanced wires, in which currents are flowing, *attract* one another when the currents flow in the *same* direction, repel one another when the currents flow in opposite directions, and these reactions begin and end with the closing and opening of the circuit. This experiment shows in the simplest manner why we are compelled to postulate some kind of *motion* associated with the wire. We cannot *see* the current; we see only identical wires whose ends are joined to zinc and copper terminals. A mere reversal of the ends of one of these wires reverses the direction of the force between the wires; in other words the "ends" of the wires are not identical. Alternatively we may simply turn one wire round through 180°; the result is the same as before. Now the only way a straight wire can acquire a "head" and a "tail" is for it to move, or for the direction of motion of something else to be determined by it. It has in technical terms acquired *vectorial* properties. The idea is of course implied in Oersted's experiments; it is merely made more explicit in Ampère's. Moreover it not only shows that something's motion is directed by the wire, but also that there need be no more than *one* such stuff to explain all the phenomena.

The epoch-making character of this discovery was made explicit by the title Ampère gave to his paper of 1822, *Experi-*

*ments on the New Electrodynamical Phenomena*; a title which he elaborated in a note, in which the following passage occurs: " I have determined to use the word *electrodynamic* in order to unite under a common name all these phenomena, and in particular to designate those which I have observed between two voltaic conductors. It expresses their true character, that of being produced by electricity in motion: while the electric attractions and repulsions, which have been known for a long time, are *electrostatic* phenomena produced by the unequal distribution of electricity at rest in the bodies in which they are observed."

To make this point clear beyond the shadow of a doubt, namely that the magnetic effects follow from the mere *motion* of electricity, Ampère showed that the effect is exactly reproduced if a current from the *same pile* is sent successively through the parallel wires free to move at right angles to their axes. There can thus be no question of the reaction of *charges on the pile*; moreover it is to be noted that the effect is altered inappreciably even when the reacting wires are at a great distance from the pile—a circumstance which suggested to Ampère the possible application of the phenomenon to the transmission of messages between distant stations, electric telegraphy in fact. Finally it must not be overlooked that the law of reaction between *currents* is *opposite* to that of electric charges, for it is currents flowing from *like* poles which *attract* one another.

The many respects in which a current of electricity resembles a magnet naturally suggested that the former might share with the latter the power of induction and also that of orientating itself with respect to the magnetic poles of the earth. The former as we have seen was first demonstrated by Arago for a straight conductor, but on the advice of Ampère he repeated the experiment using a conducting helix, or *solenoid*, as Ampère called it, with much greater effect. The second property was much more difficult to bring about owing to the weak magnetic force of the earth, but Ampère ultimately succeeded by means of his ingenious use of mercury contacts into which the ends of moving conductors could be made to dip.

It is thus clear that a current-bearing conductor has all the properties of an intact magnet. It cannot of course be divided like a magnet, since this would involve breaking the circuit. To the physicists of the period under review came therefore the challenge to prove *either* that a current-bearing conductor has the exact form of an assemblage of molecular magnets laid side by side along its axis with all their poles pointing in the same direction, *or* that the axis of every magnet is in effect the axis of an assemblage of circular current elements which encircle it and whose planes are all perpendicular to the axis—an assemblage called by Ampère an " electrodynamic solenoid ". Which of these two is correct? We are here faced once again by the sort of problem exemplified for the first time by the status of the earth in space, and, on a level of much greater complexity, by the rival theories of light. It would take too long to set out all the arguments on both sides: arguments which were sufficiently cogent to lead such eminent thinkers as Wollaston and Berzelius to adopt the former. In fact Ampère seems at first to have been almost alone in at once recognising the superiority of the latter. From the standpoint of the modern philosophy of science, as has already been illustrated with reference to the problem of gravitation, we no longer ask which of the two electromagnetic theories is the " correct " one. But we can decisively say that although the theory which gives to the magnetic element a more ultimate character does explain many of the phenomena, yet there is, as in the case of light, an *experimentum crucis*, which *while not proving the alternative theory* " right " does rule out the former as inadequate. This experiment was carried out by Faraday in September 1821: he found that when a wire connected to one pole of a battery was allowed to hang in a bowl of mercury connected to the other pole, so that the submerged end was free to move, then on placing a magnet vertically within the mercury, the submerged end of the wire at once began to describe a circle about the magnet, and *continued to do so for as long as the current passed*. Now it is of the essence of magnets that all their movements are brought about by *couples*, consequently *continuous* movement is excluded. And it is easy to see why this must be so in the light of modern mechanical views: continuous motion must involve the performance of *work*, and it is clear that a magnet cannot supply the energy for this work. This was not *fully* understood at the time; but there was no doubt that the theory which involved *something* in motion—the electric current—was more acceptable as an explanation of the continuous motion of the magnet than one in which the sole causative agent was an assemblage of magnet-couples.

That Ampère should thus have been able to reduce the whole range of electromagnetic phenomena then known—the reaction between the electric current and the magnet, the mutual reaction of currents and of magnets, and the reaction of both with the earth to the form of classical mechanics, thus earning for himself the title of founder of electrodynamics, might be regarded as a sufficient triumph for one man. But the greater triumph is yet to tell, namely the deduction of the quantitative law from which all these reactions might be calculated for any assigned conditions. The problem far surpassed in difficulty the simple law verified (by Coulomb, and earlier, though unpublished, by Cavendish) for electric charges, and by Coulomb for magnetic poles; for not only does the force vary with change of the *distance* between the reacting circuits but also with their directions relative to one another. Here in fact was the one peculiarity of the new branch of dynamics, namely that the force acts at right angles to the line joining the reacting elements, which prevented its ready acceptance by physicists.

To solve the problem Ampère had to call in the aid of the calculus—a mathematical weapon much extended in its range and sharpened in its application since its invention by Newton had rendered possible the solution of the somewhat similar problem of a planet's motion, constantly changing in magnitude and direction. The analysis Ampère brought to bear was long and involved, but the expedient he ultimately contrived to solve a problem whose direct solution would probably have surpassed the mathematical resources of his day is one of such beauty and wide application in electric science as to justify a brief outline of it.

The *analytical* nature of the infinitesimal calculus is well shown by the problem under consideration. The only procedure which can yield any information as to the reaction between circuits is to consider two *elements*, that is, regions of two neighbouring circuits whose length is considered as smaller than any *assignable* quantity. To fix the *directions* of these elements with respect to one another it is necessary, following the method derived from Descartes, to fix the direction of each with respect to the same three mutually perpendicular axes of space. The only other variable is the distance between these elements, that is, the length of the straight line joining them. The problem has now been *analysed* into the relations between the functions of two variables, distance and direction; but the nature of the *functions*, that is whether for instance it is the square or the cube of the distance which is involved, is as yet unknown. *Direct* appeal to experiment is impossible, since elements of circuits are ideal constructions of the human mind, having no real existence except as parts within a whole. The usual procedure in such a case is to integrate, or " add up ", all the elements between certain defined values of the variables—a process which is possible only if the *functions* of these variables are known or can be expressed as a function of something else which is known or can be guessed from experiment. Ampère started out on this line of attack, but could make little headway. Now, as every schoolboy knows, there are often two ways of solving problems, the austere way of calculating from the data what the answer must be, and the quicker way, often calling for much ingenuity, of looking up the answer and working backwards. This is what in effect Ampère did. It is in fact a special form of analysis peculiarly applicable to physical problems; for in physical problems there is often one answer which can be stated with certainty and reproduced experimentally, namely the answer " nought ", when all the mutually opposing factors cancel out. Ampère described his use of the method in the following words: " It consists in proving by experiment that the movable parts of the conductors are, in certain cases, exactly in equilibrium under the action of equal forces or equal moments of rotation, whatever may otherwise be the form of the movable part, and to seek directly by calculation what must be the value of the reaction between two infinitely small portions such that the equilibrium may in fact be independent of the form of the moving part." He had then no longer to express the reaction in terms of a relation between unknown functions of two variables, but to find such a relation as would *in given circumstances* be equal to nothing. He was able by beautifully contrived experiments to find three such cases. The general nature of these cases may be derived from a consideration of one of them, in which he showed that the actions on a movable straight conductor of two fixed conductors, equal in length and placed at the same distance from it, the one straight and the other twisted into any shape whatever, are nevertheless equal. This implies that the total action, taken right round a closed circuit, no matter what the shape of the circuit may be, is zero; also that the force due to any element of circuit may be resolved into two components—one, the effective one, at right angles to the movable conductor, and another perpendicular to the former—a further justification of the term *electrodynamics*, seeing that this is precisely the law found by Newton for ordinary forces and known as the " parallelogram law "

With these results Ampère was able to determine the unknown functions and write the equations from which the whole theory of electrodynamics has been derived. It is little wonder that in reviewing the subject many years later, Maxwell was to write: " The whole theory and experiment seems as if it had leaped full grown and full armed from the brain of the Newton of electricity. It is perfect in form, and unassailable in accuracy, and it is summed up in a formula from which all the phenomena may be deduced, and which must always remain the cardinal formula of electrodynamics."

## SOURCES FOR CHAPTER XXI

Magie, *op. cit.*, Gen. Bibl.

A. M. Ampère, *Receuil d' Observations electrodynamiques* (Paris, 1822). *Théorie Mathématique de Phenomènes electrodynamiques uniquement déduites de l'Expérience* (2nd ed., Paris, 1883).

CHAPTER XXII

# FROM COLOURED FLAMES TO STAR STUFF

THE story of the generalisation of " radiation " is full of moral value for the historian of science. Like many moral stories it begins in humble surroundings; for Joseph Fraunhofer was the son of a poor glazier and spent his early years in assisting his father. So skilled did he become that before he was twenty-five he had become superintendent of the glass melting and grinding processes at the neighbouring optical institute; at thirty-one he took charge of the institute which thereafter removed to Munich.

The importance of these details of his life lies in the fact that it was solely for the purpose of comparing the refractive indices of different kinds of glass for light of different colours, that he made the experiments that are the starting point of the mighty structure of modern spectroscopy. The necessity for this investigation lay not in any attempt to clarify theory but in the hope of improving achromatic objectives. Once again a purely *technological* end was the means of opening up a completely new and astoundingly fertile branch of physics.

The first recorded experiment was one in which the light from an oil lamp was refracted through a flint glass prism and viewed by means of a telescope. With the magnification effected by the latter Fraunhofer saw what Newton failed to do, namely a line in the yellow region shining more brightly than the neighbouring part of the spectrum. And now follows a further moral. For though Fraunhofer at once recognised the usefulness of this line for the determination of refractive indices of different glasses, *he did not rest content with this discovery.* Is this not one mark of distinction between the technician and the man of science, that the latter is always searching behind the particular instance for the general law? So it was at any rate with Fraunhofer; for having satisfied himself of the existence of precisely similar lines in the light from tallow and many other forms of " fire " he " wished to find out ", as he tells us, " whether a similar bright line could be seen in the spectrum of sunlight as in the spectrum of lamplight ", and he found " instead of this, an almost countless number of strong and feeble vertical lines, which however were darker than the other parts of the spectrum ". In further experiments he found the bright line again in the flames of, hydrogen, alcohol, and sulphur. Moreover he added physical to the astronomical evidence that the planet Venus shines with light reflected from the sun, for in its light appeared many of the dark lines seen only in sunlight.

A report of these investigations appeared in 1814-15, but the presence of dark lines in sunlight had already been observed by Wollaston in 1802. Why did this ingenious man carry out no further investigations of this interesting phenomenon? Is it not significant that Wollaston's interest had a strongly chemical bias—there was not the urge to apply the discovery to any practical end? How else can we account for his naïve theory that the five most prominent lines were the natural boundaries of the " four simple colours " ?

The moral thus appears complete; but not so the story. For it has to be admitted that Fraunhofer himself completely failed to account for the dark solar lines or the bright line seemingly universal in terrestrial flames. It was unfortunate for him that sodium compounds are so widely dispersed as to be almost universal, thereby hiding the fact that the line is a specific test for sodium vapour. Was it the lack of chemical knowledge which prevented Fraunhofer from examining flames coloured for instance by saltpetre or copper salts? Or was he the victim of his own narrow aim, now perfectly fulfilled in the discovery of precise landmarks, for the comparison of refractive indices? Perhaps we are too prone to look for a *single* cause for the trends in scientific advance; more likely is it that in addition to the factors already mentioned was the fact that his publications aroused no interest among physicists, who were too preoccupied with theoretical disputes. Be this as it may, the attack had to be taken up from many other directions before the Fraunhofer lines were explained or the basis of chemical spectroscopy established. This in no way detracts from the greatness of Fraunhofer's achievement, for without the bubble-free optical blanks which he first showed how to make, little further progress would have been possible.

In a book such as this, where we are concerned more with principles than a plain historical record, it will be impossible to do justice to the numerous discoveries which enlarged the boundaries of the spectrum and brought to mankind the realisation that the physical world is filled with unseen agencies as numerous and varied in their powers as any imagined in a spiritual world. It must suffice to recall that by 1859—the year of the first paper by Bunsen and Kirchhoff on spectral analysis—the simple visual spectrum had been pushed beyond the red by William Herschel and beyond the violet by Wollaston and Ritter; the emission of bright lines by hot gases had been observed by Fox Talbot; the selective absorption of light passing through coloured substances by Brewster. Moreover W. A. Miller and Foucault had independently shown the exact coincidence of the yellow lines of sodium with the dark $D$ lines of the solar spectrum; and the possibility of the latter's being due to selective absorption by the solar atmosphere had in fact been seriously considered. The mere fact that these various phenomena had been observed by so many investigators—and many more not mentioned here—seemed to lead to confused notions as to their cause. One of the most influential factors in bringing about this wide interest was undoubtedly the introduction of photography which, originating in the classical observation of the darkening of silver chloride by Scheele in 1777, was first used for making permanent records of spectra by J. W. Draper about 1840. Draper it was who came nearest to the complete solution of the problem of spectra. Born in 1811 at St. Helens in Lancashire, he was one of the first students of the newly-founded University of London, and later migrated to the U.S.A. where he filled Chairs of Chemistry and Physiology. In his most important memoir of 1847 he drew attention to the fact previously overlooked that all solids, at least, become incandescent at the same temperature; that the infra red rays are emitted at lower temperatures, and that as the temperature rises rays of increasing refrangibility are added to the emission *continuously*.

Owing to the absence of any flame devoid of glowing solid particles he mistakenly announced that the spectra of gases *resemble* those of solids, but may have bright lines in addition.

In 1854 began one of those partnerships, comparatively rare in the history of science, between two men of complementary characters, which has been the means of clarifying realms of knowledge hitherto confused by the very multitude of separate workers cultivating them. Kirchhoff was a cautious, speculator and one of the founders of mathematical physics in its post-Kantian, critical form; Bunsen, whom he met at Heidelberg in 1854, was on the other hand a daring and resourceful experimenter, whose work on the derivatives of the deadly substance, cacodyl, was one of the foundation stones of organic chemistry. It seems that the first purpose of the two investigators was to establish beyond a doubt the, as yet, uncertain belief that the vapour of every chemical element emits, when sufficiently hot, a light which can be resolved by a prism into a characteristic set of bright lines—two yellow ones in the case of sodium, a red one in that of lithium, and so on. What had so far prevented any final decision on this point was the absence of any flame free from glowing solids which produce a continuous " rainbow " spectrum. It was for this purpose that Bunsen devised the superbly simple burner which bears his name. The strongest testimony to its excellence is the fact that ninety years after its invention it is still the standard heating unit of the chemist and is the basis of the majority of those domestic appliances—fires, ovens, incandescent lighting—which have done so much, and could do so much more, to lighten labour and reduce dirt and fumes. There is a moral in this story too; for it shows how untenable is the view of those enthusiasts who see in every scientific advance a direct response to social needs: the precise opposite is here the case. Though to give them their due it must be admitted that Bunsen's burner would have had small chance to see the light of day had not Murdoch and others developed the distillation of coal to lighten the nights of London and other cities forty years before.

With this new weapon Bunsen and Kirchhoff quickly established that most fertile application of physics to chemistry—spectrum analysis. The results for chemistry were literally spectacular. Bunsen and Kirchhoff themselves discovered rubidium and caesium in the Dürkheim mineral water—metals which would never have been distinguished from potassium except, as was the case with Plattner, as a discrepancy in the numerical result of analysis.[1] Crookes followed with thallium; then came the numerous rare earths. Most sensational of all was the announcement by Janssen and Lockyer independently of a line hitherto unnoticed in the sun's spectrum corresponding to no known terrestrial element. So reliable was the new spectrum analysis that this element, unseen as such by human eye, was appropriately called "helium". Before the close of the century Ramsay had obtained measurable quantities from various terrestrial sources.

Valuable as has been this application to chemistry, of far more fundamental importance was the explanation of the Fraunhofer dark lines and the alignment of the principle thus revealed with one previously restricted to the transmission of heat. This explanation had in fact been described by Kirchhoff in a paper published a year earlier than the joint paper: " In order to test ", he writes, " in the most direct manner possible the frequently asserted fact of the coincidence of the sodium lines with the lines $D$ of Fraunhofer, I obtained a tolerably bright solar spectrum, and brought a flame coloured by sodium in front of the slit. I then saw the dark lines $D$ (of the solar spectrum) change into bright ones." So far Kirchhoff had merely confirmed " the frequently asserted fact "; thus, he continues, " in order to find out the extent to which the intensity of the solar spectrum could be increased without impairing the distinctness of the sodium lines, I allowed the full sunshine to shine through the sodium flame and to my astonishment I saw that the dark lines $D$ appeared with an extraordinary degree of clearness." Thus the *darkness* of the lines could be increased by passing the sunlight through sodium vapour. Now when an effect is *increased* by altering a given factor in the experiment, it is natural to suppose that this factor is the *cause*, or a part of the cause, of the phenomenon. He there-

---

[1] The atomic weight of caesium is twice as great as that of potassium, so that the assumption that only potassium is present inevitably led to a defect in the total weight in the elements accounted for.

fore felt driven to the conclusion that ordinary sunlight has already passed through sodium vapour before reaching the earth, presumably in the sun. The next step was to make a terrestrial model of the supposed action, namely a continuous spectrum of a lump of lime glowing in the oxy-hydrogen flame. As soon as an alcohol flame highly charged with salt was interposed between the lime and the prism, the $D$ lines appeared. Repeating the experiment with strontium and potassium, he obtained similar results. Lastly he introduced lithium into the alcohol flame, and obtained a dark line in the same position as the bright red line of this element, *although the solar spectrum contained no such dark line.* From all this he concluded that " a coloured flame, in whose spectrum bright sharp lines occur, so weakens rays of the colour of these lines, if they pass through it, that dark lines appear in place of the bright ones, whenever a source of light of sufficient intensity, in whose spectrum these lines are otherwise absent, is brought behind the flame." The single phenomenon of the $D$ line had thus been generalised, but no physical explanation had been put forward. At the conclusion of the paper he noted however that if instead of an alcohol flame he used a bunsen flame, then however little salt it might contain, the *bright* sodium lines alone showed themselves. This provided a clue to the mechanism of the phenomenon. In order to understand how Kirchhoff was able to develop this hint, we must turn aside for a moment to a discovery of great importance in the transmission of heat.

Although Lavoisier (p. 202) had given his blessing to a widespread belief in the existence of an actual substance whose transfer constituted the addition or subtraction of heat, he never supported the common man's belief in the existence of a separate " cold " substance. Towards the end of the eighteenth century however, Pictet noticed that a thermometer placed at the focus of a concave mirror showed a fall in temperature when a block of ice was placed at the focus of a similar mirror opposite the former. This was an awkward fact: it was difficult to see how the " caloric " in a thermometer suddenly became aware of the altered situation and began to leave home and to fill up the region around the ice which had become deficient. The solution of this problem was a principle of cardinal importance in physics, too often rather lightly passed over in elementary work: this was the conception due to Prévost (1791) of a "movable equilibrium of temperatures"—surely a singularly far-seeing conception at a time when the essentially *dynamic* nature of most physico-chemical equilibria (see later p. 295 on the phase rule and law of mass action) was still to be formulated. That heat could travel at great speeds in a similar manner to light, as well as being slowly conducted or convected by particles of matter, had set Newton speculating on the possibility of an *ether* whose vibrations might be the means of its transmission in the form of "radiation"; but with the rise of the *caloric* theory radiation had sunk into the background, probably through the absence of instruments sufficiently delicate to put it on to a quantitative basis. But Prévost boldly put forward the proposition that all bodies are perpetually emitting caloric in straight lines, and that the rate of emission is proportional to the quantity of caloric in the body—a measure of which is given by the temperature. Nor does this process cease when all bodies in a region have reached the same temperature—a state of affairs which on the contrary simply *means* that emission and absorption by every body are equal. With this ingenious hypothesis it is easy to explain the behaviour of the thermometer referred to above without having recourse to any separate "frigoric". It is a state of affairs which would naturally ensue from the fact that less caloric is radiated from ice than from neighbouring bodies such as the thermometer; hence the latter *lose* caloric on balance and become colder until a new equilibrium point is reached, when emission is once again equal to absorption. Though Prévost was firmly convinced that "caloric" was a "specific fluid", he nevertheless added, "if those who believe otherwise substitute waves for an emission, they may be able, perhaps, to adapt to their opinion the explanations which I give for phenomena of this class."

Among those who "believed otherwise" were Balfour Stewart in Edinburgh and Kirchhoff himself. A new weapon had meanwhile been forged by Nobili and Melloni—this was the thermopile, whose action depended on the multiplication of the minute differences of potential, developed by maintaining two metallic junctions at different temperatures; Nobili's share of the work being in the perfection of a galvanometer sufficiently sensitive to measure the minute currents generated. The thermo-electric effect itself had been first observed by Seebeck, and had played an immensely important part in the development of current measurement, since no other source of potential difference was sufficiently constant and reproducible to enable G.S. Ohm to demonstrate the law of current flow which is the indispensable rule of every working electrician.

By means of the thermopile Melloni had been able to demonstrate that radiant heat resembles light precisely in being reflected, refracted, and polarised. When the power of this new weapon was brought to his notice by Forbes, Balfour Stewart set to work to compare the emissive and absorptive powers of a wide range of substances in various conditions, using lamp black (assumed absolutely non-reflective) as the standard of reference. From the results he obtained he was able to conclude that " the absorption of a plate equals its radiation, and that for every description of heat ", that is to say, for each distinct wave length.

Kirchhoff independently developed the same conception, and by means of a mathematical argument of great elegance deduced Balfour Stewart's law as a necessary consequence of the " law of equivalence of heat and work ". The argument was based on two concepts, that of " heat enclosure " (here we recognise the idea of a *thermally* isolated system) and the " black body ". " We imagine ", he writes, " the enclosure to be composed, wholly or in great part, of bodies which, for infinitely small thickness, completely absorb all rays which fall upon them . . . A black body, in this sense of the word, must have the same refractive index as the medium in which the radiation takes place; then there will be no reflection at its surface, and all incident rays will be wholly absorbed." From these " units of thought " Kirchhoff deduced (first for black bodies) the law, " The ratio between the emissive and the absorptive power is the same for all bodies at the same temperature ", the equivalent of Balfour Stewart's law quoted above. Kirchhoff pointed out that though this involved Draper's observation that all bodies *begin* to emit waves of the same wave length at the same temperature, the *intensity* of rays of a certain wave length emitted by different bodies at the same temperature may be very different. "A body that remains perfectly transparent at the highest temperature would never become incandescent. Into a ring of platinum wire of about 5 mm. diameter I introduced some phosphate of soda and heated it in the non-luminous flame of the Bunsen burner. The salt melted and formed a fluid lens and remained perfectly clear; but it emitted no light, while the platinum ring in contact with it radiated the most brilliant light."

It is now easy to understand the genesis of the Fraunhofer and other dark lines. For if, as is the case, a gas radiates light of a certain definite wave length, it must according to Kirchhoff's law also *absorb* radiation of the same wave length. Moreover, if it is at a lower temperature than the source from which it is receiving radiation corresponding to a given band of wave lengths, it will absorb more of each wave length than it radiates; for this state is, according to the theory of exchanges, just what constitutes difference of temperature. The result will be relative darkness in those regions of the continuous spectrum which would otherwise have been formed by the source. Kirchhoff at once recognised that the Fraunhofer lines in the solar spectrum were caused by selective absorption by vapours, in the relatively cool outer layers of the sun, of certain wave lengths from the continuous band emitted from the glowing semi-solid interior. It therefore remained only to identify the lines by causing samples of the terrestrial elements to radiate, after which it could be asserted with some confidence what terrestrial elements are present also in the sun. With the improvement of refracting telescopes, optical glass, and photographic methods, it became possible to extend this method to the analysis of the stars and even nebulae. Moreover the use of the spectrometer is not confined to the determination of the *composition* of the celestial bodies, for in 1842 Doppler demonstrated that just as sound changes in pitch (frequency) when the emitting body is moving relatively to the observer, so, if light is also, as was generally believed at that time, a periodic phenomenon, its *colour* (which is correlated with frequency) should vary in a similar way. Actually owing to the enormous velocity of light, no motion of a body can be swift enough to cause change of colour but only minute shifts in the spectral lines. Over twenty years had to pass before Huggins in 1868, and Vogel in 1871, were able to detect a difference in the wave lengths of lines observed on the approaching and receding limb of the sun. In the hands of modern astronomers the spectroscope has given information concerning the motion, age, and life history of the stars, and the rotation of nebulae, and has recently led to that somewhat fantastic hypothesis of the expanding universe. From this hypothesis have been drawn the most far-reaching consequences as to the origin and destiny of the cosmos, but in applying a purely physical hypothesis to problems which ultimately lie outside the realm of purely physical concepts, too little attention has been paid to the danger of building such an imposing superstructure on the assumption of only one possible *interpretation* of a certain physical fact. For while it is undoubtedly true that the shift of the spectrum of a nebula is positively correlated with its distance from the earth, and while it is equally true that we know of no other cause of such a shift than a high velocity of recession of the emitting body, it ought to be admitted (which it generally has not been) that other causes are conceivable or at least not inherently impossible. The subject is however of great fascination and admirably illustrates that pervasive connectedness between the sublime and the trivial which the progress of science so consistently reveals.

The concepts of "heat enclosure" and "black body radiation" (often called "full radiation") have played a part in the development of physics comparable to that played by the "reversible cycle" (see p. 278). The importance of the black body is due to the fact that the behaviour of such a substance is unaffected by its chemical constitution and depends only on the temperature of the enclosure; thus we have a unit of measurement for radiation corresponding in some measure to the thermodynamic temperature scale (see p. 283) whose importance is of the same character, namely that it is independent of matter *as such*.

Space does not permit of our following up the fascinating implications of this concept; to give a mere catalogue of names and discoveries in so difficult a subject would be futile. Suffice it to say that step by step, not without the appearance of discrepan-

features, progress was made to a stage where it appeared possible to *calculate* the distribution of energy in a full radiator merely as a function of the temperature; for it must be clearly understood that a full radiator emits energy of all possible wave lengths between limits fixed by the temperature, but in no case is the energy *equally distributed* between these wave lengths. The radiation of the sun, for instance, spreads from the infra red through visible to ultra violet; but the greatest proportion of the energy of clear sky sunlight is in the orange-red. Sunlight analysed in the stratosphere would show a striking shift of this maximum towards the violet; the difference is due to the selective absorption of the lower wave lengths brought about by our atmosphere, without which absorption no living cell as we know it could long survive.

Now when attempts were made, from very general assumptions as to the dynamical character of the " ether ", to deduce the distribution of energy in a full radiator, results were obtained which were completely at variance with the results of experiment. Once again, as so often in the history of science, apparent failure was the fore-runner of resounding triumph. This triumph was of course the recognition of the " atomicity " of energy exchange demonstrated by Max Planck in the early years of this century. As part of the contemporary battlefield it lies outside the scope of this book.

The third approach to the relation between absorption and radiation was made by G. G. Stokes, Lucasian Professor of Mathematics at Cambridge, and like his great forbear, Isaac Newton, a man of great resource in experiment. Whereas Balfour Stewart approached the problem from the infra red, Kirchhoff from the visible Fraunhofer lines, Stokes was prompted to his formulation of the relationship by a phenomenon first noticed by J. F. W. Herschel, namely that an aqueous solution of quinine sulphate, though perfectly colourless when viewed by transmitted light and in light from most parts of the spectrum, yet when held in such a position that the blue-violet, and in particular the ultra violet, region of the spectrum fell upon it, glowed with a bright blue light. A similar phenomenon had been noticed by Brewster in the case of an alcoholic solution of chlorophyll, which however glows with a red light. Both these investigators had made important contributions to the theory of polarised light, so they naturally expected that these new colours were the result of some form of polarisation such as occurs in thin flakes of tourmaline; but the emitted light was *not* polarised. Stokes' study of the problem illustrates the necessity of logical analysis and courageous acceptance of the inference, even when that inference appears absurd in the light of contemporary knowledge. This is what he himself wrote:

> The firm conviction which he [Stokes] felt that two portions of light were not distinguishable as to their nature, otherwise than by refrangibility and state of polarisation, left him but few hypotheses to choose between respecting the explanation of the phenomenon. In fact having regarded it at first as an axiom that dispersed light of any particular refrangibility could only have arisen from light of the same refrangibility contained in the incident beam, he was led by necessity to adopt hypotheses of so artificial a character as to render them wholly improbable. He was thus compelled to adopt the other alternative, namely to suppose that in the process of internal dispersion the refrangibility of light had been changed. Startling as such a supposition might appear at first sight, the ease with which it accounted for the whole phenomenon was such as already to produce a strong probability of its truth.

Thus by *rejecting*, for a particular case, what Newton had with such labour and clarity demonstrated to be true for dispersion in general, Stokes was able to explain this phenomenon of " internal dispersion ". To avoid confusion, this term, which is really misleading, as it is not a refraction effect but rather a molecular one, has been replaced by that of *fluorescence*. From this result, namely that certain substances are capable of absorbing radiation of one wave length and emitting it at a *longer* wave length, Stokes developed the resonance theory of absorption. It is well known that if a tuning fork of a certain frequency is struck, it may cause such vigorous " sympathetic " vibrations in *some* neighbouring objects as to cause them to emit sounds. The effect is far greater in those objects which, if struck, would emit the same note as the tuning fork, that is, those which have the same natural period of oscillation. Stokes conceived the atoms of sodium, for instance, to have similarly a natural period of oscillation. When therefore radiation of a band of frequencies containing this frequency passes through a mass of sodium vapour, the energy is, as it were, drained out of the incident beam in just that frequency, it being transferred to the sodium atoms to increase the amplitude of their oscillations; consequently the beam will be relatively poorer in that frequency and the Fraunhofer dark line will appear. In the case of fluorescent substances absorption will occur in one frequency, but the molecular constitution of the substance is such that the energy thus absorbed by resonance is " degraded " into a lower frequency before emission. In a similar way phosphorescent substances emit the absorbed energy at a reduced rate, so are able to continue emission for some time after absorption has ceased. The subsequent history of fluorescence was spectacular; for in seeking for a persistence of fluorescence in uranium salts after removal of the exciting light, Becquerel discovered that these salts emitted rays even when no previous excitation had occurred; hence was born the science of radioactivity.

Stokes' paper on fluorescence was published in 1852; the resonance theory of the Fraunhofer lines was never actually published but had been put forward in the course of conversation with Lord Kelvin, who later claimed for Stokes priority over Kirchhoff. Stokes himself replied as follows: " I have never attempted to claim for myself any part of Kirchhoff's admirable discovery, and cannot help thinking that some of my friends have been over zealous in my cause "—a generosity of spirit not universal even among " seekers after truth ".

## SOURCES FOR CHAPTER XXII

Magie, *op. cit.*, Gen. Bibl.

Fuller citations are given in *The Laws of Radiation and Absorption* (Harper's Scientific Memoirs, New York, 1901).

CHAPTER XXIII

# THE " GO " OF THINGS

Towards the end of the last chapter a new term' was surreptitiously introduced—I say " surreptitiously " because it does not occur in the works of either Kirchhoff[1] or Stokes, in the elucidation of which I have used it. The word is " energy ". A modern writer attempting to condense and paraphrase the classical memoirs on radiation is constantly driven towards the use of this term, which he well knows became current coin in scientific transactions only some years *after* the theory of radiation in a quite general form had been firmly established. This experience is perhaps not without its significance for the history of science; it shows that physics was breaking through its material bonds into the realm which now appears the more fundamental. To put it another way, the reign of *mere* " stuff ", which was the product of the Newtonian revolution, was coming to an end. Though the recognition of the essential association of motion (force) with matter was the triumph of Newtonian physics, the accent was all along on *matter*; as to the force, "*hypotheses non fingo*". But with the development of the physics of radiation the emphasis is reversed—the preoccupation is not so much with the emitters as with *what is emitted and propagated*. Nor was this the beginning of a change in emphasis; it goes back to a revolution no less important than the Newtonian revolution, that is the Industrial Revolution, which was the result of the availability and mobility of prime *movers*. But it was essential to be able to compare the *powers* of these prime movers. James Watt, the honest, accurate, practical-minded genius, knew that his engine would sell only if compared with the existing prime mover—the horse—so he saddled us with a unit which contains the inconvenient factors 5 and 11 ! As we shall see in the sequel, the development of one

[1] Kirchhoff was however using the *concept* under the name *vis viva*.

of the major branches of modern physics sprang from an attempt to set up a theoretical standard for the " motive power " of heat.

We should be wrong however to suppose that interest in power and energy—the emphasis was first on the former, though the distinction was necessarily blurred—dated only from the social need for a standard. The germs of scientific advance can usually be traced for centuries before the time that they become empirically developed; the social environment does however seem to determine which of these germs shall spring into active growth.

The distinction between " stuff " ($ὕλη$) and the " go " ($ἐνέργεια$), whereby it develops towards the perfection ($ἐντελέχεια$) proper to its kind, was the essence of the metaphysics of Aristotle who, wiser than the mechanistic materialist of the eighteenth century, saw that the " go " and the " stuff " are only different aspects of the same thing—*mere* matter would be literally *non*-entity, *mere* form being found nowhere in nature, being indeed God Himself, the " Form of Forms "—" pure act " in the suggestive phrase of St. Thomas Aquinas.

Unfortunately Aristotle suffered, as many a young man has since, the inevitable paralysis due to a brilliant and persuasive teacher. Aristotle was an outstanding biologist, and his metaphysics of inward drive towards a dimly perceived end was no idle dreaming, but a profound intuition based on observation (see later Ch. XXV). But, being compelled with the amiable optimism of the Greeks to frame a complete account of the workings of nature, he took over from his master the worst features of a " physics " based purely on geometrical imagery—a one-sided view, which the shifting of a few blocks of stone might have served to correct, and thus started the building of a mountain of rubbish, beneath which the true genius of this remarkable man was hidden until recent times. So the " germ " of energy remained frozen in a museum of Platonic forms, which determined the " proper place " of every natural object; which restricted change to a mere return to the *status quo*. One might have expected that a clearer idea of the " go " of things would have been reached by the more inventive and practical Alexandrians, and in fact this was implicit in Archimedes' distinction between force and moment of force. But that great genius seems to have

FIG. 29. Early mining machines. A toothed wheel (A) on a horse-driven axle engages in a crown wheel (C)—the first method of changing the direction of the axis of rotation.

regarded his own inventions as playthings, of which he was rather ashamed.

The above reference to St. Thomas will show that the idea of action—closely related to the notion of physical energy—was the mainspring of that last vigorous movement of the Middle Ages; a period which modern historical research has shown to be much more fertile in invention than was previously realized. But although there was no dearth of mechanisms, theory was restricted to meteoric flashes of genius such as Buridan and Nicholas of Cusa, both of whom in rather different ways recognised the cramping artificiality of the Aristotelian physics. There was still however a gulf between the people who worked the machine and those who speculated upon the nature of motion itself. The brightest of all these, Leonardo da Vinci, not only invented, at any rate in his own mind, many mechanical devices which were actually *created* only much later, but shows by some of his almost illegible notes that he had a clear idea of what he called the " potential lever ", that is, the perpendicular distance from the axis on to the line of action of the force. He was not alone in this: both Ubaldi and Stevinus reached the same conclusion by different routes. Leonardo had the misfortune to be born just too soon. The growth of trade, the accumulation of capital, the rise of city-states and later of nation-states, the consequent growth of armies with an increased need for good communications, weapons and supplies, had begun; but it had not yet reached full stature. Otherwise it is almost certain that instead of dying with the greater part of his inventions unknown for three centuries, his claim to universal fame resting chiefly on *The Last Supper* and *Mona Lisa*, he would, like the Duke of Plaza Toro, have turned himself into a limited company and endowed the " Leonardo Laboratory " at the University of Florence.

A more important step towards the formation of the concept of work was taken by Stevinus who based his approach to the problem of vectors on the fact that an endless chain hung over a triangular support with unequal sides does not, *cannot*, move. For if it did, since nothing in the arrangement has been changed by the movement, it would move for ever. Therefore a long chain (greater weight) on a gentle incline is equal to a short chain (less weight) on a steeper incline. Here in the denial of perpetual motion is the positing of the problem of work; and in the fiction of the endless chain is the first step towards its quantitative solution.

No further advance in the development of our ideas of the " go " of things was possible until the laws of motion had been clearly enunciated by Galileo. Newton was born at the time when astronomy was reaching its climax—his formulation of the laws of motion therefore took the form of expressing the " go " of things (Newton's " quantity of motion ") as a function of *time*. In this, though in very little else in mechanics, Descartes agreed with Newton; but, less cautious than Newton, he stated it to be the *only* fundamental quantity and therefore fixed in amount. From this it seemed to follow that the " go " of any system depended *solely* on the quantity of motion in it. This result, the argument for which was couched in theological terms, having a strongly Aristotelian flavour, was attacked by Leibniz. The article of the *Acta Eruditorum* (1686) in which this attack was made well repays reading, as revealing the hopeless confusion which reigned in regard to such terms as *force, motive force, body, quantity of motion*. Leibniz distinguishes, which Descartes did not, between *motive force* and *quantity of motion*, but in referring to *body* fails to distinguish between *mass* and *weight*, and finally confuses, as Mach pointed out, the *measure* of *effective force*, and the question as to what is *conserved*. To make this latter point clear it will be necessary to summarise Leibniz's argument very briefly.

Leibniz starts from two assumptions, namely that (1) in absence of disturbing forces, a body in falling acquires a " force " sufficient to raise it to the height from which it had fallen, and (2) as much " force " is used to raise a " body, weighing one pound to the height of four ells, as to raise a body $B$ weighing four pounds to the height of one ell ". Here he uses the term " force " in two different senses, namely (1) *momentum* and (2) *work*, neither of them equivalent to Newton's definition as that which is proportional to the *change* of momentum—the one accepted to-day. He then proceeds to argue that the body $A$ in falling four ells has acquired the same " force " as the body $B$ has in falling one ell, and this because *each has now sufficient force to return it to its original height.* But since, as Galileo had shown, the velocity acquired by *A* is only *twice* that acquired by *B*, the quantity of motion gained by *A* is only *half* that gained by *B*. From which Leibniz quite correctly concluded that " there is a great difference between motive force and quantity of motion ", and quite dogmatically continued that " from this it appears in what way the force should be estimated from the quantity of the effect which it is able to produce; for example, from the height to which it can lift a heavy body of known magnitude and nature, not from the velocity that it can impress upon the body ". This assertion is dogmatic in the sense that the definition of " force " is quite arbitrary; it is neither better nor worse than Descartes'. The " simple demonstration " was entirely irrelevant to the " nature of things " and was concerned merely with a choice of definitions; it was certainly no basis for the patronising manner in which Leibniz asserted that " Descartes, by an error common to great men, became a little too confident ".

The more one reads of the squabbles between the Cartesians and the Leibnizians the more one realises the greatness of Newton, whose *Principia* appeared the year after the above article. The truth seems to be that the continental thinkers had failed to rid themselves of metaphysical crudities deriving from theological preoccupations. Newton was just as much engaged on theological issues as they were, but he had the admirable English faculty of keeping the two realms apart. It may be admitted that this delimitation cannot be permanent if the human race is to avoid mental disintegration, but it was essential, *at that stage of development of physics*, that clear and unambiguous concepts and definitions should alone be enunciated even if they were later discovered to be inadequate or based on over-simplified assumptions.

That the whole dispute was entirely metaphysical and consequently futile *from the point of view of mechanics* was argued with clarity by D'Alembert in his *Traité de Dynamique* (1743). "All that we see distinctly in the motion of a body is that the body traverses a certain distance and that it takes a certain time to traverse that distance. It is from this one idea that all the principles of mechanics should be drawn . . . I have entirely excluded forces inherent in bodies in motion, obscure and metaphysical entities which can only cast shadows on a science that is in itself clear." Here we have the ruthless abstraction which foreshadows the criticism applied a century later by Ernst Mach to the whole Newtonian conceptual scheme. D'Alembert is undoubtedly right in rejecting "forces inherent in bodies in motion" which is a relic of animism and is entirely irrelevant to the science of *mechanics*. But though he shows clearly the cause of the long and futile controversy concerning the measure of "forces" by "quantity of motion" (proportional to velocity) or *vis viva* (proportional to *square* of velocity) he did not perhaps sufficiently realise the growing importance of the concept of work.

It is perhaps not without significance that this *concept of work* was clearly defined first in relation to purely *statical*[1] problems. This was done by Jean Bernoulli in a letter written in 1717 to the French mathematician, Varignon, who was so struck by its generality and beauty that he set it forth in one of his own books, *Nouvelle Méchanique ou Statique*. Put briefly, the proposition amounts to the statement that if in any system of forces in equilibrium one of them is given a very small displacement, then all the remainder will adjust themselves so that the algebraic sum of all the products of the several forces and their respective displacements *in the direction of their lines of action* will be zero. This is the principle now known as the *principle of virtual work*, though to add to the modern reader's confusion Bernoulli called these products—the virtual "works" of the forces—their "energies". We might now without error call them changes of *potential* energy, but the correlative concept of *kinetic* energy was not introduced until nearly a hundred years later.

Meanwhile Lagrange had derived the whole of his superb *Méchanique Analytique* (1788) by the application of analysis to the principle of virtual work in such a way that the problems of mechanics, which Newton had reduced to problems of *geometry*,

---

[1] One notable exception to this was the invention by Huygens (1673) of the pendulum clock. The problem of finding the centre of oscillation of a compound pendulum was solved by assuming that no one of the component masses could have its centre of gravity raised above the common centre of gravity without the application of a force external to the "force" inherent in the system. Here the concept of potential energy is implicit. The principle became the basis of the further development of rotational dynamics. The stimulus to this study was of course the hope of using the clock for the determination of longitude at sea.

could be posited in the form of equations (i.e. relations between co-ordinates) whose solution involved only mathematical skill. This had a profound influence on the development of physics. It subsequently caused a striking acceleration of progress and an extension of vision into unexpected fields; but it did so at the expense of cutting much thought adrift from the " solid ground of nature " in the aim to achieve merely formal elegance and consistency. We shall return to this subject in the next chapter.

Meanwhile all this had been achieved *without any clear idea of the nature of energy having been formulated*. Why was this? It is at least plausible to suppose that it was just about this time that the steam engine passed from the pumping to the rotary stage—previously mechanics was largely concerned with the problems of *structures*—bridges and fortifications. It is significant that in the Bakerian Lecture in which Young confirmed the falsity of measuring *force* by the product of mass and velocity squared, proposing the term *energy* for the latter quantity, he refers to Smeaton, an early pioneer of the steam engine, as one who had raised the question. It is perhaps not an exaggeration to suggest that the formally mechanical concept of kinetic energy might never have been reached or at any rate been widely used, had it not been forced upon man's attention by the need to find a relation between work and *heat*.

We have already seen how the quantitative basis of the science of heat had been laid by Black and Lavoisier before the close of the eighteenth century. It was a practical engineer, Benjamin Thompson (Count Rumford) who in 1798 first suggested the inadequacy of the caloric hypothesis. His rough but ingenious experiments carried out with a specially *bluntened* borer on brass castings in a cannon foundry showed without a doubt that heat was not a " subtle fluid " extruded from the pores of material bodies, but was in some way provoked by friction. Rumford's clear understanding of what was at issue is shown by his subsequent demonstration that the capacity for heat of the brass *turnings* did not materially differ from that of a block of the same weight. There could be no question of a " loss " of something previously held within the interstices. In Rumford's own words " it is hardly necessary to add that anything which any *insulated* body,

T

or system of bodies, can continue to furnish without *limitation*, cannot possibly be a *material substance*; and it appears to me to be extremely difficult, if not quite impossible, to form any distinct idea of anything capable of being excited and communicated in the manner the Heat was excited and communicated in these Experiments, except it be MOTION ". The creation of a new " climate of opinion " is evidenced by Davy's youthful experiments " with a peculiar mechanism " whereby ice was melted without addition of heat. The fact that Davy's optimism concerning the complete exclusion of heat invalidates the demonstration, does not alter the interesting fact that he thought it worth attempting. Moreover at the end of the description of this and similar experiments he expresses the opinion that heat is " the power which prevents the actual contact of the corpuscles of bodies ", adding less happily that " it may with propriety be called the repulsive motion ".

We must at this stage clearly distinguish what is at issue. The supporters of *both* hypotheses regarded motion as constitutive of the phenomena of heat, but whereas at the end of the eighteenth century the prevailing fashion was to regard any such motion as being in fact the motion of particles of the hypothetical " fluid ", *caloric*, the view being urged by Rumford and Davy was in the orthodox line of physics from William of Occam, namely that caloric is *superfluous*, the heat of a body being *nothing more* than the motion of the particles of matter. This was not a new *hypothesis*; it had been stated confusedly by Boyle (who unfortunately ascribed weight to the " corpuscles of fire "), with a fair degree of precision by Hooke, and in mathematical terms by Daniel Bernoulli in 1738, in a work that laid the basis of modern hydrodynamics. What was new was the attempt, qualitatively successful, to demonstrate it by direct experiment. But physics thrives neither on hypothesis nor on qualitative demonstration, but on quantitative law; and for a comprehensive demonstration of the quantitative interconvertibility of heat and mechanical motion it had to wait nearly half a century.

It is interesting to speculate why this was so. One reason perhaps was the fact that during this period the best intellects in physics had all they could manage in establishing **the undulatory** theory of the propagation of light, the laws of magnetism and of electrodynamics. Another reason was the necessity for the establishing of *detailed* quantitative relationships between heat and matter, which were carried out by Dalton and Gay-Lussac (thermal expansion of gases and pressure of vapours) and by Dulong and Petit (atomic mass and specific heat). During this period also fundamental observations were made on the functional relationship between temperature and aggregational state (Cagniard de la Tour 1822, Faraday 1823). All these advances were essential for the full understanding of the theory of the steam engine —Cagniard de la Tour expressly mentions this in his paper— but the obstinacy with which the caloric theory was retained is evidenced by the long absence of any attempt to investigate the possible relation between heat supplied and mechanical work obtainable, and the fact that when the attempt was made it was argued in terms of a flow of caloric.

It has been pointed out by Pledge (*Science Since* 1500) how well the history of steam power illustrates the difference between the British and French intellectual tempers. In Britain the invention of the piston engine in its modern form was quickly recognised as being a major event in the history of technology; but it was in France, where the spread of steam power was much slower, that the first analysis of the *physical* relation between heat and power was undertaken.

The small memoir of Sadi Carnot, a product of the Ecole Polytechnique, and captain of military engineers, was published in 1824 under the title *Reflexions sur la Puissance Motrice du Feu*. Nowhere in the history of science is the close relation of technological progress and enlargement of theory more closely demonstrated. At the very beginning of this epoch-making memoir the question is asked, " What happens, in fact, in a steam engine at work? " The analysis which follows is *expressed* entirely in terms of the conception of the *flow* of caloric—doubtless under the influence of the highly suggestive *Théorie Analytique de la Chaleur* published by Fourier two years previously. The caloric flows *with* the steam; it is stated categorically that " the production of motive power in the steam engine is therefore not *due to a real consumption of the caloric*, but to its transfer from a hotter to a colder body—that is to say, to the *re-establishment of its equilibrium* . . ." (italics mine). It is clear that Carnot is here misled by the universal practice at that time of over-stretching the analogy between " palpable " and " impalpable " fluids. Fourier's *Théorie* was faultless in so far as it relates only to the " flowiness " of heat; Carnot's result was completely erroneous in so far as it was concerned with the question, " *How is work obtained from heat?* " If this is so, why is Carnot's memoir regarded as one of the masterpieces of scientific reasoning? The answer amounts to the fact that Carnot was asking several other questions at the same time in which the actual characteristics of heat were irrelevant; and for the purpose of answering these questions he devised an analytical weapon—the reversible cycle of operations —which has been well described (by Pledge) as the essential " unit of thought " for an immense branch of physics and physical chemistry—a branch commonly known as " thermodynamics " or " energetics ".

What then were these questions? With commendable clarity he enunciates them himself. First, " Is the motive power of heat invariable in quantity or does it vary with the agent which one uses to obtain it . . .? " To answer this he employs only well attested facts, for instance that " wherever there is a difference of temperature the production of motive power is possible ", and vice versa. But in order to *isolate* this action he is compelled to carry through the operations in such a way that if *they were reversed in sequence, the original state of affairs would be re-established.* In this way the incalculable factors of *internal* work and *dissipation* of heat are avoided. No actual engine works in this way; but the postulation of such an " engine " provides a *unit* or norm by which all such processes can be compared. The answer to the question is stated as follows: " The maximum motive power resulting from the use of steam is also the maximum motive power which can be obtained by any other means."

The second question is little more than a corollary to the first. " What is the meaning of the word maximum? How can we know that this maximum is reached . . .? " The answer follows at once from the introduction of the reversible cycle, namely that " the necessary condition of the maximum is that in bodies used to obtain the motive power of heat, no change of temperature occurs which is not due to a change in volume ".

While Carnot was assuming that to assert the loss of heat as such during the reversible production of work " would overthrow the whole theory of heat of which it is the foundation ",[1] James Prescott Joule was growing up in the invigorating atmosphere of his father's Manchester brewery. He was, to be sure, only six years old when Carnot's memoir was published, but his concern for the utilisation of physical " forces " began at so early an age that his first publication on this subject was accepted by the Royal Society when he was only twenty-two. This first paper was concerned with the relation between heat and *electricity*. In it he shows that the heat obtainable from the electric current produced by dissolving a known weight of zinc in a Daniell cell is approximately equal to the heat obtainable by dissolving the same weight of zinc in sulphuric acid. This was a most ingenious extension of the problem considered by Rumford and Carnot—far too ingenious for his contemporaries who remained quite unmoved. Obtuse as were the British men of science, it is some comfort to know that the Germans were even worse; for the first philosophical attempt to clarify the confusion surrounding the various " forces " was actually refused by the editor of *Poggendorf's Annalen*. Its subsequent publication by Liebig in 1842 evoked so little interest that the author was requested not to send any more. The author was Robert Mayer, who, when acting as ship's surgeon, had his attention drawn to the subject by the fact that blood taken from members of the crew in the tropics was redder than usual. This implied less oxidation and prompted the question "What do we understand by 'forces'?" And how are different forces related to each other? His answer, though not free from the metaphysical obscurity which D'Alembert had been at pains to remove, and vitiated by the retention of the word " force " when Young's term " energy " would have made the distinction clearer, at any rate pointed to the paramount necessity of *asking* the question and of the possibility of fruitfully applying the answer to the case of

[1] His notebook, which did not become public till much later, shows that he ultimately corrected this error. " D'après quelques idées que je me suis formées sur la théorie de la chaleur, la production d'une unite de puissance motrice nécessite la *déstruction* de 2,70 unites de chaleur (quoted by Pledge *Science Since* 1500, italics mine).

the relation between work and heat. His analogies were also most suggestive, as when he compares a train to a " distilling apparatus; the heat applied under the boiler passes off as motion and this is deposited again as heat at the axles of the wheels ". Moreover he gave the first calculation of the ratio of work done to heat produced, on the incompletely verified assumption that no heat is lost or gained when the volume of a gas changes without the performance of *external* work.

Mayer's far greater compatriot Helmholtz shared with him the distinction of being rejected by Poggendorf. In his great memoir, *The Conservation of Force*, published by the Physical Society of Berlin in 1847, Helmholtz set forth the first generalised account of the modern principle of the *conservation of energy*. The introduction at least should be read by every student of physics. Critical physics has of course progressed far beyond Helmholtz, but uncritical physics, of which there is still a great deal, is a generation behind him. The only weakness in the memoir is that, despite the derivation of the well-known *energy equation*, which led to the use of $\frac{1}{2}mv^2$ for the *vis-viva*, the term " Force " is used far too loosely. The memoir of the Danish physicist, Colding, published also in 1842, was likewise without influence. It is a most remarkable paradox that in an atmosphere so apathetic Mayer, Joule, and Helmholtz continued the struggle to convince their contemporaries of the quantitative equivalence of heat and work, and this in spite of the absence of any *adequate* experimental basis. For though Joule's later results were of an unparalleled exactitude, his earlier results varied between 700 and 1000 ft.-lbs. per B.Th.U. The establishment of the modern dynamical theory of heat thus repeats the history of the earlier atomic theory; both provide convincing evidence of the driving power of scientific *faith* which is able to remove mountains of experimental inconsistencies. To this faith Joule testified in the statement that he was " satisfied that the grand agencies of nature are by the Creator's *fiat* indestructible; and that whatever mechanical force is expended, an exact equivalent of heat is *always* obtained." At this time (1843) Joule evidently did not realise the implications of Carnot's discoveries—if indeed he had ever heard of them—for

THE "GO" OF THINGS 281

he summarised the results of his experiments to determine the ratio of the " force " used to turn a " magneto-electric machine " and the heat generated in the words: " the quantity of heat capable of increasing the temperature of a pound of water by one degree of Fahrenheit's scale is equal to, and *may be converted into,* a mechanical force capable of raising 838 lbs. to the perpendicular height of one foot."

It will not be necessary to examine Joule's work on the mechanical equivalent in detail; it belongs rather to the history of *technique* and in any case has been admirably set out in easily accessible memoirs. It will be sufficient to emphasise that Joule's views on the interconvertibility of natural forces equalled Mayer's in comprehensiveness; and he added to this speculative insight an equal comprehensiveness of experimental methods, such as were suggested by the march of mechanical engineering among whose latest triumphs he found his vocation. To the electro-chemical, magneto-electric, and the better known purely mechanical (paddle wheel) experiments, he added one in which the heat generated by driving water through capillaries was measured. And he repeated with greater precision Gay-Lussac's experiment in which was demonstrated within the then permissible experimental errors, that no heat is lost or gained when a gas expands into an empty vessel so that the mean pressure throughout the apparatus remains constant.

This unambiguous demonstration of the precise relation between " destroyed " work (friction) and heat production sealed the fate of caloric and enabled Clausius to correct the error in Carnot's principle.

Great as were the contributions of Joule and Carnot, Clausius seems to stand out as the key figure in this great revolution in thought. Starting his career at the age of twenty-eight as Professor of Physics at the Royal Artillery and Engineering School in Berlin, he occupied chairs at Zürich, Würzburg, and Bonn. His attitude to the problem is clearly shown by the title and introduction of his great memoir (1850) *Ueber die bewegende Kraft der Wärme* (On the Motive Power of Heat): " Since heat was first used as a motive power in the steam engine, thereby suggesting from practice that a certain quantity of work may be treated as equivalent to the heat needed to produce it, it was natural to assume also in theory a definite relation between a quantity of heat and the work which in any possible way can be produced by it . . ." How was this relation to be sought? After paying tribute to Carnot and Clapeyron, who had translated Carnot's verbal analysis into symbolic form based on the indicator diagram familiar to engineers, he criticised their assumptions that the quantity of heat, being a substance, remains unchanged. The results of Joule on the contrary seemed to prove that the quantity of heat produced by friction is proportional to the mechanical work done. These facts and others " support the view that heat is not a substance but consists in the motion of the least parts of bodies ". If this view is correct, it is admissible to apply to heat the general mechanical principle that a motion may be transformed into work, and in such a manner that the loss of *vis-viva* is proportional to the work accomplished. These facts " almost compel us to accept the equivalence between heat and work, on the modified hypothesis that the accomplishment of work requires not only a change in the distribution of heat, but also an actual *consumption* of heat " (italics mine). When this change has been made then the Clapeyron diagram shows how the amount of work obtainable in a perfect engine depends only on the temperature relations of cylinder and condenser; and the amount of work so obtained is in fact equivalent to the amount of heat " lost ". This result amounts to a law of nature additional to that of Joule. The latter, the first law of thermodynamics, asserts that *whenever* a conversion of heat into work or vice-versa occurs, the quantities bear a constant relation to one another; the former—called the second law of thermodynamics—asserts the *conditions* under which the maximum conversion of heat into work may take place. Clausius drew from the second law a corollary which has been used and abused with great vigour by all sorts and conditions of men desirous of proving both that the universe was and was not created in time—an example of that vicious extrapolation so fashionable in our time. The corollary itself, regarded as a physical principle is unexceptionable and has valuable applications to such apparently remote problems as the forecasting of the movements of depressions. For a *reversible* transfer of heat

Carnot had shown that $(Q_1 - Q_2)/Q_1 = (T_1 - T_2)/T_1$, which may be written in the form $Q_2/Q_1 = T_2/T_1$ or $Q_2/T_2 = Q_1/T_1$. In other words in a *reversible* process the function $Q/T$ undergoes no change. To this function Clausius gave the name " entropy ". In the vast majority of processes that are going on around us however it happens that a quantity of heat passes from a body at one temperature to a body at a lower temperature, without the performance of any work. The gain of entropy of the body receiving the heat is bound to be greater than the loss of entropy of the body losing heat; therefore the entropy of the whole system, as Clausius said, tends to a maximum. A physical interpretation of this will appear after we have studied the kinetic theory.

While Clausius was engaged on these matters, a young Scotsman, William Thomson, had been quite unconsciously dogging his steps. Like Clausius he had read Clapeyron's ingenious development of Carnot's principle and, while engaged on that reprehensible habit of nosing round the second-hand bookstalls in Paris, came upon a copy of the almost entirely neglected original memoir of Carnot. He was intrigued with the great possibilities of this little work—too intrigued, as it turned out—for he failed to recognise the error in it. Thomson, as his ultimate career proved, had the mind of a precision engineer; among the chief of his services to science was the great triumph of the successful laying of the Atlantic cable, the invention of a reliable mariner's compass, and the complete overhaul and establishment of accurate and reproducible standards of measurement in electricity. But at this time the subject of heat claimed his interest; and he must soon have realised that the further progress of that branch of science must be seriously hampered by the absence of any but *empirical* standards of temperature, in which the unit was perforce based on the behaviour of some individual substance. If it be argued that the same was then true of the whole of dynamics, since the units of mass, length, and time were similarly bound to *material* standards, it must be admitted that this was indeed the case, but that, as subsequent investigations have shown, the uncertainty involved is of the order of the reciprocal of the velocity of light and therefore of no significance on the terrestrial scale. Fortunately the robust euclidean materialists of the Victorian age were quite unconscious of any such arbitrariness in the fundamental units.

Thomson had the best possible preparation for an investigation of this kind—a year under the rigorous discipline of Regnault's laboratory. This great physicist had shown that the gas thermometer was as near a perfect instrument as human wit was likely to devise; but Thomson recognised that even so " the scale actually adopted is *an arbitrary series of numbered points of reference sufficiently close for the requirements of practical thermometry* " (italics his). What was wanted was an *absolute* scale, and Carnot's principle provided the basis for it, for " a unit of heat descending from a body $A$ at a temperature $T°$ of the scale, to a body $B$ at a temperature $(T-1)°$, would give out the same mechanical effect, whatever be the number $T$. This may justly be termed an absolute scale, since its characteristic is quite independent of the physical properties of any specific substance." In his first exposition (1848) of this proposed scale Thomson had no qualms about Carnot's belief in the indestructibility of heat—" the conversion of heat (or caloric) into mechanical effect is probably impossible, certainly undiscovered "—consequently the first scale worked out by Steele under his supervision departed far from the gas scale. This is all the more surprising seeing that he was one of the few British physicists who showed any interest in the work of Joule, with whom he had been in correspondence before the publication of this scale. In 1851 however he had made the great renunciation and published what is commonly regarded as the most comprehensive account of the relation between heat and dynamics—and it is to be noticed that in it he fully admitted Clausius' priority in the establishment of the second law. The new absolute scale being re-calculated on the basis of the corrected Carnot function was found to depart but little from the nitrogen gas thermometer used at ordinary temperatures. We can only refer in passing to the fact that it was to test this correspondence that Joule and Thomson carried out the well-known experiment on the elasticity of real gases, whose result exerted so profound an effect on the attempts then being made to reduce these gases to the liquid state. The scale is commonly known as the " Kelvin " scale, Thomson being ultimately raised to the peerage as Baron Kelvin of Largs.

Henceforth the ghost of caloric ceased to haunt scientific laboratories and to pervert the minds of even such vigorous innovators as Kelvin. The established interconversion of heat and work in strict numerical proportion was the foundation of all further studies of the " go " of things. But the task, even in this limited field was not ended; for it left entirely unsolved the problem of the actual nature of heat, and of the absence of a complete reciprocity between heat and work. Both speculations, such as those of Bernoulli and Davy, even when backed as in the former case by mathematical argument, seem to have been ignored or regarded as aberrations. The boldest and most comprehensive of all was submitted in 1845 to the Royal Society, and after a careful and conscientious examination by two referees was filed in the archives as being according to one of them " mere nonsense ", and by both considered unworthy of the time it would take to read. The paper *On the physics of media that are composed of free and elastic molecules in a state of motion* was disinterred by Lord Rayleigh in 1892 and published with a sympathetic and generous tribute. The author, J. J. Waterston, had however by then passed on to a realm where priorities of publication presumably mean little. Waterston's approach to the problem marked a notable advance on that of Bernoulli and Herapath in that he clearly recognised that the temperature of a mass of gas is measured by the *mean kinetic energy* of the particles, since all facts of experience point to the necessity for postulating the existence of *unequal* velocities among the particles. On this assumption and that of the perfect elasticity of the particles he was able, by the application of the classical laws of motion of Galileo and Newton, to derive Boyle's law, Charles' law, and the then recently discovered diffusion law due to Graham.

Twelve years later a paper on very similar lines by Joule, which had already appeared in outline in another journal some years earlier, was published in the *Philosophical Magazine*. It would be interesting to know what Waterston's referees thought when Joule's " nonsense " was published. Joule went a little further than Waterston and made the first calculation of the velocity of a molecule at a particular temperature. On the other hand he does not make it clear that the assumption of uniform and constant velocity is a simplification which is unlikely to be true. In the same year Clausius, with his usual felicity of title, published his famous memoir *Ueber die Art der Bewegung, welche wir Wärme nennen*—a declaration of faith that the random and ceaseless rush of the molecules is not the *cause* of heat but *is* heat. He covered the same ground as had Waterston and Joule, but he realised, as they had not, the importance of the exceedingly small distance traversed by a molecule between " collisions ".

In the year 1860, then, a most interesting problem had been clearly posited. Its investigation had produced most promising results, but the principal concept involved—the kinetic energy of the molecules of a gas—could not be expressed without ambiguity. Waterston had realised the necessity for admitting the existence of a great number of velocities in a mass of gas under any given conditions, and had therefore expressed his result as functions of the *mean* velocity. Joule had tacitly, and Clausius overtly, assumed that the velocities were all the same. Now it might be thought that the velocity of hydrogen calculated by Joule was in fact, though not stated by him, the *mean* velocity as postulated by Waterston. But the question at once arises " What mean? " For the calculation involves the equation $c^2 = 3\ pv/mn$ where $c$ is the " velocity "; that is the square root of the arithmetic mean of the sum of the *squares* of all the separate velocities—a quantity which is demonstrably not equal to their simple arithmetic mean. The problem was to show that this " root mean square " function followed from postulates involving no assumptions other than those of Newtonian dynamics. But the method of carrying it out involved the application of the technique hitherto restricted to the solution of problems in games of chance (Pascal); in random observations of physical quantities (Gauss, Laplace, and Poisson, who together founded the mathematical theory of probability, i.e. the probability of a particular value of a given quantity lying between given limits); and human mass characteristics (Quetelet). To apply this to a problem in particle dynamics demanded a " new star of the first magnitude in the firmament of mathematical physics ". That star had already risen—James Clerk Maxwell.

Maxwell was not altogether unprepared; he had just been

awarded the Adams Prize for an essay on the nature of Saturn's rings, in which he had proved that the apparent stability of these rotating objects was possible only if they consisted of a swarm of separate particles. He approached the problem of the distribution of the velocities of a swarm of particles of a gas in a purely formal manner—a manner which was characteristic of all his work and which disconcerted those who regarded physical " models " as not only essential to fertile thought, as did also Maxwell himself, but as either " right " or " wrong ", which he did not.

" In order to lay the foundations of such investigations on strict mechanical principles I shall demonstrate the laws of motion of an indefinite number of small, hard, and perfectly elastic spheres acting on one another only on impact. If the properties of such a system of bodies are found to correspond to those of gases an important physical analogy will be established, which may lead to more knowledge of the properties of matter."

The molecules in this case are to be regarded as Newtonian particles whose gravitational properties are negligible compared with their elasticity.

But " instead of saying that the particles are hard, spherical and elastic, we may if we please say that the particles are centres of force, of which the action is insensible except at a certain small distance, when it suddenly appears as a repulsive force of very great intensity. It is evident that either assumption will lead to the same results."

This alternative assumption was not original. First systematically developed by Boscovitch in 1745 and 1759, it had greatly influenced many physicists. What *is* original is that Maxwell does not appear to be interested in the question as to which assumption is " true ". For the problem in hand it suffices to discover whether the deduced consequences of the *formal* postulates fit the known facts; if so, we can proceed to deduce further as yet unknown " facts ", that is, we can use the hypothesis as a *guide* to the gaining of further knowledge which might never have been suspected in its absence. The mathematical problem is " to find the average number of particles whose velocities lie between given limits, after a great number of collisions among a great number of particles " By rigorous *mechanical* argument

Maxwell discovered that " the velocities are distributed among the particles according to the same law as the errors are distributed among the observations in the theory of the ' method of least squares '." Thus no matter what change a gas is made to undergo, the velocities of the " particles " composing it will ultimately reach a steady *distribution*, though the velocities of individuals are constantly altering, and further that the mean kinetic energy is proportional to the (square) of the " root mean square " velocity. Unfortunately Maxwell's first proof (1859) of this result depended on the solution of a " functional equation " in which the three velocity components of any molecule were *assumed* to be independent.

With this capital achievement as a starting point, Maxwell proceeded to open up huge vistas of new knowledge of gases into which we cannot enter, suffice it to mention one example of the " unsuspected " mentioned above, namely viscosity. Maxwell showed that viscosity, like other properties which are, so to speak, " transported " across a space by the opposed flow of two streams of molecules endowed with some determining factor (in this case linear momentum) to different degrees, is proportional to both the number of molecules in unit volume and their mean free path. But these two quantities are inversely proportional to one another. Hence change of pressure, which at constant temperature depends only on the former quantity, will not affect the viscosity—a hardly credible forecast, but one quickly verified by experiment.

We must here introduce another great figure who worked in friendly rivalry—the Viennese physicist, Boltzmann. Both he and Maxwell independently attacked the problem of the partition of energy between the various possible motions of a body. The special interest of this problem was the possibility of its bearing on the " molecules " being point-centres of force; for obviously such an entity could possess no kinetic energy save in respect of simple translatory motion. Both Maxwell and Boltzmann discovered that, in the more general case of a rigid body, the ratio of translational to other (e.g. rotational) energy must be constant and equal to unity—the so-called equipartition law. Boltzmann also gave a method of expressing the partition in terms of the number of " degrees of freedom " of the molecule, that is the number of co-ordinates necessary to determine the energy-state of a body—three for a " point ", six for a rigid body. It was easy to show, following a suggestion of Clausius, that the ratio of the specific heats of a gas, measured at constant pressure and constant volume respectively, would equal $1+2/n$. Attempts to verify this completely failed: no gas gave a value of 1.67 corresponding to $n=3$, and the majority of gases gave 1.40 corresponding to $n=5$. It must be remembered that all this was going on only a few years after Canizzaro's resuscitation of Avogadro's theory of the distinction between atoms and molecules of elementary gases, and many years *before* the use of Kundt and Warburg's method for measuring specific heats for such gases as mercury. The equipartition law was for the time being saved by Boltzmann's suggestion that the molecules of the simpler elementary gases might be conceived as a pair of " points " rigidly connected by a single axis—one of the six degrees of freedom would thus vanish. At the end of the century verification came dramatically with the discovery of five gases whose atoms were so inert as to be unable to combine even with their own atoms, and all have a specific heat ratio of 1.67.

Boltzmann carried the analysis of the distribution law much further than had Maxwell, but even he made the assumption of " molecular chaos "—an expression implying that the mode of choice of a statistical group of molecules has no influence on the distribution.

We now have to enquire how the kinetic theory of gases was related to the " go " of things in general, and, on the other hand, to the " unilateral " conversion of mechanical work into heat. To the solution of the first problem the botanist, Robert Brown, had in 1828 furnished a clue, when he observed the ceaseless random motion of *dead* pollen grains suspended in liquid and free from any external disturbance, such as draughts or convection currents. At first he thought he had discovered the " primitive molecules " constituent of living things; but after having observed the same motion in particles of inorganic origin he concluded that the " primitive molecules " must be common to all matter. It is doubtful to what extent Brown realised the nature of the phenomenon he had observed; the modern explanation—that the particles are driven hither and thither by the *unbalanced* impacts of much smaller particles—was given by Delsaux in 1877, and fairly accurate measurements of the mean displacements were made by Gouy in 1888. It was Einstein however who, in a classical series of papers dating from 1905, *proved* that " according to molecular kinetic theory of heat, bodies of microscopically visible size suspended in a liquid will perform movements of such magnitude that they can be easily observed in a microscope on account of the molecular motion of heat ". It must be understood that this prophesy was based on exact calculations, and was entirely independent of the already observed " Brownian movement " of which Einstein had heard such vague accounts as to raise doubts in his mind as to whether it was really the same thing. Moreover, he went further, and actually calculated the size of the molecules. By 1914 Perrin and his associates in Paris had verified this prediction, and made an estimate of the number of molecules in a cubic centimetre—a quantity soon to be confirmed by many independent methods, thus giving to the molecule a physical reality in spite of the impossibility of its ever being seen.

Maxwell himself supplied a philosophic answer to the second question. Useful work may be regarded as the proportion of *ordered* motion available in a disorderly crowd of particles. Similarly " heat " cannot flow from a " cold " body to a " hotter " one, because *on the whole* the kinetic energy of the former is less than that of the latter, hence there are *fewer* fast moving molecules in the former, therefore *on the whole* fast moving molecules will tend to move from the hotter to the colder body. But, said Maxwell, an intelligent demon might operate a trap door in such a way as to allow only the relatively *fast* molecules to move into the *hot* gas, and an equivalent number of relatively *slow* ones to move into the *cold*. The change of mean kinetic energy, that is " heat ", would then be from cold to hot. The device of the " sorting demon " thus reveals that the second law of thermodynamics is a purely *statistical* law, that is, it applies only to the *average* behaviour of large numbers. There is nothing *dynamically* impossible in the passage of heat from a cold mass to a relatively hotter one—it is merely vastly less probable than that a pack of cards by continual shuffling should ultimately resume the order in which it was sent out by the makers.

So much for the " go " of things in general. But there is a special "go" which has from the earliest times been recognised as " different "; this is chemical " affinity "—a word said to have been used first in this sense by Albertus Magnus. The chequered history of chemical affinity is but another example of failure to distinguish between *quantity* and *intensity*—made peculiarly difficult in this case seeing that *quantity* may in certain circumstances modify *intensity*.

So long as alchemy ruled the day, the " go " of a chemical process was conceived solely in terms of intensity—the philosopher's stone being as it were pure intensity, which, as Bacon remarked, was to turn a " sea of quicksilver into gold ". Despite his inability to rid himself entirely of the alchemical miasma, Boyle on the other hand said, " I have long thought that in divers cases the quantity of a menstruum may much more considerably compensate its want of strength than chemists are commonly aware of." In the eighteenth century the two views had their adherents, an " affinity table " being constructed by Geoffroy (1718) and given greater precision by Bergman (1775) and Beaumé (1773); Wenzel on the other hand pointing out that concentration could not be neglected in the order of affinities (1777). It was however above all Berthollet who, in his *Essai de Statique Chimique* (p. 229), returned to Boyle's hint and illustrated it by many examples. Berzelius recognised that Berthollet had confused the possibility of influencing the course and amount of chemical reaction with the *composition of the product*. This judicious reminder prevented Berthollet's work being entirely lost sight of, and as a result chiefly of the work of Wilhelmy (1850) on the slow, virtually irreversible, hydrolysis of cane sugar, and of Berthelot and St. Gilles (1862-3) on the *reversible* action between alcohol and acids, Guldberg and Waage were able to enunciate in 1864 the law of Mass Action, one of the foundation stones of physical chemistry. The essence of this law is that the rate at which a chemical reaction proceeds is, at constant temperature, proportional to the product of the active masses of the constituents. If the resultants are capable of mutual interaction, the same will be true of this action also. Hence an equilibrium mixture will ultimately result. The constants of proportion of the two actions will in general be different from one another, hence the equilibrium may rest on either side of mere equality. Any other condition, such as the nature of neighbouring surfaces, which affects the rate will in general affect both rates to the same extent; hence the equilibrium state, though established more or less quickly, will be unaltered. The definition of active mass—at first regarded as the number of molecules per unit volume—has turned out to be something of a much more subtle kind; the interpretation of the word " active " has been the business of much recent research.

It now remains to be shown why an idea, superficially purely chemical, has been introduced at the end of a chapter dealing with the concept of energy. The reason lies in the restriction placed on the law of mass action, namely that it is not necessarily true except at constant *temperature*. The explanation of this is easily seen by reference to the second law of thermodynamics: if the reaction is accompanied by any heat change, addition of heat must favour the action normally proceeding with the absorption of heat (endothermic), otherwise it would be possible to extract an unlimited amount of heat from the reaction (Le Chatelier, 1888, a special case of which is Lenz's law, p. 301). Thus the equilibrium point would be shifted. Now the constant in the law of mass action (the ratio of the two *velocity* constants referred to above) is a quantitative measure of the " affinity " of the reaction, hence this affinity must be influenced by the heat relations of the reaction. At first Berthelot and, independently, Thomson taught that the *total* heat evolved or absorbed by equivalent quantities of reactants gives a direct measure of the affinity. It was Helmholtz who, on the basis of extensive studies by Horstmann, pointed out that not the change in *total* energy but *free* energy is the true measure of affinity, the change in free energy being defined[1] as the equivalent of the work done when the system changes at constant temperature, and also reversibly, so that no internal dissipation occurs. The reason

---

[1] The reader is warned that the term " free energy " is used by Lewis and Randall (*vide inf.*) to define a different function.

why the " free " energy is in general different from the total heat exchange of the reactants is that the process *may* involve transfer of *heat* to or from the *surroundings*. But this has nothing to do with the *work* the system may be able to do.

This was in 1882; five years later with the publication of the first number of the *Zeitschrift für Physikalische Chemie*, physical chemistry celebrated its birth as a separate discipline. This was a fitting culmination of a crowded decade; for during those fateful years that versatile genius, Van't Hoff, had effected the great generalisation whereby not only equilibria between chemical reactants but also the equilibrium between a solution and its pure solvent (osmotic pressure) were shown to be of the same fundamental *form*; and this form was none other than that of the formulation of the second law of thermodynamics established by Clapeyron and Clausius (p. 282). The number of the *Zeitschrift*, in which appeared Van't Hoff's generalisation of the empirical laws of Raoult for solutions, contained also the explanation of the apparent anomalies presented by electrolytes. This explanation, due to Arrhenius, was the theory that electrolytes contain a proportion of the molecules of the solute in the form of charged particles called " ions ". This at once provided the basis for an extension of the thermodynamic laws—now because of their generalisation better known as the laws of energetics—to the elucidation of the electromotive force developed between electrodes and electrolytes. Once again Helmholtz (and later Nernst) was the leader.

Behind all this feverish activity was Wilhelm Ostwald. No one more than he realised the importance of this young branch of chemistry in which the form of the energy relationship was studied in abstraction from the properties of individual chemical substances. Skilled alike in research, in teaching, in administration, and in publicity, he smoothed the path for the marriage of science to industry in Germany—a union fruitful in the vast range of amenities which distinguishes our age, but creator also of the colossal resources without which the world catastrophes of 1914 and 1939 could never have occurred.

Meanwhile there was living in respected obscurity in Yale University a man whose name is mentioned with reverence in every book of the history of science and in every text book on physical chemistry. But apart from stating the fact that he discovered a principle of great importance called the " phase rule " and expressed it in terms which hardly any of his contemporaries could understand, these books are content to allow Josiah Gibbs to remain a Great Name. This is a pity; since although the reading of Gibbs' original memoirs demands mathematical facility of a fairly high order, the direction of thought can be followed with great profit by anyone capable of grasping the simple outline of thermodynamics given in this chapter.

It will be remembered that it is from the indicator diagram showing the relation between pressure and volume of a mass of steam in a cylinder that Carnot's principle was put in a form from which many unsuspected results could be drawn by mathematical manipulation. Gibbs' first contribution to knowledge was to show that far wider consequences could be drawn from curves showing the relation between volume and *entropy*. Though, as he himself pointed out, " the use of a method involving the notion of entropy, the very existence of which depends upon the second law of thermodynamics will doubtless seem to many far fetched. . . This inconvenience is perhaps more than counterbalanced by the advantages of a method which makes the second law of thermodynamics so prominent." His great powers of geometrical imagery enabled him to introduce a third variable, energy, whereby thermodynamical *surfaces* instead of curves were generated. By the analytical manipulation of these surfaces Gibbs was able to break entirely new ground in the application of thermodynamics. The nature of this novelty is put very well by Mr. J. G. Crowther (*Famous American Men of Science II*—Pelican Book). " The founders of thermodynamics ", he writes, " . . . were not interested in the private life of the steam, the internal relations between its different portions. They assumed the states were the same all through. The problem before a chemist is quite different. He is interested primarily in the actions inside a flask, not in the work that can be got out of steam inside a cylinder." In other words, whereas the engineer is concerned with the energetic changes of a changeless substance, the chemist seeks to

which does not seem to be mentioned in the ordinary text books, came nearest to an original creation. Here he showed how to determine the most general mechanical relationships (which, since they involved an indefinitely large number of " degrees of freedom ", were necessarily statistical) from which the statistical *thermo*dynamics of Maxwell and Boltzmann could be deduced as special cases. This achievement was made possible by his creation (1880) of a special type of vector algebra, in which he drew upon the " multiple algebra " of Hamilton and more particularly Grassmann—the latter's being a wonderful generalisation for dealing with manifolds of points ordered in any number of dimensions.

Gibbs' sole recorded speech to the Faculty of Yale closed a discussion of the relative importance of mathematics and language with the words: " Mathematics is a language." No one knew better than he how to make it speak with the tongues of angels.

## SOURCES FOR CHAPTER XXIII

E. Mach, *op. cit.*, Gen. Bibl., for Stevinus and L. da Vinci.

Magie, *op. cit.*, Gen. Bibl., for Leibniz, Descartes, d'Alembert, Rumford, Joule, Clausius, Helmholtz, W. Thomson, Maxwell.

J. J. Waterston, *Collected Works* (Edinburgh, 1928).

R. Glazebrook, *James Clerk Maxwell and Modern Physics* (London, 1896).

G. N. Lewis and M. Randall, *Thermodynamics and the Free Energy of Chemical Substances* (New York, 1923). This is not a historical work, but it puts the new views in an extremely clear light.

J. Van't Hoff ⎱ *The Foundation of the Theory of Dilute Solutions*,
S. Arrhenius ⎰ Alembic Reprint No. 19.

J. W. Gibbs, *Collected Works* (London, 1928). Gibbs' works make very difficult reading, but the *Heterogeneous Equilibrium* is largely free from mathematical symbolism.

J. G. Crowther, *Famous American Men of Science: Edison and Gibbs*, (Pelican Books); an attractive account, but rather biassed in outlook.

## CHAPTER XXIV

## THE PARTICULAR "GO"—FROM PARTICLES TO FIELDS

THE year 1860—in which Maxwell's proof of the distribution law of molecular velocities was published—may be taken as marking the complete acceptance of the principle that the work that can be got out of any mechanical system depends in the last resort on its kinetic energy or its potentiality for acquiring kinetic energy—to mark which a concept called "potential energy" was introduced. Further, that the heat content of any body is the sum total of the kinetic and potential energies of its constituent molecules, from which however useful work can be gained only in so far as these molecules are prevented from increasing the *random* motion of those of another body. This, and the many consequences both for physics and chemistry which could be deduced from it, was a great achievement. For one thing it could be applied with equal cogency to electrical phenomena; for, as Joule had shown, the mechanical equivalent of heat was the same whether the heat was produced by "degrading" electricity direct in a conductor or indirectly by means of a motor and paddle wheel. The latter method was used with great precision by Rowland. Sound offered no difficulties; Joule with characteristic thoroughness, allowed for the kinetic energy "lost" as sound from his paddle wheels. But light, and radiation generally, although, like sound, propagated with finite velocity, remained outside this otherwise all-embracing scheme. It is true that many mathematical physicists were taxing their own ingenuity and other people's credulity in devising *mechanical* "models" of an oscillatory character to account for light, which would consequently involve something equivalent to energy in the undulating medium; but that is as far as anyone could go. In

1865 however the situation was transformed with dramatic suddenness, and in a most unexpected fashion, by the publication of Maxwell's paper, *A Dynamical Theory of the Electromagnetic Field*. There was no attempt to account for radiation; Maxwell's object was to discover a theory which would account more consistently and more convincingly for the forces of attraction and repulsion between magnetised and electrified bodies than would the existing theory of Weber and Neumann. This latter theory assumed that the forces were due to *particles* either stationary or in relative motion, capable of acting on each other at a distance—it extended, in fact, the Newtonian gravitational conception to the new " forces " of magnetism and electricity. Maxwell however " preferred to seek an explanation of the facts in another direction, by supposing them to be produced by actions which *go on in the surrounding medium as well as in the excited bodies, ...*" (italics mine). Now this is just what the undulatory theory did to explain light. When therefore we speak of the electromagnetic theory of light as a consequence of Maxwell's procedure, we should remember that it is not so much that light has been deduced from electricity and magnetism as that electric and magnetic phenomena are explained as special characteristics of a medium, or at any rate of a *field*; which is precisely the concept found to be essential for the explanation of certain properties—interference and polarisation—of light.

This revolutionary shift of emphasis from particle to field was not originated by Maxwell but borrowed—with the most respectful acknowledgments—from one who had played no part in the hub-bub which had been going on since Ampère had founded electrodynamics—Michael Faraday. Faraday was debarred from joining in the hub-bub for the sufficient reason that he knew no mathematics: in the eight hundred odd pages of his *Experimental Researches* there is not a single mathematical formula. Yet, as the late Sir J. J. Thomson remarked, " he showed by the way he handled his conception of the lines of force that he was a born geometer." Faraday once said that he hadn't time to make money; presumably he hadn't time to learn mathematics; he was, like Bach, too busy using them ready made. While the continental mathematicians were laying down more and more elegant roads in the wrong direction, Faraday passed his time exploring the country to find the right direction. His explorations, the careful description of which fill two large volumes, brought him a view of many of the secrets of nature, but two above all compelled him to regard the *space* (see later, p. 302) surrounding a magnetised or electrified body as altered in some way, which may be most accurately pictured by supposing it traversed by actual lines of force having the characteristics of tension and mutual repulsion whereby they may, unlike those of the gravitational field as pictured by Newton, follow *curved* paths. These discoveries were specific inductive capacity and electromagnetic induction. It will repay us to take a swift view of these well-known phenomena.

The discovery of electromagnetic induction was the result of Faraday's conviction, amounting almost to an obsession, that natural phenomena are complementary—a kind of generalised law of action and reaction—and also to his skilful use of analogy. Now it was well known that static electric charges are always accompanied by induced charges on bodies in their neighbourhood, and that every *current* is accompanied by a magnetic field. Surely then it should be possible to detect an electric current induced by a current flowing in a neighbouring conductor, and also a current in a conductor in the neighbourhood of a magnetic field? Unfortunately Faraday was met at the outset by an unexpected difficulty. It will be remembered that a magnetic field is produced whenever a current flows—it matters not whether the current is steady or not. It was reasonable to expect a similar result when seeking an induced *current*; but no matter how strong the current, how long or intimately wound together the conductors, how thorough the insulation of the two circuits from one another, not a sign of an induced current was to be seen. It is however the privilege of genius to allow faith to ignore facts; Faraday *knew* that there must be an effect. He used a battery with ten times the number of plates previously employed, and " when the contact was made, there was a sudden and very slight effect at the galvanometer, and there was also a similar slight effect when the contact with the battery was broken "; moreover the movement of the needle at " break " was in the opposite direction to that at " make ".[1] But while the current was steady there was no movement at all—even when the current was strong enough to heat up the helix. It is probable that Faraday's final success was due to the nature of the result when testing the effect of a *magnet* on a conductor: only when the magnet was *moved*, so as to *change* the strength of the field, was any result obtained. His account of these experiments—the care, the perseverance, the seizing of the essential in the minute effects obtained—make fascinating reading. In the end he had shown that *whenever the magnetic* field in the neighbourhood of a conductor is changing, whether by means of a simple or electro-magnet, an electric current tends to flow in the *conductor*. Moreover he gave an exact description of the *direction* of flow in relation to the direction and polarity of the field; but he took near a couple of pages to do it. By regarding it as a special case of *electrodynamic* action and reaction Lenz was able to condense it into one sentence.

The discovery of electromagnetic induction made possible the electric age in which we are now living—if the sub-atomic age has not yet dawned. But for our immediate purpose its importance lies in the stimulus it gave to Faraday's belief in the existence of " physical lines of magnetic force ". At first he was inclined to think that the momentary appearance of a current in a conductor, placed near a current-bearing conductor, was due to the *matter* of the former being thrown into an abnormal state, which he called " electrotonic ". Later he abandoned this view for one in which the intersection of the " magnetic curves " surrounding the current were regarded as the cause. Then came his discovery of specific inductive capacity.

It is well known that if a charged body is brought near to one previously known to be uncharged, a charge is *induced* on the latter for as long as the *inducing* charge is in the neighbourhood. The question Faraday wished to answer was whether the *magnitude* of the induced charge depended on the nature of the " dielectric ",

---

[1] Working quite independently of Faraday, Henry, in America, noticed that the spark formed when a single circuit is broken depends on the length and shape of the circuit. " I can account ", he says, " for these phenomena only by supposing the long wire to become charged with electricity which by its reaction on itself projects a spark when the connection is broken ". He thus anticipated Faraday in the discovery of *self*-induction.

or medium (usually air) between the two bodies. The contrast between Faraday's point of view and that of the founders of electricity is shown by the fact that the question was presumed to have been settled by Coulomb, who had shown that a clean wire and one coated with shellac acquired the same induced charge. Faraday's comment on this was: " Charge depends on induction; and if induction is related to the particles of the surrounding dielectric, then it is related to *all* the particles of that dielectric inclosed by the surrounding conductors,' and not merely to the few situated next to the charged body." He is clearly looking for a proof of his growing conviction that electricity, as it were, *resides in* the medium, and is not, as was then supposed, merely shot through it. In a beautifully contrived electrostatic balance he showed that the balance is at once upset by lowering an insulated sheet of shellac between the inducing charge and either component of the balance. The difficulty of the experiment is due to the rarity of a completely non-conducting material: shellac and sulphur were the substances finally chosen. The result had, of course, the most profound effect on the design of condensers.

Faraday's final statement concerning the existence of physical lines of magnetic force was published in 1851. It is a masterly piece of natural philosophy comparable with the *Queries* that conclude Newton's *Opticks*. In reviewing the various types of " lines of force " by means of which physical phenomena may be more readily comprehended, he contrasts those of gravitation, in which " no effect which sustains the idea of an independent or physical line of force is presented to us " (he would probably change his mind were he alive to-day) with those of the sun, in whose case " rays (*which are lines of force*) pass across the intermediate space; but then we may affect these lines by different media applied to them in their course" (italics mine). The case is similar with static electric charges (as shown by the phenomena of specific inductive capacity) and far more evident in the case of the current. But the magnet is the supreme example, since " it presents a system of forces perfect in itself and able therefore to exist by its own mutual relations ", for instance the absolute equality of its opposite poles. The existence of *physical* lines is evidenced by their being commonly curved, and by the fact that when a wire moves through them a current is generated in it. " The mere fact of motion cannot have produced this current: there must have been a state or condition around the magnet and sustained by it, within the range of which the wire was placed; and this state shows the physical constitution of the lines of magnetic force." In the last paragraph he rises to the heights of the Newton of the *General Scholium*—heights from which Maxwell must have viewed the Promised Land:

> What this state is, or upon what it depends, cannot as yet be declared. It may depend upon the aether, as a ray of light does, and an association has already been shown between light and magnetism . . . Whether it of necessity requires matter for its sustentation will depend upon what is understood by the term matter. If that is to be confined to ponderable or gravitating substance, then matter is not essential to the physical lines of magnetic force any more than to a ray of light or heat; but if in the assumption of an aether we admit it to be a species of matter, then the lines of force may depend upon some function of it. Experimentally, mere space is magnetic; but then the idea of such mere space must include that of the aether when one is talking on that belief; or if thereafter any other conception of the state or condition of space rise up, it must be admitted into the view of that which just now in relation to experiment is called mere space. On the other hand it is I think an established fact, that ponderable matter is not essential to the existence of physical lines of magnetic force.

In this great passage he refers to the fact that an association had already been shown between magnetism and light. The experiment had been described in *Phil. Trans.* for 1846 and was a model of scientific method. In the path of a beam of polarised light was placed a plate of borosilicate lead glass (manufactured by Faraday himself—a tribute to his astounding versatility); the glass itself lay across the lines of force of a powerful electromagnet. By adjusting the eyepiece of the analyser the polarised beam could be extinguished. When however the closing of the circuit energised the magnet, the light immediately became visible; on opening the circuit it at once disappeared. That this was in fact due to the rotation of the plane of polarisation was shown by rotating the analyser, when the light disappeared; and the *direction of rotation depended on the direction of the magnetic field.* This was virtually a proof of the existence of a *magnetic* factor in light; but since Faraday had already convinced himself that these "two forces [electricity and magnetism] are convertible into each other in a greater or smaller degree", it can further be regarded as establishing an electromagnetic factor in light. But it was a hint only; to *prove* that light *is* an electromagnetic disturbance demanded the special type of genius which could deduce the laws of motion of Faraday's lines of force; this triumph belongs to Maxwell alone. To him we therefore return.

It was said above (p. 299) that Maxwell was under a great obligation to Faraday for supplying the then revolutionary concept of a *field* of action in contradistinction to an assemblage of particles. This statement requires two qualifications: first, that on the one hand Faraday never quite rid himself of the preoccupation with chains of polarised *particles* filling space—the view being strengthened by his discovery of the laws of electrolysis, which seemed to him to be the expression of an "axis of power" consisting of charged particles ("ions", as he was the first to call them) which were so to speak shot out at the electrodes; second, that Maxwell was at first very much under the influence of Kelvin. In trying to recreate Maxwell's thought processes we are fortunate in possessing the many informal letters in which he invited Kelvin's advice. Though his great memoir did not appear till 1865, already in 1854 he is asking Thomson, as he then was, how by reading to "get a little insight into the subject" of electricity. The answer was evidently satisfactory, for in the same year he wrote again that in the light of Faraday's and Kelvin's "lines of force" he was proposing to attack the problem of reducing the theory of electricity to a dynamical form by "considering magnetic polarisation as a property of a magnetic field or space and developing the geometrical ideas according to this view". "Polarisation" expresses for him the fact that at a point of space the south pole of a small magnet is attracted in a certain direction with a certain force. He then defines the polarisation of a curve in space as the integral of the polarisation at any point resolved along it, multiplied by the element of the curve. There was nothing strikingly new about this procedure for *magnetic* forces; what Maxwell was proposing however was to use this integral as determinative of any *electric* change in the field.

The first fruits of this new procedure were announced in two papers read at Cambridge in December 1855 and February 1856. In these papers he clearly envisaged the task which was to reach complete fruition only ten years later, namely of taking Faraday's *facts*, and without any presupposition as to the physical nature of electricity, expressing them in precise mathematical form, and thence by purely mathematical reasoning deducing some of the well-established interactions between electric currents and magnets. The method adopted by Maxwell marks a critical transition in modern physics from an attempt to explain natural phenomena as the results of interactions between units supposed known (such as the " hard, massy, particles " of Newton, or the " imponderable fluids " of the eighteenth century) to one in which the *form* of the relationship only is sought in analogy, the " model " being discarded as soon as its usefulness had been exhausted. The meaning of this will appear more clearly in the sequel.

In the paper under review Maxwell recalled Thomson's comparison of the electric current with the motion of a fluid, not, be it noted, a vaguely conceived " imponderable " fluid of purely hypothetical character, but a fluid behaving according to the well-established quantitative *mechanical* laws involving pressure, energy, velocity, and the like. Precision had been given to this notion by the postulation of an ideal " unit of thought "—the vortex filament—the number and distribution of which determine the motion of a *closed mass* of fluid. The axes of such tubes became for Maxwell the lines of force suggested by Faraday. By recognising the correspondence of velocity to current and difference of fluid pressure to difference of potential, he was able to deduce the properties of steady currents. In the second part of the paper he extends the analogy to include the case of *changing* currents—the changes in the density of the tubes of fluid being used as an expression of Faraday's " electrotonic state ".

Years later this youthful forging of new weapons was replaced by a more detailed " model " of the electromagnetic field. In his paper on physical lines of force (1862) he supposed the lines of magnetic force to be in *rotation* about their axes, the angular velocity of each tube being proportional to the force along it. From this he was able to deduce the experimental facts of the magnetic field, namely that the *pressure* due to these rotating filaments is proportional to the square of the angular velocity and is accompanied by a *tension* along their axes—the very characters assigned by Faraday to his lines of force. Now the value of any hypothesis is its power to embrace phenomena which have not been taken into account in its creation. So far the picture is of a purely magnetic field; moreover there is a latent contradiction in a system of contiguous rotating tubes, namely that they cannot rotate in the same direction. To rectify this Maxwell introduced subsidiary tubes of relatively small diameter, like the idle wheels which maintain gear wheels in motion in the same direction (Fig. 30). This shift to make the original system work turned out to be its most important part. So long as the " magnetic " tubes rotate

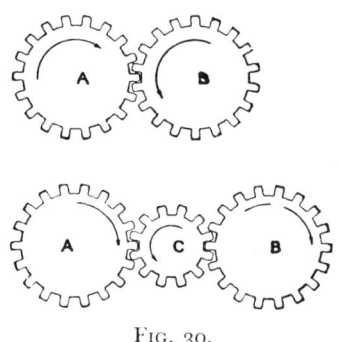

FIG. 30.

with uniform angular velocity, the subsidiary tubes appear to be redundant as constitutive of the magnetic field. But let a difference in the angular velocity of contiguous tubes appear and the idle tubes no longer remain idle, but are translated bodily, while rotating, between the magnetic tubes whose *axes* are fixed. *And this is just what happens to the Faraday lines of* electric *force*, which manifest themselves only when the distribution of magnetic force is *changing*. Moreover there is an even more novel inference. The change in the angular velocity of the magnetic tubes causes a translation of the electric tubes or particles as they may be thought of. Now the distinction between conductors and insulators lies just in this, that in the former the " particles " of electricity are free to move from one molecule of the conductor to the next, but in the latter such movement is resisted. The medium thus behaves like an elastic body, yielding until the pressure is balanced by the internal stress; if the pressure is increased until the stress is overcome (discharge) the release of the pressure will set in operation an oscillatory motion, that is the discharge will produce a wave-like disturbance in the medium. The velocity of propagation of this disturbance must necessarily be a function of the electrical properties of the medium. When Maxwell came to make the calculation from the then accepted value of the ratio of the electromagnetic to the electrostatic units of current for " empty " space, he found it to be in strikingly close accord with the supposed velocity of light. Here, as the late Sir James Jeans pointed out, Maxwell had just the opposite kind of luck to that which may have been one of the factors in deferring Newton's publication of the law of the moon's gravity; for the two quantities which were in such close accord with one another were in fact separately in error by a factor two hundred times as great as the small discrepancy between them. In any case this paper created comparatively little interest—the ether could not really consist of cog-wheels, so the results deducible from the supposition that it did were hardly to be taken seriously.

In 1865 however as has been noted, came the final paper on the *Dynamical Theory of the Electromagnetic Field*. " The theory I propose may therefore be called a theory of the *Electromagnetic Field*, because it has to do with the space in the neighbourhood of the electric or magnetic bodies, and it may be called a *Dynamical Theory* because it assumes that in that space there is matter in motion, by which the observed electromagnetic phenomena are produced " (italics in original). Arguing from the recognised transmission of *energy* in the form of light and heat at an immense speed across vast distances, he supposes that there exists some medium in which energy is thus stored, in part as actual motion, in part as *potential* energy of elastic deformation. So much had been assumed by many, from Fresnel onwards, though the concept of *stored* energy had not been fully clarified. But Maxwell's earlier investigations had given strong evidence for the view that the movements of the same medium would be apprehended by us as the phenomena of magnetism and electricity. In the case of currents the energy is dissipated as heat, that is, the movement of the medium is communicated as the random movement of the molecules of matter; in the case of insulators there is a displacement or polarisation which represents potential energy, but is without further effect. The charging of an insulator thus involves arrested current flow.

Thus then we are led to the conception of a complicated mechanism capable of a vast variety of motion, but at the same time so connected that the motion of one part depends, according to definite relations, on the motion of other parts, these motions being communicated by forces arising from the relative displacement of the connected parts, in virtue of their elasticity. Such a mechanism must be subject to the general laws of Dynamics, and we ought to be able to work out all the consequences of its motion, provided we know the form of the relations between the motions of the parts.

The last sentence gives the clue to the nature of Maxwell's aim and achievement. The " cog wheels " showed that a plausible connexion between magnetism and electricity could be conceived, that the connectedness fills all space and suggests the " form of the relations between the motions of the parts ". They have done their job; so they are not even mentioned. What did Maxwell put in their place? At first tubes and cells of incompressible fluid; but even they are not mentioned now. What then? The answer is *nothing*. It is here that Maxwell's *method* breaks away from tradition. He is not concerned with what electricity " is "; a distinguished French savant said of his *Treatise on Electricity and Magnetism* that he understood all of it except what Maxwell meant by a body charged with electricity! He is not concerned, as another great Frenchman, Henri Poincaré, has reminded us, with *discovering* a mechanical theory of the electromagnetic field, but only with showing that such a theory is *possible*; that is that the whole universe of " Matter and Motion " is *conformable* to one generalised abstract scheme. How was this possible? To answer this question it will be necessary to make a somewhat lengthy digression on the history of the " field " concept—at first only implicit, becoming explicit only with Faraday and Maxwell.

Two strands of thought must be followed; one of which, namely the generalisation of mechanics, we have already alluded to (p. 274), the other being the principle of least action. As far back as the first century of our era we find Hero of Alexandria proving that the path taken by a reflected ray of light is the shortest possible. Fermat, seeking to extend this result to the case of refraction, found that the path followed by the ray is the one which takes the least *time*. Leibniz noted that Hero's " law " is a special case of Fermat's, namely that in which the velocities of incidence and rebound are equal. Following on this the great mathematicians of the seventeenth and eighteenth centuries investigated a number of problems in mechanics of a similar nature, of which the catenary and the brachistochrone may be cited as examples. But all these were isolated problems. More general were what is known as D'Alembert's principle and Gauss's principle of least constraint which are concerned with the possibility of predicting the behaviour of a system of forces *which are so to say interlocked*—in D'Alembert's principle the interlocking is such as to sustain equilibrium; in Gauss's no such restriction is imposed. These principles, valuable as they have proved for the solution of mechanical problems intractable by the classical Newtonian methods, still retained some of the latter's " earthiness "—its dependence on geometrical diagrams. At the beginning of Lagrange's immortal *Mécanique Analytique* the reader is appraised of the author's intention to rid mechanics of its dependence on diagrams. The path had been cleared for him by Euler who resolved curvilinear forces into tangential and normal components, and more completely by Maclaurin, in whose *Complete System of Fluxions* all forces are resolved along three mutually perpendicular Cartesian axes, whereby the geometrical (spatial) characteristics of mechanical problems could be expressed in the form of relations between symbols (i.e. algebraic equations). For dynamical problems Lagrange started from D'Alembert's principle that when forces are impressed upon the several masses of a connected system they will undergo *accelerations* which give a measure of the *effective* forces; these accelerations may be written as second order differential equations with respect to the co-ordinates of the masses. But for infinitesimal displacements the *impressed* forces and the *effective* forces must be in equilibrium; and the summation of the product of the force-pairs and the respective displacements is, by the principle of virtual work (p. 274), zero. This is the fundamental equation of dynamics.

As such it is of very little use; it is necessary to find some function which would determine the action of the constraints, which in the fundamental equation has been left completely unspecified. This function was derived from the Principle of Least Action.

The ingenious demonstrations of Hero and Fermat, referred to above, were felt to be special cases of a more general principle, namely that Nature works in the most economical manner to achieve her aims. This dogma—for dogma it is—derives from the uncritical acceptance of Aristotelian teleology, namely that every natural object is the best that could be designed for the specific purpose in view. Aristotle presumably had not time to count the fifty million odd eggs of the Sunfish or the astronomical number of pollen grains from every pine tree. Curiously enough it met less widespread recognition on the physical side, for though Lucretius sang the conservation principles in immortal verse, the theories of alchemy and of phlogiston for long flourished only by the ignoring or distortion of these principles. From Descartes however came the *neo*-Aristotelian dogma of the conservation of momentum, from which in all probability came the " metaphysical " confusion which D'Alembert was the first to clear up (p. 273). It cannot be too often emphasised that the seventeenth and early eighteenth centuries were the heyday of theology; so that when Maupertius propounded the Principle of Least Action in 1747, namely that in any mechanical action the product of mass, velocity, and space traversed is a minimum, his sole *reason* was that such an economy accorded best with the wisdom of the Creator. It is interesting to note that Euler, who took up the matter, gave the same reason, namely the assimilation of purpose and means, for the adoption of a least action principle; but for his " action " he chose a dynamically well-founded function—the space integral of the velocity for constant mass. Euler clearly recognised that the principle of least action holds good only for " conservative " systems (in which politicians may see an omen)—systems in which the *total* energy (work and *vis viva* in contemporary terminology) remains constant. It was in this form that the principle was applied by Lagrange.

The problem before Lagrange was to find an expression for any dynamical system such that the actual process, as defined by the analytical form of the second law of motion described above, would at the same time be performed with the expenditure of *least action*. The problem was of a kind[1] entirely different from those which the mathematicians of the seventeenth century had had to face and for which they had to anticipate the methods of the differential calculus, ultimately codified by Leibniz. There it was merely a matter of finding values for the variables which would give a minimum (or maximum) value for the function. Here it is a case of *finding a function* such that its value under certain stated conditions may be a minimum. The brachistochrone mentioned above provides a convenient example of the nature of the problem; this problem is to find the path between two points in a vertical plane such that a freely falling body reaches the second point in the least *time*. Now a path in space referred to given axes can be characterised by an equation, that is, a function of the variables represented by those axes. The problem is therefore to find the *function* which of all possible functions shall make the time value a minimum, the possible functions being restricted by the well-known function expressing the fact that the vertical rate of fall is itself proportional to the time. This specific problem could be solved by a geometrical method, but the *general* problem of the variation of functions demanded an entirely new technique invented by Lagrange and called the " calculus of variations ". By means of this he was able to deduce three equations of the form

$$\frac{d}{dt}\left[\frac{dL}{dq}\right] - \frac{dL}{d\dot{q}} = Q$$

where $L$ (the so-called " Lagrangian function ") represents the difference between the potential and kinetic energies of the system, $Q$ the external force tending to increase the value of $q$, which in turn stands for any one of the co-ordinates $x$, $y$, $z$, of Euclidean space. There remained for Lagrange only to point out that the equations were of purely *general* character, $q$ standing not necessarily for the space co-ordinates, but for any " degrees

[1] Newton was an exception. In 1696, returning after a day's work at the Mint, he found a " challenge problem " (that of the brachistochrone mentioned below) awaiting him. He solved it the same evening. The solution, sent anonymously to its propounder, John Bernoulli, drew the famous reply, " the lion is recognised by his claw ".

of freedom " (i.e. possible changes of any kind, subject to the quantity $Q$, which likewise need not be a " force ").

At last we see where this long digression is leading us. Reference to the paragraph quoted on p. 308 from Maxwell's final paper on the electromagnetic field will make this clear: Lagrange's equations comprise those " general laws of dynamics " from which Maxwell hoped to work out the behaviour of the field. The original Lagrangian equations had in fact been put into a more manageable form by W. R. Hamilton, whose statement of the principle of Least Action is so far regarded as the final form that it is now generally known as " Hamilton's Principle ". The problem was further complicated by the fact that magnetic and electric forces do not act in constant directions. To allow for this Maxwell had to make use of that specialised branch of mathematics, the vector calculus, which had been applied at first by Laplace, and later by Hamilton, to give greater generality to the equations of Lagrange. Consideration of the process by which Maxwell showed that the equations revealing the structure of the electromagnetic field were of precisely the same *form* as the generalised dynamical equations lies entirely beyond the scope of this book; all that concerns us is its place in the growth of physical ideas. No one recognised the crisis in physical ideas created by Maxwell better than Poincaré. In comparing the methods of one of the " classical " physicists with Maxwell's, he says:

> Behind the matter of which our senses are aware and which is made known to us by experiment, such a thinker will expect to see another kind of matter— the only true matter, in his opinion—which will no longer have anything but purely geometrical qualities, and the atoms of which will be mathematical points subject to the laws of dynamics alone. And yet he will try and represent to himself by an unconscious contradiction, these invisible and colourless atoms, and therefore try and bring them as close as possible to ordinary matter.
>
> Then only will he be thoroughly satisfied, and he will then imagine that he has penetrated the secret of the universe. Even if the satisfaction is fallacious, it is none the less difficult to give it up . . .
>
> Maxwell does not give a mechanical explanation of electricity and magnetism; he confines himself to showing that such an explanation is possible . . .
>
> The same spirit is found throughout all his work. He throws into relief the essential, i.e. what is common to all theories; everything that suits only a particular theory is passed over almost in silence. The reader therefore finds himself in the presence of form nearly devoid of matter, which at first he is

tempted to take as a fugitive and unassailable phantom. But the efforts he is thus compelled to make force him to think, and eventually he sees that there is often something rather artificial in the theoretical " aggregates " which he once admired.

Maxwell, perhaps unconsciously, was formulating the modern aim of theoretical physics, which may be broadly expressed as follows:—Given certain special *formal* characteristics of the continuum, to discover more *general formal* characteristics by which the former may be determined. In its latest phase it has reached the logical conclusion which is that of reducing physics to a set of orders of connectedness to whose individual symbols can be assigned *no* physical meaning. Although this method has achieved startling triumphs (e.g. in foretelling the discovery of the meson) Professor Max Born speaks for several of the leading mathematical physicists in expressing his belief that this must end in self-stultification. Had such a doctrine been universal in 1890 it is doubtful whether the X-ray would ever have been discovered. It was *not* Maxwell's way; on the contrary it was said of him as a young man that " although somewhat deficient in analysis, it appeared impossible for him to think wrongly on any *physical* subject ". It was his power of visualisation of the *form* of hydrodynamic action which enabled him to visualise the *form* of connexion between electricity and magnetism. On the other hand he avoided the error of supposing that there need be any *pseudo*-material fluid to express this form. As the formal characteristics became clearer in his mind, the material trappings, which were at best redundant and at worst contradictory, were shed one by one. In this respect he showed himself superior to Kelvin, from whom he at first " poached ", as he called it, so much. In an age of the triumph of mechanical engineering, Kelvin remained at heart an engineer; at the flights of Maxwell's unearthly genius he could and did only regretfully shake his head.

Nor was Kelvin alone in this. To most of his contemporaries, especially on the continent, Maxwell's theory was a beautiful piece of imaginative virtuosity, but quite out of touch with the " realities " of physics. And so it might have remained had not one of its quite unexpected corollaries been the oscillatory nature of electromagnetic action. Light was of course accepted as oscillatory; the velocity of electromagnetic disturbance as calculated by Maxwell turned out to be the same as that of light—a striking coincidence, but no one had ever met waves having electromagnetic characteristics. How difficult it may be for even the most powerful intellects to break through the crust of tradition is shown by the fact that Kelvin had already in 1853 given a mathematical proof, based on the principle of conservation of energy, that the *discharge* of a Leyden jar would in certain cases be oscillatory in character. It was moreover from his theory of " electric images " that Maxwell admitted that he intended to " poach ". But Kelvin could not bring himself to take seriously the possibility that the oscillatory *discharge* might be only, as it were, the outward and visible sign of an oscillatory displacement in the surrounding " ether ". Helmholtz was more open-minded and induced his pupil, Heinrich Hertz, to investigate the matter, though rather in the hope that he would be unsuccessful, as Helmholtz had managed to deduce Maxwell's equations from assumptions which debarred the possibility of the propagation of waves of finite velocity.

Hertz first announced the success of his attempts in the *Annalen der Physik* in 1887. As a source he had used an induction coil whose secondary circuit included a pair of rods whose capacity was increased by fixing metal spheres to their ends. To detect the presence of oscillatory currents (" electric waves ") in the neighbourhood he constructed a wire hoop of such dimensions as would cause it to be " in tune " with the waves, i.e. to constitute a resonator. With a train of sparks passing between the spheres of the radiator, he moved about the room with the detector. At certain positions sparks were clearly seen within the gap; positions determined by the fact that reflection from the walls of the room was causing the phenomenon well known in waves on a fluid surface, of stationary waves. We may marvel at the simplicity of his apparatus and at the certainty of the results; but within two years, working with waves of ten times the frequency produced by an improved apparatus, he was able to report results of an astonishing precision. Using a parabolic cylindrical reflector to concentrate the primary waves he detected stationary nodes at various points along the wave axis; with a plane sheet of metal placed normally to the axis he demonstrated rectilinear propagation, and then by rotation of the sheet equiangular reflection with polarisation. Finally by interposing a prism of pitch one and a half metres high in the path of the waves he showed that the latter, freely transmitted by non-conductors, are nevertheless refracted. In describing the results of reflection he drew attention to the fact that since only *electrical* phenomena were being detected, it was to be supposed that a magnetic *effect* would be observable in a plane perpendicular to that in which the electric was polarised; and this he was able to detect by using a low frequency source.

The title of Hertz's memoir was *Ueber Strahlen elektrischer Kraft*. He summarised the result in the following words: " We have applied the term rays of electric force to the phenomena which we have investigated. We may perhaps further designate them as rays of light of very great wave length. The experiments described appear to me at any rate eminently adapted to remove any doubt as to the identity of light, radiant heat, and electromagnetic wave action." History has completely endorsed his claim. But just as Maxwell's life was brought to a tragically early end ten years before the complete realisation of his theory of the pervasive nature of electromagnetic action, so with even greater haste was Hertz's young spirit called away from this mortal stage. Thus those keen eyes, whose perception of the varying intensity of a hardly visible spark had confirmed the greatest physical advance since Newton, never glimpsed even from afar the Promised Land.

The rest of the story of Matter and Motion—the rapid extension of the Hertzian waves down and up, especially up, the scale of frequency; the atomisation of energy; the collapse of matter into systems of energy or, if you prefer it, the materialisation of energy constellations; the particulate manifestation of wave trains, the oscillatory specification of material particles; the final—but is it final?—apotheosis of the nature of the physical world in a *Walpurgis* dance of symbols with no physical meaning —all this falls within the lifetime of men of science still with us, many of whom have told the story in a series of books to which the reader may turn in order to gain his knowledge at first hand from these great thinkers themselves. It is the object of this outline to bring before the modern reader some of the thoughts of the great minds of the past as they have contributed to the growth of scientific ideas. It would be contrary to the spirit in which this task has been undertaken to attempt to " bring the story up to date ". We are still too close to the landscape of ideas to judge of it as a whole. Above all we cannot yet assess with certainty the, at present quite plausible, expectation that the power of *physical* analysis may not be nearly exhausted. The sceptical reader will of course remind us that much the same thing was being said about 1890—just before the discovery of X-rays showed that a new epoch of revelation was just *beginning*. But to-day there is a difference in principle—there always is when one has an axe to grind! For only recently has the physicist's probe been comparable with the object of his seeking. If in seeking a piece of Dresden china in a dark cabinet the only available implement is a heavy walking-stick, it is perhaps wiser to leave the shepherdess in the dark. I do not say that physics *has* of necessity reached the end of its long road, but the possibility at least provides a good excuse for closing an outline which has already been drawn out long enough.

## APPENDIX TO CHAPTER XXIV

The Lagrangian function, L, is what is now called a *potential function*. The growth of the idea of *potential*—one of the most fertile in science—has taken place in several fields, a fact which may easily lead to confusion. The term *potential* seems first to have been used in a memoir of George Green published in 1828 but which suffered a fate similar to that of Waterston's until 1845. The *idea* is implicit in Lagrange's analytical device ($F_x = -\partial V/\partial x$, in modern notation, where $F_x$ is the $\times$-component of the force at the point $x$, $y$, $z$, and V is the " potential " at that point) for the calculation of the effects of central forces. In 1784 Laplace generalised the concept to include " fields " other than gravitational, in particular those involving continuous fluid motion, and showed that

$$\frac{\partial^2 V}{\partial x^2} + \frac{\partial^2 V}{\partial y^2} + \frac{\partial^2 V}{\partial z^2} = 0$$

in empty space. Poisson extended the application of this equation to space occupied by gravitating matter of average density $\rho$—zero in this case being replaced by $-4\pi\rho$. In the paper in which he coined the name Green applied the idea of " potential " to the electromagnetic field; but here the matter is complicated by the fact that the lines of force of such fields are in general not straight lines, so a new algebra was required to simplify the calculations. The hint of this was provided by Hamilton who restated the fundamental field equation in the form:

$$\mathbf{F} = -\left(i\frac{d\psi}{dx} + j\frac{d\psi}{dy} + k\frac{d\psi}{dz}\right)$$

Where $\psi$ is the potential at a point $x, y, z$ of the field and $i, j, k$ are unit vectors (called " right versors " by Hamilton and recognised as operators) having the property that although $i^2 = j^2 = k^2 = ijk = -1$ and $ij = k, jk = i, ki = j$, $ji$ is *not* equal to $ij$ but to *minus ij*, etc.; that is, the commutative axiom of multiplication which gives consistent results in " ordinary " algebra must be given up in the algebra of " quaternions ", as Hamilton called it. This cardinal discovery was given a lukewarm reception by his contemporaries for several reasons, among which the charitable will give most weight to his inconvenient notation. The idea, though not the notation, was however essential to Maxwell, who introduced two further vectorial functions. The modern form of vector analysis owes most to Grassmann (p. 297). The part of the above expression which on being applied to the potential $\psi$ gives the negative value of the force, is known as the " Hamilton operator ". It must not be confused with Hamilton's principle (1834-5) which was his generalisation of Lagrange's equations in the form of a " stationary " principle. By " stationary " is meant the fact that the value of the function (here a time integral) does not vary for paths " infinitely " near to one another. The actual value may be a maximum or minimum.

## SOURCES FOR CHAPTER XXIV

Magie, *op. cit.*, Gen. Bibl., for Maxwell, Faraday, Hertz.

Faraday's *Experimental Researches in Electricity* (London, 1849) should be consulted as models of the experimental method practised by the " Prince of Experimenters ".

*The Origin of Clerk-Maxwell's Electric Ideas* (Cambridge, 1937). This is largely mathematical, but contains many interesting insights into Maxwell's way of thought.

H. Poincaré, *Science and Hypothesis*.

# PART TWO
# NATURE AND LIFE

## CHAPTER XXV
## WHAT IS LIFE?

At the beginning of this book we saw that for those thinkers who first began to ponder on the nature of things, the problem of life hardly existed. The world itself was pictured as a living thing whose parts also were living in the sense that they " grew " out of whatever elements happened to seize the particular thinker's fancy, and were in turn subject to corruption. The problem of the distinction between the living and the non-living had therefore to be *created*. When and how did this distinction arise? It is, I believe, impossible to point to anyone in whose thought the distinction became explicit. There can on the other hand be no question that it was *implicit* in the teaching of the Atomists, in that the atoms were unambiguously described as eternal and immutable in shape and size. Growth and decay were now pictured as a concourse and dispersal of units themselves simple and " senseless ". The state of " being alive " was maintained by the inspiration and expiration of " soul " or " fire " atoms; but these latter differed in no way from the atoms composing non-living bodies except in being rounder and smoother, that is to say, in respect of purely *geometrical* characteristics. Here then we see the traditional attitude reversed. Nature is regarded as essentially chaotic and lifeless; life is now the state that has to be explained. Only two characteristics of the living were attributed to the atoms, namely continuous motion in a straight line with occasionally an unexplained swerving away from the straight and narrow path.

The great merit of atomism was its uncompromising consistency: no " souls ", no intervention of gods, " for methinks no golden chain let down from heaven above the races of mortal things . . . but the same earth bare them as now feeds them from herself ", and again, " nature replenishes one thing out of another and does not suffer anything to be begotten before she has been recruited by the death of some other." The words are those of the Latin poet, Lucretius Carus, in whose superb poem *De Rerum Natura* we have the fullest exposition of atomism, and one which, scholars are generally agreed, probably departs little from the original teaching of Demokritos, amplified as it was by Epikuros in the fourth century B.C.

Majestic in its simplicity and admirable in its freedom from question-begging, the atomic theory was nevertheless an almost sterile system in the ancient world. Not until it was linked to seventeenth-century dynamics and the experimental method could it have more than a formal value. The impulse to a fertile characterisation of the nature of living things came from quite another quarter, namely the study of the human body in the aim to advance the art of medicine. We have already (p. 12) had occasion to refer to the origin of Greek medicine and its influence on natural philosophy. While Demokritos was theorising on the structure of the cosmos, the disciples of Hippokrates were exploring the microcosm, as the human body was wont to be called. One of these physicians was the father of Aristotle, whose encyclopaedic works on animals show that the sober reliance on observation taught in the medical schools must be wedded to the speculative reason if a just idea of nature is to be achieved.

The works of Aristotle have come down to us by two paths. The original Greek texts were translated into Syriac by pious members of the Nestorian Christian church in Asia Minor, and with the spread of Islam these Syriac versions were further translated into Arabic by order of the Caliph al-Mamun (A.D. 813-33), son of the notorious Harun-al-Rashd. Later, when the Saracen was thrown back from Western Europe by Christendom, Toledo became a centre for the translation of Arabic versions into Latin. Among those working at Toledo was one Michael who had probably travelled there from the lowlands of Scotland, and we know that before 1217 he had completed the translation of the books known in their Latin form as *Historia Animalium, De Partibus Animalium*, and *De Generatione Animalium*. The original Greek texts were recovered a little later and were found to be far from complete; but advantage has been taken of these two largely independent sources to reconstruct a text as near as possible to the original text of Aristotle himself. It must be further remembered that even Aristotle's " text " consisted largely of notes for lectures; Aristotle's works are not literary masterpieces as Plato's are, though there is reason to believe that he wrote in dialogue form when still a member of Plato's Academy.

It is of course quite impossible to survey even superficially all Aristotle's writings—in this place we shall consider only those which may help us to understand what was his *idea of a living organism*. In later chapters we shall make further use of his teachings, for with all his faults he is the fountain-head of more ideas in the science of biology than anyone before or since.

It is not to be supposed that Aristotle carried out his biological researches with the immediate object of advancing knowledge of the human body. Just as his philosophy is unintelligible to anyone completely ignorant of biology, so his biological *method* can be grasped only when it is realised that his *aim* is ultimately philosophical. Towards the end of the First Book of *The Parts of Animals* in which he works out the implications of the store of detailed knowledge he has gathered together in the *History* (the Greek word ἱστορια means a " reasoned survey ", not a chronologial record) occurs a famous passage which expresses his aim and attitude to nature. Having mentioned that he has already treated of " things divine " (in the *Metaphysics*) he goes on to say " We must not betake ourselves to the consideration of the meaner animals with a bad grace, as though we were children; since in all natural things there is somewhat of the marvellous. There is a story which tells how some visitors once wished to visit Herakleitos and when they came to call and saw him in the kitchen, warming himself at the stove, they hesitated; but Herakleitos said: ' Come in; don't be afraid; there are gods even here.' In like manner we ought not to hesitate or to be abashed, but boldly to enter upon our researches concerning animals of every sort and kind, knowing that in not one of them is Nature or Beauty lacking. I add ' Beauty ', because in the works of Nature purpose and not accident is predominant; and the purpose or end for the sake of which these works have been constructed or formed has its place among what is beautiful."

This strikes the keynote of his researches. If we are to understand the works of nature—and our aim is to reveal the "somewhat of the marvellous"—we must discover the *purpose* of every part for itself and in relation to the whole. A thing is *known* when we know what it is *for*. Aristotle has of course been accused of having failed to achieve *science* just because he started at the wrong end. Of this more later. But it cannot be too often urged that one of his principal claims to genius was that he had the intellectual courage, and for a pupil of Plato this must have been great—to *insist* that " we ought first to take the phenomena that are observed in each group, and then go on to state their causes. This applies just as much to the subject of the process of formation: for here too we ought surely to begin with things as they are actually observed to be when completed." Thus though we ought always to be mindful of *purpose* in *interpretation*, the impartial observation of what *is* must come first. Aristotle occasionally lapsed from this ideal, but much less often than his critics of the sixteenth and seventeenth century believed. The most famous case is that of the small shark *Mustelus*, which Aristotle described

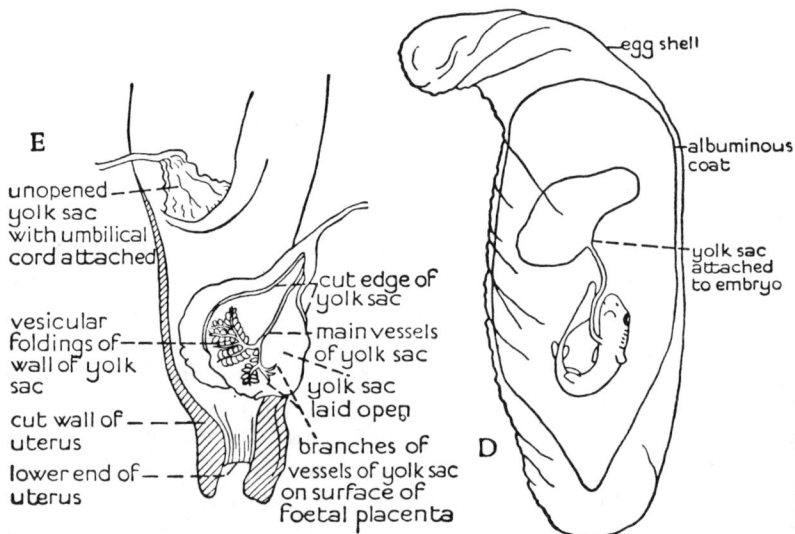

Fig. 31. The yolk-sac placenta of the viviparous shark *Mustelus*.

as carrying its living embryos within its body, whereby they derived nutriment direct from the mother. This was faithfully recorded by the " dogmatic systematist ", but regarded as fantastic by modern naturalists until its truth was completely established by Müller in the early years of the nineteenth century. On his own evidence Aristotle gathered much information from fishermen, hunters, and other men, the success of whose enterprise depends on accurate observation. On the other hand where, owing to the religious ban against dissection of the human body, observation was difficult or impossible, Aristotle often interpreted the facts to suit his preconceived hypothesis of what the purpose " ought " to be. The outstanding example of this is his insistence on the function of the brain being to " cool " the blood.

In this place we are concerned chiefly with his views on the nature of life; and it is from his most famous zoological work, *The Generation of Animals*, that we may learn most on this score. In strict accord with the facts of observation—for there were no microscopes in those days—Aristotle asserts that " among the bloodless animals, while some groups have both male and female and hence generate offspring which are identical in kind with their parents, there are other groups which, although they generate, do not generate offspring identical with their parents. Such are the creatures which come into being not as the result of the copulation of living animals, but out of putrescent soil and out of residues " [excrements?]. These forms which are " spontaneously generated " are always of the nature of larvae, never " perfect " animals. Plants also may be formed " when either the soil or certain parts in plants become putrescent, since some of them do not take shape independently on their own, but grow upon other trees, as for instance the mistletoe does ".

Aristotle has therefore been driven by observation from the crude view that even the " blooded " organisms, including man himself, might at some previous time have been formed out of the earth. On the other hand he does not make the mistake of regarding the " life " of one group as being in any way different from that of the other.

In his theory of the *mechanism* of spontaneous generation

Aristotle goes beyond the facts—goes back to the dogmatic basis of all his theoretical natural philosophy, the four elements, or rather their highly refined correlate, the αἰθήρ, which fills the celestial realm of the stars. For though the " matter " of which organisms are made is composed of the four elements, varied as to proportions in the different parts of the body, the " form " is imparted by a substance—πνεῦμα—analogous to αἰθήρ. " Animals and plants are formed in the earth and in the water because in earth water is present, and in water πνεῦμα is present, so that in a way all things are full of soul; and that is why they quickly take shape once it has been enclosed. Now it gets enclosed as the liquids containing corporeal matter become heated and there is formed as it were a frothy bubble." It is important to note that πνεῦμα is neither material air nor on the other hand soul itself. It is however the *material* cause through which Soul acts as *motive* (usually called *efficient* by the commentators) cause.

The reference to " Soul " explains why Aristotle's treatise *De Anima* is usually studied in conjunction with his purely biological works; for him living things are unintelligible in complete abstraction from the idea of Soul. This fact is of first importance for the history of biology. In its wider acceptation it gave rise to the endless (or is the end at last in sight?) wrangles between " vitalists " and " mechanists "; in a narrower application it set the problem—now happily resolved—of the relative importance of male and female in the reproduction of the higher animals. It may be best to set Aristotle's views on these subjects in the context of a summary of his answer to the question, " What is Life? "

First it is necessary to repeat that without the study of at least the guiding ideas of Aristotle's philosophy, confusion and distortion are bound to arise in the reader's mind. Should he consider, as many leading biologists are once more considering, that Aristotle was not merely a muddle-headed and prejudiced dogmatist, but a painstaking genius who, just because he was *not* prejudiced as we are by centuries of theoretical construction, saw the problems of life in their naked clarity, he will be well advised to make the effort to become acquainted with that philosophy. Thus armed against taking a superficial view of Aristotle's rich imagery he will perhaps come to see how much we have lost, as well as gained, by a long dissociation from his concrete method.

To sum up then. For Aristotle organisms on a higher level or, rather, on a hierarchy of higher levels, illustrate the method by which the *whole* of nature is to be understood, namely, as progressive realisations of *ends* at first present only as *possibilities*. In each organism, then, we must seek the *end* ($\tau\epsilon\lambda os$ = " final cause ") to which its activity (motive or efficient cause) is tending. But since there can be no *end* completely devoid of substance, we must also seek the *material* cause in which the end is *potential*. Finally, realisation of the end must proceed through one path rather than another, the actual path taken being determined by the *formal* cause. And this method applies to the parts as much as to the wholes: of each organ we must ask, what is it *for*? How does it contribute to the end?

In those organisms nearest to " matter " (that is, stuff in which potency is nearly all, form barely present) the $\pi\nu\epsilon\hat{v}\mu a$ present in matter is sufficient to produce the degree of organisation adequate to the end. But for the higher levels, including Man, " fetation " must intervene, in which the " coction " (or " cooking ") of matter may proceed whereby the residues inessential to the form may be removed. And here we meet with the *particular* hypothesis which caused no little dissension in the later centuries not only to biologists but also to feminists. For Aristotle unambiguously regards the male as supplying that little something—the motive cause through which soul acts—the female hasn't got. The conclusion he draws from this, namely that woman is imperfect man, has been regarded as the sort of thing to be expected from a slave-owning member of the privileged class. Perhaps too much has been made of it. After all Aristotle is speaking biologically; and modern comparative studies do point to the fact that for the majority of eggs the stimulus of the sperm is a determining factor for cleavage in more ways than one. Though whether Aristotle would regard woman as imperfect Epsom salts or hot needles (whose stimulus is sufficient in some cases) is doubtful!

It is impossible to give anything like a complete evaluation of his theory of life. It suffers like all Greek science from attempting too much; it is over-simplified by reason of the absence of knowledge of the complexity of the " lower " organisms which only the microscope could reveal. A more serious fault is its acceptance of the hierarchy of nature as already established, with the implication that every part has an end, thus ruling out the possibility of its being a " dead end " of a series that has been transformed. This opens the way to all sorts of special pleading and blindness to alternative—and often more adequate—explanations. But whatever its faults, it is a superb expression of one characteristic of biological thought which, in the hands of a naturalist like Aristotle, might have led to little obstruction. But there were, after Theophrastos,[1] no naturalists like Aristotle; only imitators, with results we shall see in the sequel.

To understand the further development of the idea of life, it is necessary to take some account of the progress of Greek medicine; for of biology as a separate interest apart from its application to the healing art, there was after Aristotle virtually none. The history of medicine shows no such clear-cut problem as that which characterised astronomy. The removal of the guiding hand of Hippokrates (p. 12) seems to have been followed by a period of stagnation—a state likely to arise in any branch of knowledge relying too exclusively on *mere* observation without the stimulus of theory. The wise and liberal art of Hippokrates, relying so much on the personality and " intuition " of the physician, and the *vis medicatrix naturae*, degenerated in the hands of lesser men into a stereotyped " system " of cures, thus gaining for itself the label " dogmatic ". Reaction came in the new centre of learning at Alexandria. Here professional curiosity overcoming religious awe, the human body was first systematically dissected, and in consequence human anatomy, beyond that of the skeleton and muscles, was born. This was largely the work of two outstanding men, Herophilos of Chalcedon (fl. B.C. 300, hence a contemporary of Euclid) and Erasistratos of Chios (fl. 280). Both of these men

---

[1] The work of Theophrastos (B.C. 372-287) on plants was the only notable exception. Comparable in respect of range and precision with the works on animals by his master, the book extends the Aristotelian idea of potency into the world of plants. For these organisms display in an even more marked degree that " we should observe these things not for what they are but for what they are becoming "—a point particularly well brought out by the nature of fruit and seed, and by the possibility of growing a whole plant from a cutting.

broke away from the dogmatic " rational " medicine of the Hippokratic school; but whereas Herophilos retained the doctrine of the " harmony of the humours " as the basis of health, Erasistratos repudiated any pseudo-chemical theory at all—even to the extent of denying the value of drugs.

Probably the most important discovery of the Alexandrian school—initiated by Herophilos—was that of the true function of the nervous system. It will be recalled that the worst blunder of Aristotle was the dogma that the heart is the seat of the emotions, the brain being a subsidiary organ to temper the emotions by " cooling " the blood. Herophilos correctly regarded the brain as the centre of the nervous system, and as such the seat of the intelligence. The matter was pushed further by Erasistratos who, applying the comparative method to the brains of animals, rightly deduced that the convolutions of the cerebrum, which he distinguished from the cerebellum, are a measure of the animal's intelligence. It seems almost certain that by actual experiment he anticipated by 2,000 years Charles Bell's discovery (see p. 454) of the sensory and motor functions of the dorsal and ventral roots respectively of the spinal nerves. The self-restraint characteristic of modern science at its best was, however, lacking in these great anatomists, who could not forbear to regard the nerves as hollow *tubes* conducting a special *fluid* whereby action could be transmitted from point to point. Erasistratos particularly was obsessed by the purely mechanical nature of the phenomena of life propagated into every part by three sets of vessels, veins to carry blood, nerves an invisible fluid, arteries *air*. We see here the baneful influence of classical atomism, in which every kind of action was " explained " by a variation in the purely *spatial* characteristics of its co-operating particles. Nevertheless, although ill-founded, the theory of Erasistratos was of great importance for the history of biology; its importance lying in its *comprehensiveness*. It comprises the first attempt at a systematic approach to the problem of " how life works "; it was in fact the first *physiology*.

Erasistratos' physiology was based on his view that the " vital spirit " (a kind of $\pi\nu\epsilon\hat{v}\mu a$) was drawn by the lungs from the air, whence borne along by the *arteries* it permeates every part of the body. In its wanderings it passes through the brain, whence, modified into a more subtle πνεῦμα, the "animal spirits", it is distributed to the body through the hollow nerves. Connecting arteries to veins he believed there existed "synastomoses" through which, in *diseased* states only, blood was forced from the veins into the arteries, thereby impeding the movement of the πνεῦμα.

The affinities of this theory to atomism are clear enough; what is not so clear is why Erasistratos, who completed the discovery of the lacteals begun by Herophilos, ignores the part played by food in the maintenance of life. Or rather, why he had no *theory* as to this part; for his *pathology* and *therapy* were frankly based on the assumption that all disease is ultimately *mal*-nutrition.

For us the physiology of Erasistratos is of importance mainly as the source of the system of Galen, which dominated European medical thought for more than thirteen hundred years. Galen, born in Pergamos in A.D. 129, is one of the most difficult writers to assess justly. It is easy to dismiss him as of negligible importance in the history of science since he made few original observations, and modified the existing theory in only one way, and that not for the better. He was indeed a proficient dissector, but his system of *human* anatomy is retrograde, in that the nearest relative of man to be dissected by him was the Barbary Ape; and he was not always careful to avoid drawing detailed conclusions from the one to the other, though he warns other anatomists of this very danger. What then gave him his influence, equalled only by that of Aristotle and Ptolemy? It is generally agreed that his renown rests chiefly on his judicious eclecticism, which made of him a *conservator* of knowledge at a time when not only was little new being *added* to published knowledge, but much that had existed was being lost. It will be noted that he lived in the heyday of Roman Imperialism, when theory and speculation were at a discount, and the practice of medicine had sunk to the level of a mechanical routine. Into this unimaginative world came this well-educated Greek, possessed of a ready—too ready, for he wrote 150 medical works alone—flow of language, encyclopaedic learning, endless patience, and a mind receptive to the human touch, despite his enormous literary output. As one of his recent translators, the late Dr. A. J. Brock, so admirably puts it, " He combated the tendency . . . to reduce medicine to the science of finding a label for each patient, and then treating, not the patient, but the label." For us to-day the cardinal importance of his works is that they form, as Aristotle's did for early Greek natural philosophy, our chief source for the history of Greek medicine.

By the time that Galen was browbeating with his pen every medical author except Hippokrates, for whom he had an unbounded admiration, the schools had grown to at least five. Since these cause great confusion to students of medical history, it seems desirable to pass them briefly in review at this stage. Besides the *Dogmatists* already mentioned (p. 328) there had arisen in the first Alexandrian period (fourth century B.C.) as the culmination of Herophilos' rejection of dogma, the *Empiricists*.[1] From Alexandria there came to Rome in the first century B.C., Asklepiades, in the medical tradition of Erasistratos. The latent atomism of Erasistratos became patent in Asklepiades, for was not Lucretius now singing its praises in immortal verse? Asklepiades' pupil, Themiso, was the first of the *Methodist* school, which sought a middle position between the dogmatic rationalism of the " classical " school of Hippokrates and the equally dogmatic denial of dogma by the Empiricists. From the Methodists came the age-old weakness of attaching too much importance to mere *names of diseases*, but among them was one outstanding figure, Soranos of Ephesos, whose modernity of outlook is well exemplified by his injunction to mothers to refrain from putting a child to the breast every time it cries—a lesson hardly yet learned after 2,000 years! Lastly came the *Pneumatists* founded by Athenaeus of Attaleia. Their name betokens a veneration for the all-pervading πνεῦμα, showing the influence of the Stoic philosophers, and in turn teaching the early Christians to call the Holy " Ghost " the Lord and Giver of *Life*.

In the small space that we can devote to a more detailed study

[1] Highly important as a school of thought until comparatively modern times. As A. J. Brock has pointed out, Locke, the founder of modern empirical (i.e. observational) philosophy, himself a physician, was the intimate of Sydenham, a medical empiric in the best sense of the word.

of Galen, we can do no more than glance at those aspects of his teaching on which he laid greatest stress, and which, in some measure at any rate, bore the stamp of originality. A quotation from his most characteristic work *On the Natural Faculties* will illustrate the first of these. " We say that animals are governed at once by their soul and by their nature (φῦσις), and plants by their nature alone, and that growth and nutrition are the effects of nature, not of soul." Here is the essence of his teaching, namely that living things have a *nature* of their own and are not merely singular combinations of blindly operating atoms. Moreover, this nature is characterised by growth and nutrition. The purpose of his treatise is to enquire " from what faculties these effects themselves . . . take their origin ". In an ingenious passage, unfortunately too long to quote, he distinguishes *organic growth* from mere increase in size, likening the latter to the balloons made by children from pigs' bladders which may be progressively increased in volume, but only by attenuation of thickness. How then can *growth* occur? By the alteration of substance, for " no one will suppose that bread represents a kind of meeting-place for bone, flesh, nerve, and all the other parts,[1] and that each of these subsequently becomes separated in the body and goes to join its own kind; before any separation can take place, the bread obviously becomes blood; (at any rate if a man takes no other food for a prolonged period he will have blood enclosed within his veins all the same)." Galen is here (though he did not know it) staking out a claim for the study of *chemistry*—the science which establishes the conditions of alteration of substances—to be an essential propaedeutic to physiology. His own solution of the problem was a stage in advance of that of Erasistratos, but shows the inevitable weakness of Greek natural science in its failure to recognise the need for simplification of the problem at the risk of temporary distortion. For him—and for more than a millenium thereafter—the problem is " solved " in terms of " spirits " and " faculties ". The food in the gut is changed into blood by the " blood-making faculty " of the portal vein. On reaching the

---

[1] Possibly a reference to the teaching of Anaxagoras (B.C. 460 ca.) who seems to have recognised the importance of this very problem without being able to give any kind of solution in unambiguous trems.

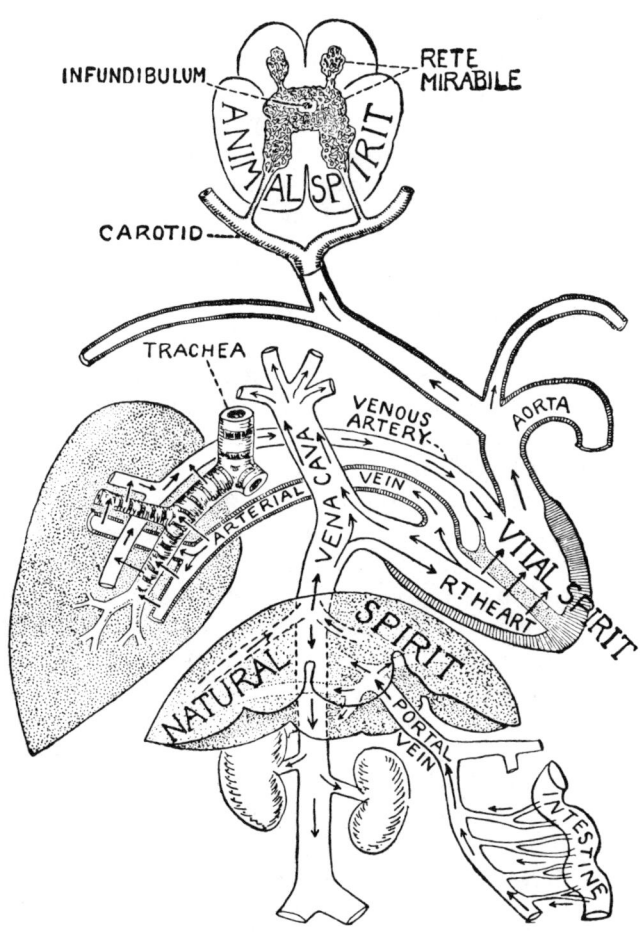

Fig. 32. The course of the "spirits" according to Galen. "Natural spirit" is introduced by the liver. Only one lung is shown.

liver it receives the " natural spirit ". In the heart it is further enlivened by the influx from the lungs, via the arteries, of the " vital spirit "; and that part of it which reaches the brain is there changed into the " animal spirit ", which passes through the hollow nerves to every part of the body. Galen admitted that the arteries were not filled with *air*, as had been previously taught—his scientific spirit, aided by a callousness characteristic of the age, is shown by his opening, between two ligatures, the pulmonary artery of a live dog—but stated that air is absorbed by the superficial arteries, the lungs being used rather as exhaust pipes.

Parallel with his account of nutrition and assimilation was his re-affirmation of the doctrine of the four " humours "—sanguine, phlegmatic, yellow bile, black bile (melancholic)—a doctrine absurd enough in its literal application to the explanation of disease by the acquisition of a surfeit of any one humour from a particular food, but rendered immortal by its being linked with the problem of *temperament,* as shown in the expressions commonly used to-day. Moreover, the germ of truth in it is made clearer by our recognition of the chemical basis of character, in so far as it may be modified by the tilting of the balance in favour of one or other of the secretions of the ductless glands. Galen knew, too, of the profound influence of mental states on physical well-being, as is evidenced by his skilful treatment of a young woman in love with a famous dancer.

Galen is seen at his best in his theory of the action of the kidneys. Having first demonstrated by experiments, as ingenious as they were callous, that urine passes *from* the kidneys *via* the ureters to the bladder, he proceeds to argue at inordinate length and persistently *ad hominem* that the whole process is no mere mechanical separation, but an " attraction " by the kidney for the urinary matter in the blood. If we are inclined to scoff at his naïve *physical* analogy of the " attraction " of the lodestone for iron, let us remember that we are still far from the whole explanation of the mechanism whereby the kidney pumps urine out of the blood *against* the osmotic gradient. To his credit he had recognised what the kidney is *for,* even if he stumbled hopelessly in trying to explain how it performs its vital task.

## WHAT IS LIFE?

It is, I believe, fair to say that Galen's service to biological thought was that of sensing the right questions to ask, his disservice in asking them in the wrong terms, or in trying, as in his famous " discovery " that blood percolates through the septum of the heart from the right to the left ventricle, to take a short cut to the answer. Vitalist and teleologist he was, at a time when such a view was perhaps the only profitable one to take. He can hardly be blamed for the idolatrous acceptance of his views by men who had every opportunity of correcting his errors.

With the break-up of the Roman Empire the biological thought of Greece decayed almost more completely in Western Europe than any other part of the Greek culture. One of the earliest evidences of the practice of anatomy in Europe (the so-called *Anatomia Cophonis*) was probably written about the time of the battle of Hastings—that is about the time of the opening up of communication between the Norman world of Western Europe and the world of Islam, particularly through the " passage " of Sicily. This island, one of the earliest colonies founded during the westward movement of Hellas, seems to have retained in some measure more of the Hellenic tradition than any other of the western outposts. Even the language was a corrupt Greek. Although the tradition of liberal culture of the city of Salernum is now known to have been legendary, yet we have indisputable evidence that dissection was carried on in a spirit of unprejudiced observation, little distorted by Galenic errors of interpretation. Several " demonstrations " are known whose terseness and clarity fall not so far short of modern manuals of practical anatomy. Of any text book there is no evidence until the appearance of a Latin translation by Constantine the African, " Monk of Monte Cassino ", of an Arabic summary of the medical knowledge of Islam written by Ali Ibn el Abbas, probably a Persian court physician. This was perhaps the first book on Eastern medicine (there was yet no Western worthy of the name) to enter Europe. It was followed shortly afterwards by the earlier works of Rhazes, which were considered of sufficient importance four hundred years later for Vesalius (p. 336) to write his graduation dissertation on some of them—and later still by the most influential of all, the Arab, Ibn-Sina's (Avicenna) *Canon of Medicine*, which remained the " Bible " of medical doctrine until Galen's *original* works were translated into Latin by the famous English physician, Linacre, in 1523.

With the founding of many universities and hospitals in the eleventh century, the subject of medicine was cultivated in many countries besides Italy. But with the flood of Arabic literature the standard of medical science fell below that of the crude but vigorous study at Salernum. The professor was wont to sit above the dissecting table, which was served by one or more barbers, and from the eminence of this *cathedra* hurl platitudes and errors alike, mostly culled from the book before him. We have already seen how Paracelsus (p. 165) reacted to this custom; but a more effective opposition was started by Andreas Vesalius who revolutionised the study of anatomy, giving to it the form in which it has appeared ever since. Having made his mark as a brilliant and not very discreet *prosector* in Paris, he was called to the Chair at Padua, whence in 1543 came his greatest book, *De Humani Corporis Fabrica*. Even a cursory examination of this book, actually published at Basel, reveals a completely new point of view. The title page shows Vesalius himself speaking to a packed theatre, not from the eminence of the *cathedra* but beside the cadaver which he himself is demonstrating. Behind the table is an erect skeleton for all to see, since for Vesalius the purpose of anatomy was the understanding of the *living* body; and he was quick to point out that though in the process of dissection the bones are reached *last*, nevertheless the fabric of the human body must be viewed in relation to that framework which is the foundation of its strength, the means of its motion, and which in many ways dictates the course of vessels and nerves. Near the table sits a figure with pencil and paper; this figure, some have thought, represents his countryman, Calcar, who had come to Venice to study under Titian. The presence of Calcar, if it is he, is perhaps explained by turning the pages of the *Fabrica*; for thus are brought to view the magnificent plates revealing the structures laid bare by dissection. Anatomy and the plastic arts had recently drawn together to their mutual benefit. One has only to compare the muscular athletes of Michael Angelo (1500 ca.) with the delicate but flat-bodied figures of the early masters to realise that anatomy

PLATE V

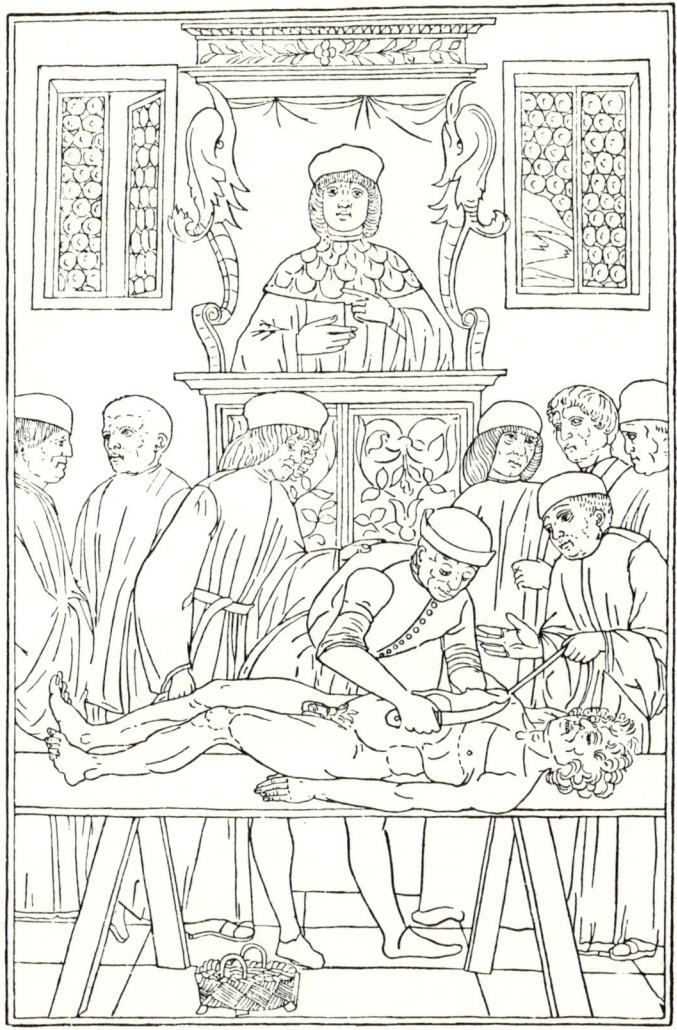

Dissection scene from the *Fasciculo di Medicina*, Venice, 1493. It represents an academic anatomy at Padua University. The Professor, in academic dress, reads his text from his "chair", while a menial performs the dissection, directed by a "demonstrator" in cap and gown.

(and perspective) had made great strides before Vesalius. This fact and the great improvement in the making of wood blocks undoubtedly gave Vesalius a great advantage. In fact there is evidence that his book sold in comparatively small numbers, whereas a condensed version (the so-called *Epitome*) issued separately together with the plates, was a best seller. Be that as it may, there is no doubt that anatomy, hence the idea of the living body, was never the same after Vesalius had breathed life into the dead members. The importance of the *Fabrica* for the history of science lies not so much in its correction of the errors of Galen (which task Vesalius by no means completed) as in the fact that in the words of the latest commentators of Vesalius, Dr. C. Singer and Mr. C. Rabin, " it opens a new scene as with the quick rise of a curtain, for it is suddenly, essentially, and brilliantly modern —modern in appearance, modern in outlook, modern in method, modern in its art and in its technique. And modern, too, in what it omits, no less than what it includes. It is a startling apparition in the very midst of that imitative world of the revival of classical learning which still pervades its language. To appreciate at all the magnificence of the great adventure of the *Fabrica* we must know something of its starting-point. " The starting point was an earlier work, published in 1538 without a title, but generally known thereafter as the *Tabulae anatomicae sex*. It consisted merely of six annotated drawings. Of these, three rather inferior ones were of the skeleton and were stated by Vesalius to be the work of Calcar; technically they are little in advance of an earlier one included by Charles Estienne in his *De dissectione partium corporis humani* (1545—the plate being dated 1532). The other three were, so to say, anatomical charts of the veins and arteries drawn by Vesalius himself in a form not previously used. The authors of *A Prelude to Modern Science* regard it as highly unlikely that Calcar drew the majority of the figures of the *Fabrica*, which are for the most part much superior to the skeletons of the *Tabulae*. If it was not Calcar, it is impossible to say who it was; for though in the *Tabulae* Vesalius had shown himself to be a highly competent *draughtsman*, we have no evidence that he had such talent as the *artist* who wrought so finely in the *Fabrica*.

The last word on the problem of the swift birth of modern

observational science has by no means been said. Vesalius was not the *first*, as he himself claimed to be, to " put his hand to the business " in the matter of dissection; he was not the first to use anatomical figures; he was not the first to systematise the nomenclature of the organs—this he learnt from Sylvius in Paris; but he was nearly the first to do each of these things, perhaps actually the first to do them all simultaneously; and what is much more important, to " photograph " the body itself on to a diagram. Henceforth the disputes would not be about the *opinions* of medical authorities, but, if at all, about the *facts* of nature. Such illustrations as were used before his day—and such great men as Fernel frowned on their use—were literally illustrations. After Vesalius they became the definitive charts of every biological structure. One other point is worthy of mention, namely that he brought about the marriage of learning and craft; for it was not only the printing press and the block-maker that he called in to his aid; by observing the tools in use by craftsmen of various callings he effected great improvements in the design of instruments.

The revolution in anatomy hardly touched the Galenic physiology; such change of viewpoint as there was gave rise to a brief flowering of a chemical interpretation of life—a flowering which as we have seen produced many weeds. But with the advance of mechanical devices the subtleties and confusions of chemistry were largely ignored by the wiser thinkers, and the cardinal problem of the understanding of the living organism was recognised to be in the means of motion.

## SOURCES FOR CHAPTER XXV

Book I of Aristotle's *Parts of Animals* should be read to get an insight into his method and idea of *cause*. (Loeb trans., 1937).

The brief sketch of Graeco-Roman Medicine given above is based on: A. J. Brock's *Greek Medicine* (Library of Greek Thought, London, 1929); J. L. Heiberg's *Science and Mathematics in Classical Antiquity* (London, 1922).

Aristotle's views on the nature of life may be gathered from *The Generation of Animals*, which contains a full contents-summary (London, 1943).

The Introduction to A. J. Brock's Loeb translation of Galen's

# WHAT IS LIFE?

*On the Natural Faculties* is a valuable picture of the man himself and of his place in the history of science (London, 1916).

The General Introduction to W. H. S. Jones's Loeb translation of Hippokrates (several vols.—1st, London, 1923) provides a guide to historical method in the study of the history of medicine.

Recent works on the medieval tradition and the birth of modern medicine are:

G. W. Corner, *Anatomical Texts of the Earlier Middle Ages* (Washington, 1927).

H. Cushing, *A Bio-Bibliography of Andreas Vesalius* (New York, 1943).

C. Singer and C. Rabin, *A Prelude to Modern Science* (the *Tabulae Anatomicae Sex* of Vesalius) Camb. 1946.

C. S. Sherrington, *The Endeavour of Jean Fernel* (Camb. 1946). Fernel was the first to use the term " physiology " in something like its modern sense. Sir Charles Sherrington makes a very careful estimate of his almost unique position in the history of science.

## CHAPTER XXVI

# ORGANISMS AS MACHINES

"As we are about to discuss the motion, action, and use of the heart and arteries, it is imperative on us first to state what has been thought of these things by others in their writings, and what has been held by the vulgar and by tradition, in order that what is true may be confirmed, and what is false set right by dissection, multiplied experience, and accurate observation."

Apart from the word "vulgar" this might be the opening paragraph of a modern text book and not of the *Exercitatio Anatomica de Motu Cordis et Sanguinis in Animalibus* of William Harvey, published in Frankfurt in 1628. The necessity of studying this work in detail has apparently nothing to do with the question "What is Life?" But though such contributions as Harvey made towards the solution of this problem are contained not in the *Exercitatio* but in his only other major work *De Generatione*, it is the former that has placed him among the immortals and altered the whole approach to the problem of life. We are not concerned therefore so much with the actual *circulation of the blood*, a fact sufficiently familiar to the modern reader, as with the method by which it was arrived at. The basis of this is stated in the above excerpt, confirmed by the famous sentence from the Dedication: "I profess both to learn and to teach anatomy, not from books but from dissections; not from the positions of philosophers but from the fabric of nature." With Harvey then the reign of Authority is closed (but see p. 346), much more so than with Vesalius. But we should stray widely from the truth if we were to suppose that the spirit of Aristotle had been exorcised. On the contrary the method of Harvey is in essence the method of Aristotle, but applied to the "cultivation of the narrow plot of ground" and with a caution and economy taught by two thousand years of error. Moreover, what turned his thoughts to

## ORGANISMS AS MACHINES

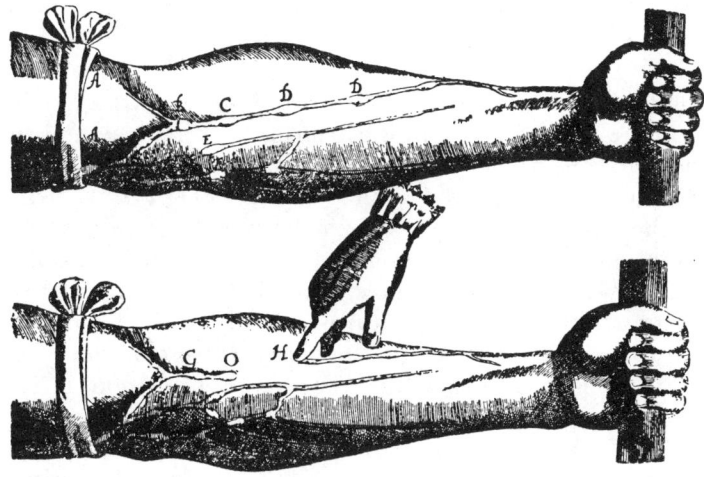

Fig. 33. Harvey's demonstration of the valves in the veins of the forearm.

the problem of the movement of the blood was the demonstration by Fabricius, his teacher at Padua, of the valves in the veins—structures which seemed to have no *purpose* on the current theory.

It is one of Harvey's great merits that he was both fully acquainted with the traditional views on the "action and use of the heart and arteries", and at the same time ready to expose the errors contained in them. This was all the more difficult, since there were many rival theories which were not even internally consistent. It will suffice for our purpose to single out, as Harvey did, the three main causes of confusion which were: (1) that the function of the arteries is to distribute air (Erasistratos) or "vital spirits" (Galen); (2) that the arterial pulse is due to an *innate* faculty of contraction in their tunics; and (3) that the right ventricle has no other function than to "nourish" the lungs. Harvey's reply to the first of these is to remind his readers of Galen's experimental proof of the existence of blood alone in the pulmonary artery, and to point out that blood is no less blood for carrying "spirits". In this reply Harvey was hampered by the confused state of contemporary chemistry in which, despite Van Helmont's work, no clear idea of a *material* gas seems to have been entertained. It is a testimony to Harvey's genius that he remained true to his ideal of not going beyond the facts. The arteries contain *blood*—no question of that, though he points out that *after death* " we find so large a quantity of blood in the veins, so little in the arteries . . . circumstances which probably led the ancients to believe that the arteries . . . contain nothing but spirits during the life of the animal ". But " whether or not the heart, besides propelling the blood . . . adds anything else to it—heat, spirit, perfection—must be inquired into by and by, and decided upon other grounds ". Exactly why, and when, did natural philosophers first throw off the ancient ambition to explain everything at once, and by concentrating on one essential create the spirit of *modern* science? The transition is clearly complete in Harvey with all his hangover of " spirits, heat, and perfection ". But the complete answer to the question lies in a volume of the history of science which has still to be written.

To the second of the cardinal errors Harvey replied with a mastery made possible by the fact that *mechanical* processes were far better understood than chemical. Though it is doubtful how much Harvey knew about the formal laws of hydrostatics, the fact that Stevinus had laid the basis of this branch of science before Harvey was born is evidence that the working of pumps and water supply was becoming a familiar matter. Harvey was thus able to regard the heart and arteries as a pump subject to the limitations of any other kind of pump. Now it had been recognised from the time of Aristotle that the heart and arteries *beat* simultaneously, from which it had always been assumed that they *expand* simultaneously. But, asks Harvey, " how can two bodies mutually connected, which are simultaneously distended, attract or draw anything from one another? " Reason makes such a supposition absurd; and observation of the living heart of a snake or a fish, especially when it is slowing down with the approach of death, makes it clear that the dilatation (diastole) of the arteries *succeeds* the *contraction* (systole) of the heart (ventricle). Moreover, the spurt of blood from a punctured artery follows its *diastole*. " The arteries dilate because they are filled like bladders or leather bottles; they are not filled because they expand like bellows."

In his onslaught against the ancients because they believed that the right ventricle is concerned only with the nutrition of the lungs, the left with obtaining " spirits " from the aorta, he turns the method of their master, Aristotle, against them with consummate skill. His main concern is with the cuspid valves between auricles and ventricles and the semilunar valves at the bases of the two great arteries. In seeking what each is *designed or*, he notes their actual structure and presumed mode of operation. Like cross-examining counsel he fires a broadside of questions at the supporters of the ancient doctrine, driving them rhetorically from corner to corner up to the famous outburst: "Above all, how can they say that the spirituous blood is sent from the pulmonary veins by the left ventricle into the lungs without any obstacle to its passage from the mitral (cuspidate) valves, when they have previously asserted that the air entered by the same vessel from the lungs into the left ventricle, and have brought forward these same mitral valves as obstacles to its retrogression? Good God! How should the mitral valves prevent the regurgitation of air and not of blood? " The diatribe is by no means finished at that point; he no longer, like Vesalius, " marvels at the fineness of the pores " in the septum of the heart which makes them invisible, but flatly denies their existence. We pass over the remainder of his objections to the final words of the introduction where he proposes " to contemplate the motion of the heart and arteries, not only in man, but in all animals that have hearts ".

It is impossible to do justice to the thoroughness with which Harvey establishes by comparative study of a great variety of animals that (1) the heart is composed of muscle whose contractions conform to those of muscle in general but, being hollow, is *lengthened* when these circular muscles contract; (2) the arteries, whether pulmonary or systemic, are filled with blood " like the fingers of a glove " whenever the ventricles contract, which they do together; (3) the blood from the right ventricle passes through the substance of the lungs and is thence carried by the pulmonary veins to the left ventricle; (4) the contraction of the two auricles is simultaneous and precedes that of the ventricles. The third of these positions was of course by no means novel; it was implicit—as Harvey shows—in Galen's account of the valves, explicit in the writings of Servetus (*Restitutio Christianismi*, 1553) and Columbus (*De Re Anatomica*, 1559). But since all the copies but three of his books were burnt with Servetus it is unlikely that Harvey had ever heard of his account. Harvey mentions Columbus, and indeed claims nothing revolutionary in urging the acceptance of the *pulmonary* circulation. " But what remains to be said upon the *quantity and source* of the blood which thus passes is of so novel and unheard-of character, that I not only fear injury to myself from the envy of a few, but I tremble lest I have mankind at large for my enemies . . ." (italics mine). All the preparatory reasonings from design and purpose were Aristotelian in character; but Aristotle would not have worried about the *quantity* of the blood; it is Harvey's particular glory to have started the custom of asking such questions in biology—to have brought the weapons of physics to aid in the revelation of living function. His concern for his reputation was little exaggerated. John Aubrey puts it on record that after the publication of the theory, Harvey's practice fell off and he was regarded as " crack-brained ". What was this unheard-of folly?

> I frequently and seriously bethought me and long revolved in my mind what might be the quantity of blood which was transmitted, in how short a time its passage might be effected, and the like; and not finding it possible that this could be supplied by the juices of the digested aliment without the veins on the one hand becoming drained, and the arteries on the other getting ruptured through the excessive charge of blood, unless the blood should somehow find its way from the arteries into the veins, and so return to the right side of the heart; I began to think whether there might not be A MOTION AS IT WERE IN A CIRCLE.

This great result is a piece of pure deduction based on circumstantial evidence. Harvey never *saw* the circulation of the blood, could only guess at the existence of the capillaries—the first recorded observation was by Malpighi in 1661. But he could and did estimate the *least possible quantity* of blood shot irreversibly from the left ventricle into the aorta in an hour. The remainder of the book is taken up with the review of a great many further arguments—the results of ligatures of varying strengths, the rapid spread of poisons and of drugs, etc., many of which are, as Harvey says, *a posteriori*, that is, more readily explained by the assumption of a circulation than by any other means. So far as the *mechanical* aspect of the circulation is concerned, we feel, on laying Harvey's book down, as we feel after reading Ampère's work, namely, that it is perfect in design and in conclusion, and that it has established for ever on a sound basis a new department of science. But living as he did at the birth of modern scientific method, being himself one of its creators, he is unable to throw off the wordy inanities of scholasticism. In the absence of chemical *knowledge* he indulges in transports of high-flown fancies. The blood having become in the body " cooled, coagulated, and, so to speak, effete; whence it returns to its sovereign, the heart . . . here it resumes its due fluidity and receives an infusion of natural heat, powerful, fervid, a kind of treasury of life, and is impregnated with spirits, and it might be said with balsam; and thence it is again dispersed; and all this [!] depends on the motion and action of the heart." This is quoted in no depreciatory spirit of the genius of Harvey, but only to show how the greatest of thinkers are apt to be the victims of their intellectual circumstances. The ideas of science grow in mutual interaction, vigorous young shoots being often entangled in old and decaying wood.

Harvey first taught this doctrine in his lectures at the Royal College of Physicians in London, probably in the year 1616. A few of the more enlightened of his colleagues persuaded him to publish the *Exercitatio*. This took place in 1628. But it was perhaps thirty years or more before the medical profession were ready to accept it. One keen spirit, however, so far from regarding the author as crack-brained, paid him the compliment of including a summary of Harvey's teaching in his own epoch-making book. This was Descartes, who admitted that to Harvey " we must give the praise for having broken the ice in this matter ". It is hardly necessary to say here that Harvey had not merely " broken the ice " but had cleared away all obstruction to just views on the subject, and in other places Descartes himself gave Harvey the undoubted credit for having made a discovery of the greatest importance to medicine. There is no doubt that, contrary to the conventional view, Descartes' activity was not restricted to metaphysics and mathematics; he relates in his less well-known book, *Le Monde*, how many hours he spent at the *abattoirs* dissecting viscera. In the Fifth Part of the *Discourse on Method* he uses Harvey's discovery (doubtless confirmed by his own observations) as an illustration of the *mechanical* nature of living beings as such. Thus he writes: " the movement I have just explained comes as necessarily from the particular disposition of the organs which can be seen in the heart by the eye, and from the heat which can be felt with the fingers, and from the nature of the blood which can be known by experiments, as does that of a clock from the force, the situation, and the shape of its balance and its wheels." Moreover he sees in the heart " one of these fires without light which I had already explained, and which I conceived to be of the same nature as that which kindles hay when it is shut up before it is dry, or which makes new wines boil when they are left to ferment on the stalk ". Thus he repeats the age-long error of supposing that the heart *alters* the blood. And worse than this, he put forward a mechanical theory to explain the movement of the heart itself (as distinct from the circulation of the blood) which ran counter to the facts as elucidated by Harvey. Professor Gilson sees in this the failure of Descartes to rid himself of the phantoms of scholasticism even in the effort to abolish them for ever.[1]

Descartes' lengthy study of this subject advanced biological knowledge not at all, but it had the profoundest effect on the growth of biological ideas, which henceforth took place mainly in an atmosphere where mechanical modes of thought were dominant.[2] As the century wore on a counter movement arose (Stahl); and so the stage was set for the battle of vitalists and mechanists, the dust of which has scarcely yet subsided. Nor has this been a merely technical matter, for Descartes did not accept the mechanical working of the circulation of the blood as a result, to which any student of nature must be driven by an impartial review of the facts; he embraced it as an articulated demonstration of the dogma which vitiated his whole philosophy. This dogma was the belief that organisms, including the human *body* differ in no way, except as regards complexity, from the

[1] The matter is too long even to summarise here, but forms a fascinating study in historical influences. The article occurs in *Etudes sur le Rôle de la Pensée Mediévale dans la Formation du Système Cartesien.*
[2] But see later, p. 385.

machines made by men; so that animals, being devoid of a rational soul, are mere *automata*. The dire consequences of this view, which set the course of European thought, cannot be discussed here. For biology the effect was at first wholly beneficial, as it promoted precision in observation and the purging away of redundant hypotheses; but even here it tended to sharpen the opposition of mechanism and vitalism into two equally untenable dogmas.

We shall consider here first the great advances in the understanding of *movement*. It will be recalled that Galen clearly distinguished the " motor " and " sensory " nerves; it was therefore natural that advances in the theory of movement should run parallel with those unravelling the nervous system. A great advance in the topography of the nervous system was made by Willis in his *Cerebri Anatome* of 1664. Although his mapping of the brain was in some respects confused, his numbering of the twelve pairs of cranial nerves is the same as every medical student uses to-day—a system which is of course one of convenience, as it masks the true relations of the nerves to the developmental units or somites of the body. In regard to the physiology of nervous *action* however the great ones of the seventeenth century forgot all their high resolves to keep within the narrow range of experience and in the absence of any concepts adequate to this exceedingly recondite subject (the electric current was to lie hidden for a century; as also the ideas of potential, and capacitance of condensers) gave imagination rein to luxuriate among the only forces they understood. Thus Descartes follows his accurate anatomico-mechanical description of the heart by the words " what is most remarkable in all this is the generation of animal spirits, which are like a very subtle wind, or rather a very pure and lively flame which, continually mounting in great abundance from the heart to the brain, flows from thence through the nerves into the muscles and gives motion to all the members ". Remarkable indeed; but less so than Willis's phantasmogoria in which the growing " science " of chemistry is called in to give greater force to the theory. As a modern historian of biology, E. Nordenskiöld, has remarked, " When a scientist of Willis's rank can seriously discuss the question whether the vital spirit can be compared with spirit of wine or hartshorn oil, it is easy to realise to what hopeless lengths the natural-scientific speculation of that age could go."

In 1679 appeared one of the great masterpieces of anatomical physiology, *De Motu Animalium*, of Borelli. Great strides had been made during the previous century by Belon (1551), Rondelet (1553-4), Coiter (1572-5), Ruini (1598), Fabricius, and of course Harvey, in discovering the parts of different animals having similar functions, and also those which have the same morphological relationship (e.g. wing of bird and fore limb of mammal), though the term " comparative anatomy " was first used in its modern sense in English by Grew in 1674. In Borelli's work the idea is applied with consummate art to show how different animals move about and also move their parts (e.g. heart); the upshot being to prove that they are—machines. Thus at the outset he says: " Just as it is customary in other (*aliis*) physico-mathematical sciences, we shall attempt to expound this science of the movement of animals from phenomena as our foundations." He is as good as his word; it is unnecessary to quote at length his straightforward description of muscles as consisting of bundles (*fasciculi*) of parallel fibres commonly joined at both ends by means of tendons to bones. And whereas the " nerves, tendons, and membranes suffer no contraction " it is of the nature of muscle to contract. He corrects the ancient false distinction between muscle and flesh, and correctly states that the contraction of muscle is effected by lateral enlargement (swelling) while the total volume remains constant. The properties of muscle having been established, he demonstrates in a series of diagrams how every movement consists of a pair of balanced lever actions, discussing the moment of perpendicular and oblique forces and of the effect of muscle sheets in which the fibres are contracting in different directions relative to the axis of the whole muscle. In addition to the simpler movements he considers problems of balance, the running of quadrupeds, the perching of birds, and the power of the stomach muscles of a turkey to crush glass marbles. It is, in fine, a text book of what Sir Arthur Keith has aptly called " The engines of the body." In the second volume he deals with the internal movements, for instance, the heart, and gives the explanation of breathing accepted to-day, namely the filling of the lungs by the elasticity of the air and the enlargement of the thoracic cavity.[1] Finally he reviews and in many ways corrects the prevailing opinions on many points of physiology, e.g. nutrition, not obviously connected with motion (see p. 430). But to the historian perhaps the most interesting part is the section which deals with the *cause* of motion of the muscles. In a writer who introduces anatomical physiology as by implication a mathematico-physical science, it is not surprising to find the " animal spirits " characterised as a " very subtle self-moving *fluid* ". Having rejected the pressure of blood or air as the motive power, his own hypothesis, extravagant as it may seem to us, is ingeniously argued and stated with scientific caution, namely that whatever it is, and we cannot see it, which is the immediate cause of the contraction of muscle fibres, it is transmitted from the brain along the hollow nerve fibres. Moreover this transmission is worked on the hydrodynamic analogy of a tube filled with liquid, from whose further end a quantity may be squeezed out by a light pressure at the near end. He even puts forward a speculative scheme whereby the brain through repeated over-all action on the near ends of the tubes " learns " by trial and error to squeeze only the appropriate tubes, as may be inferred by watching the blundering efforts of small children to walk. The *mechanism* is crude, but at least consistently linked to known—but here inappropriate—causes; the *form* is just, if over simplified, since with a little goodwill we can recognise our specific motor branches with their localised action and also the rudiments of final common path and facilitation. Borelli was one of the great pioneers of physiological anatomy; his imprint is everywhere; his office has been so universal that his name is never mentioned in the texts upon which our medical education is founded!

Application of the new mechanics to the problems of plant growth was naturally slower to come. Its chief exponent, Stephen Hales, was one of the outstanding curiosities of that remarkable age. An exemplary parish priest, punctual in the discharge of his duties to the poor, he nevertheless found time to carry out in his garden at Teddington a series of investigations, in the execu-

[1] Mayow's *De Respiratione* (1668) had included a more accurate account, as might be expected from one who was studying at the time when Boyle and Hooke were demonstrating artificial respiration by means of pumps.

tion of which he brought to bear a resource in manipulation and a critical insight in interpretation such as to give him an importance above the enthusiastic but often ill-conceived enquiries of most of the Fellows of the Royal Society. The measure of the man may be gained from the following extract from his *Vegetable Statics*, published in 1727, under the *imprimatur* of Isaac Newton. Having referred to the observation of Grew and Malpighi on the *structure* of the vessels of plants he says:

> Had they fortuned to have fallen into this statical way of enquiry, persons of their great application and sagacity had doubtless made considerable advances in the knowledge of the nature of plants. This is the only sure way to measure the several quantities of nourishment, which plants imbibe and perspire, and thereby to see what influence the different states of air have on them. This is the likeliest method to find out the sap's velocity and the force with which it is imbibed: As also to estimate the great power that nature exerts in extending and pushing forth her productions by the expansion of the sap.

This " statical way of enquiry " had been applied by him to animals about " twenty years since " in order to find the pressure of the blood in horses, dogs, and deer. At that time he had despaired of being able to discover the pressure in plants until " about seven years since, by mere accident I hit upon it, while I was endeavouring by several ways to stop the bleeding of an old stem of a vine which was cut too near the bleeding season . . . Having . . . tyed a piece of bladder over the transverse cut of the stem I found the force of the Sap did greatly extend the bladder, whence I concluded, that if a long glass tube were fixed there in the same manner, as I had before done to the Arteries of several living Animals, I should thereby obtain the real ascending force of the Sap in that Stem, which succeeded according to my expectation . . ." It is interesting to note that this country clergyman was not driven to mutilate his vines, cherry, apple and quince trees, or with complete callousness to tie down live horses for haemastatical experiments while he attended Divine service, from a disinterested desire for knowledge of the wonders of nature. "As the art of Physic has of late years been much improved by a greater knowledge of the animal oeconomy; so doubtless a farther insight into the vegetable oeconomy must needs proportionably improve our skill in Agriculture and Gardening." The last

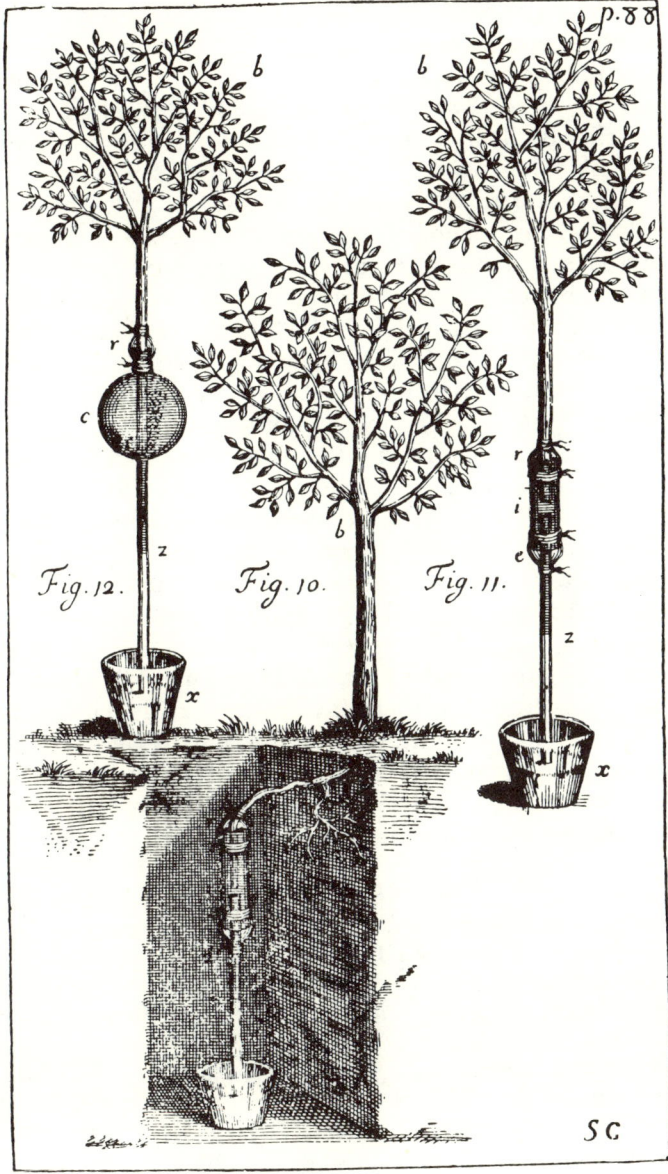

FIG. 34. Stephen Hales' experiments to show the mechanical nature of the movement of sap.

paragraph of the book might be taken as a text for a sermon on the function of agricultural research associations.

It is impossible to quote further from this absorbing book. We are not here concerned with detail. But when it is noted that he clearly distinguished root pressure as the dominant force in the " bleeding season " from the " pull " of " perspiration " as he calls it, of the active leaves; that he observed the possibility of the lateral movement of sap from one branch to another, and the presence of large quantities of air in the vessel; and when finally he rejected the plausible theory of " circulation " as unlikely; we may fairly call him the " father of plant physiology ". In his experiments to " Find in how great a proportion Air is wrought into the composition of Animal, Vegetable, and Mineral Substances ", though his observations are full and accurate, he is handicapped in his interpretation by the absence of any clear concept of the chemical individuality of the different kinds of " air ", and by too great subservience to the views of the " Illustrious Sir Isaac Newton " whom he constantly quotes. He was convinced of the great importance of "Air " in the life of plants, both by " absorption " and " generation ", but he was unable to break through the alchemistical hangover of " sulphureous acid vapours", etc., to a greater precision of statement than that "Plants very probably draw thro' their leaves some part of their nourishment from the Air ". The nature of his experiments on every conceivable animal and plant material, with no guiding hypothesis as to their " way of life ", illustrates the limitations of the " realistic " approach, for the " improvement of Agriculture and Gardening ".

## SOURCES FOR CHAPTER XXVI

W. Harvey, *Anatomical Disquisition on the Motion of the Heart and Blood in Animals* (Everyman Library).

R. Descartes, *Discourse on Method*, pp. 37-43 (Everyman Library).

J. A. Borelli, *De Motu Animalium*, Lugduni in Batavis (1685).

S. Hales, *Vegetable Staticks* (London, 1727).

Sir Charles Sherrington's *Jean Fernel* gives a very clear idea of the current views of the action of the heart and blood vessels, and towards the end of the main section of the book (pp. 140-6) he contrasts Fernel's use of facts (which he respected with the freshness of mind characteristic of the Renaissance) with Harvey's, only three generations later.

## CHAPTER XXVII

# THE MICROSCOPE AND THE ORIGIN OF LIFE

WITH Borelli and Hales the first stage of the quest for the manner of working of the living organism came to an end. But we may recall that Aristotle set the problem of life in a two-fold form—the working of the parts and the origin of life itself. To the second part of the great quest Fabricius addressed himself with such success that he is generally regarded as the founder of embryology. In his *De Generatione Animalium* (1651) Harvey confesses to using Fabricius as his informant, Aristotle as his guide: the result is a historical document rather than a scientific classic. Though scattered about through this very long, rambling, and repetitive book are passages in which the acute anatomist, whose ambition it was to " learn and teach from dissections ", was able to correct the errors of his forerunners, he is clearly too uncertain of himself to throw off his guides and thus rid himself of the temptation to wander off into metaphysical by-paths. In the *Exercitatio* Harvey applied the Aristotelian *method* to interpret stubborn *facts*; in the *De Generatione* he is constantly attempting to frame concepts from *words*, as when he quotes Aristotle as having defined the relations of seed and fruit as prior and posterior " for the fruit is that which comes from another, seed is that from which another comes "; and the egg as a " mean between the animate and the inanimate, for neither is it wholly endowed with life nor is it entirely without vitality ". When one calls to mind that *life* is the ultimate *problem,* one may easily see that such aphorisms merely obscure the immediate problem of relating the *egg* to the *hen*. Nevertheless there does emerge from this fog of words the categorical assertion that *An egg is the Common Origin of All Animals*, for " whether they arise spontaneously, or from others, or in others, or from the part or excrements of these (they) have this in common, that they are engendered from some principle adequate to this effect, and from an efficient cause inherent in the same principle. In this way therefore the primordium from which and by which they arise is inherent in every animal . . . Such primordia are the eggs of animals and the seeds of plants ". This is in effect the definition of a " germ "—whether it be egg, seed, or vegetative " bud "; it is also an implicit denial of *real* " spontaneous generation ", experimental proof of which had, as we shall see, to wait for more than two centuries.

When it comes to explaining the *origin* of the embryo his vacillation seems criminal. He is certain that neither cock nor hen is sufficient by itself; he is almost certain that their contributions are of equal potency; but he cannot bring himself to admit that the semen has any *material* effect, since " though it is a mixture of some sort that renders an egg fruitful, still this does not happen according to the predominance of the genitures, or the manner of their mixture, for it is certain, and Fabricius admits it, that the semen of the cock does *not reach the cavity* of the uterus; neither is there any trace of the egg to be discovered in the uterus immediately after intercourse . . ." (italics mine). How necessary it is to put oneself in the place of each investigator and try to see things exactly as he did! *To the unaided eye both these statements are literally true.* The union of the *sperm* with the egg is so simple a matter to *read about*, but how difficult to observe —even with a good microscope, a thing non-existent in Harvey's day! So his vacillation turns out to be the scientific restraint of one unwilling to go beyond the *observable*. Yet restraint does not go far enough to admit blank ignorance of *how* the cock's concurrence is essential to the transaction, so he looks back to Aristotle to hallow the action of some " effluvium."

On the progress of differentiation he put forward an opinion which had the most baneful influence on subsequent workers: " Now the common instrument of all vegetative operations is in the opinion of all men an internal heat or *calidum innatum* or a spirit through the whole and in that spirit a soul or faculty of a soul, the egg therefore . . . has its own operative soul which is all in the whole and all in each individual part and contains within itself a spirit or animal heat, the immediate instrument of that soul." It is doubtful what Harvey meant by all this—in another place he plainly refers to God as the cause—but we can see in it an excuse for the *entelechies* of later workers. In another place he states that " the blood is that in which the vital principle (*anima*) has its dwelling place." He bases this belief on the observation of the tiny red spot which leaps into view and again disappears in the early foetus. To him it appeared as *self-moving* blood. It takes careful preparation even to-day to render visible the containing muscular membrane whose contraction and relaxation is the cause of the " motion " of the blood.

Here then let us leave him, as Professor F. J. Cole has called him " a very puzzled man "; but before doing so let us recall that at the end of the book, though probably based on much earlier observations, is an accurate and full account of the embryology of the deer, for whose dissection at every stage he had unrivalled opportunities, as it was the custom to hunt the doe at a time when they bore within them the fruit of the autumn mating.

Nearly two centuries were to pass before the observation of the mammalian ovum on the one hand, and the subsequent development of microdissection on the other, were to transfer the question of the origin of animals from theology to biology. But a late contemporary of Harvey, Francesco Redi, called in question the Aristotelian dogma of spontaneous generation a few years after his fellow-countryman, Galileo, had the false dogma of circular motion.

In the sixteenth century so astute a man as Van Helmont could deceive himself into believing that he had brought about the spontaneous generation of mice (which even Aristotle would have doubted) by a fantastic recipe of damp flour and soot. The more usual view seems to have been that the earth had become too exhausted (cf. Lucretius) to produce anything more elaborate than worms and maggots. Redi however denied even this. What led him to deny this view in defiance of " common sense " ? One can only suppose that he applied in a particular direction that principle so generally accepted in the seventeenth century, thus expressed in his own words: " I have given myself all possible trouble and have taken the greatest care to convince myself of *facts with my own eyes by means of accurate and continued experiments before submitting them to my mind as matter for reflection* " (italics mine). Having observed that three dead snakes placed in an open box became covered with " worms ", he carefully closed the box in order to determine the fate of the latter. The fact that from the *same meat* there were produced after pupation two or three *different kinds of flies* made him suspect that the meat was not the sole factor in the generation. With glorious prodigality, such as to cause a sigh in the modern reader, he " made ready six boxes without covers " and " placed in the first two of the snakes, in the second a large pigeon, in the third two pounds of veal, in the fourth a large piece of horseflesh, in the fifth a capon, in the sixth a sheep's heart; and all became wormy in a little more than twenty-four hours. The worms, five or six days after birth, changed as usual to eggs " (*uova*—we should call them *pupae*) from which in due course there emerged a variety of flies. He noticed also the eggs from which the worms hatched. The experiments were repeated with raw and cooked flesh of the most various kinds, always with similar results.

Having considered these things [he continues] I began to believe that all worms found in meat were derived directly from the droppings of flies, and not from the putrefaction of the meat, and I was still more confirmed in this belief by having observed that before the meat grew wormy, flies had hovered over it, of the same kind as those that later bred in it. Belief would be vain without the confirmation of experiment, hence in the middle of July I put a snake, some fish, some eels of the Arno, and a slice of milk-fed (!) veal in four large wide-mouthed flasks; having well closed and sealed them, I then filled the same number of flasks in the same way, only leaving these open. It was not long before the meat and the fish, in these second vessels, became wormy and flies were seen entering and leaving at will; but in the closed flasks I did not see a worm, though many days had passed since the dead flesh had been put in them. Outside on the paper there was now and then a deposit, or a maggot that eagerly sought some crevice by which to enter and obtain nourishment.

Here is a biological experiment in modern dress, complete with a *control* set of apparatus; the " milk-fed veal " alone suggests the artistic touch of more spacious days. Moreover Redi recognised that the control was not an exact copy of the experiment as " the trial had been made with closed vessels into which the

laid the basis of minute anatomy both in plants and animals, but his excursions into comparative speculation were largely responsible for our heritage of such unsuitable terms as the " ovary " of a plant, and of others far worse, which have happily been expunged.

In our study of the search for the nature of life we must confine ourselves to a glance at the life and work of Leeuwenhoek, linen draper and later Chamberlain of Delft. This remarkable man, born in 1632, gained at an early age a state of modest affluence to enable him to devote his powers to his noble hobby—that of opening up with the aid of microscopes of his own contriving a whole new universe of living creatures, whose little worlds were raindrops, grains of dust, and the organs of larger animals.

In 1673 Oldenburg, Secretary of the Royal Society, received from de Graaf a letter introducing some observations " by a certain most ingenious person here, named Leeuwenhoek (who) has devised microscopes which far surpass those which we have hitherto seen ". The rather crude illustration of the sting of the bee and other curiosities so pleased the gentlemen of the Royal Society that they asked for more. So began that wonderful series of letters from the Father of amateur microscopists. In 1674 he describes how in a drop of water from a lake he saw " green streaks, spirally wound serpentwise . . . Among these there were besides very many little animalcules whereof some were roundish, while others a bit bigger consisted of an oval. On these last I saw two little legs near the head and two little fins at the hindmost end of the body . . . others were again green in the middle and before and behind white . . . and I judge that some of these little creatures were about 1,000 times smaller than the smallest ones I have ever seen upon the rind of a cheese, in wheaten flour, mould, and the like." In the famous Letter XVIII he describes organisms with such precision that they may be certainly identified as *Vorticella, Stylonichia, Colpidium, Monas*, and others. Bacteria were first observed in an infusion of peppercorns. Of a mass of *Spirilla* he says, " This was for me among all the marvels I have discovered in nature the most marvellous of all . . . no more pleasant sight has ever come before my eyes than these many thousands of living creatures all alive in a little drop of water." In human faeces he saw the first recorded protozoon inhabiting the gut of another animal; in the contents of the frog's rectum he saw a quite different species with a similar mode of life. In the sediment between his own teeth he observed several different kinds of bacteria. As Professor Dobell admits (from whose beautiful memoir, *Antony van Leeuwenhoek and His Little Animals* all the above facts are taken) there was not one branch of modern microbiology to which he himself turned his attention but Leeuwenhoek had already trodden the path. His influence was twofold: it was his opening up of new worlds within the known world that almost certainly led Leibniz to say in his *Monadology* (1714): " We see that a world of creatures, living beings, animals, entelechies, souls, exists in the minutest parts." Leibniz himself took no part in the fruitless controversy which raged throughout the eighteenth century, namely whether the sperm or the egg was the means whereby life was handed on. It was Leeuwenhoek's discovery of the spermatozoon in human and animal semen which provided the " evidence " for the former, whereas de Graaf's discovery of the Graafian follicles of the mammalian ovary (which he misconceived as being the equivalent of the bird's " egg ") seemed to give equally strong confirmation of the latter. What is important in regard to this controversy is not to take sides on a dilemma which arose simply as a result of the lack of a technique adequate to the problem, but to estimate the truth of the " encasement " hypothesis to which it inevitably led. For if every animal contains within itself a *perfect replica of itself*, which the enthusiasts of the infant science of microscopy were only too ready to assert, then this replica must contain a further replica, and so *in infinitum*. This was evolution in its literal sense: " There is neither absolute birth nor complete death in the precise meaning of the separation of the soul and body. What we call *births* are developments and unfoldings, and what we call *deaths* are foldings and shrinkages." Even if the world seemed a much younger thing than it does to us (Archbishop Usher in the seventeenth century had fixed the creation at B.C. 4004), even so the imagination boggles at the Universal mother Eve containing within her body a geometric and apparently infinite series of men and women in embryo.

Crude as the hypothesis was, it is not quite so unimaginable as might first appear, for had not the microscope shown that " each portion of matter may be conceived as a garden full of plants and as a pond full of fish. But every stem of the plant, every limb of the animal, every drop of sap or blood is such a garden or pond." Here Leibniz is clearly influenced by the discovery of bacteria within living bodies, of the blood corpuscles within the blood itself. So while we must reject as crude and unscientific both ovist and spermist doctrines as such, we should reverence Leibniz for his sagacity in seeing the all-in-allness, the intimate and far-flung *organisation of nature*: " Thus there is nothing arid, sterile, or dead in the universe, no chaos, no confusion, save in appearance; exactly as a pond would appear to us at a distance were we able to see only the confused movement of the swarming fish and not the fish themselves."

The views of Leibniz were popularised in a much simplified form by the great French naturalist and stylist, who could write only when caparisoned in all the elegance of silk, lace, and braided wig, Jean, Compte de Buffon. His *Histoire Naturelle* (started 1749) marks the close of an epoch in biology. In it he taught that the " seeds of life " are everywhere, and are at all times being absorbed by living things. The " residue ", as it is secreted by testis and ovary, provides the embryos of the next generation. In this way he avoided the crudities of the preformation hypothesis.

Meanwhile the more extended use of the microscope had re-opened the question of the *origin* of life. Leeuwenhoek himself was dogmatically opposed to the belief in spontaneous generation —dogmatically in that he had demonstrated the existence of *living* bacteria in infusions contained within exhausted tubes— evidently what we now know as anaerobic bacteria. But the English priest, John Needham, was not so easily satisfied. In 1748 he published a report of experiments done in conjunction with Buffon in which he sought to find out whether *microscopic organisms* could not arise spontaneously. He boiled mutton broth and sealed the flask with mastic in order to exclude the possible entrance of air-borne germs; but after a few days the broth was swarming with " animalcules ".

With knowledge of micro-organisms at such a low level there was really no *evidence* upon which to base belief one way or the other. Only to an investigator dogmatically opposed to spontaneous generation would it have appeared worth while to repeat Needham's experiments. Such a one was Lazzaro Spallanzani, one of the founders of experimental biology. A naturalist of a high order, he realised from his own observations, and from those of the Swiss, Bonnet, on the birth of aphids from virgin females and the regeneration of the parts of worms, snails, and polyps, that the problem of " generation " was an exceedingly complex one: attention must be paid to the size and nature of the organisms which appeared in the boiled infusions. In his *Observations and Experiments upon the Animalcula of Infusions* he tells us that the number of animalcules varies with the time of boiling; the longer the boiling the more numerous the animalcules. This result might have misled a less careful experimenter; but since after a few weeks all the infusions were equally populated, he rightly concluded that the length of boiling determines the degree of maceration of the seeds used. He then repeated the experiments in the following manner: " I hermetically sealed vessels with the eleven kinds of seeds mentioned before. To prevent the rarefaction of the internal air I diminished the thickness of the necks of the vessels, till they terminated in tubes almost capillary, and putting the smallest part to the blowpipe, sealed it instantaneously, so that the internal air underwent no alteration." Having assured himself that animalcules would develop in this limited supply of air he " took nine vessels with seeds, hermetically sealed. I immersed them in boiling water for half a minute. I immersed other nine for a whole minute, nine more for a minute and a half, and nine for two minutes . . . I examined the infusions and was surprised to find some of them an absolute desert; others reduced to such a solitude that but a few animalcules, like points, were seen . . . I thence concluded that the heat of boiling water for half a minute was fatal to all animalcula of the largest kind." By boiling for three quarters of an hour he deprived all the infusions of animalcules. " We are therefore induced to believe that those animalcula originate from germs there included, which for a certain time withstand the effects of heat but at length yield under it . . . "

Here, in these simple experiments, conducted with thoroughness and manipulative skill, adequately " controlled " and accurately interpreted, are the beginnings of our ideas of the thermal death point, of sterilisation and of food preservation. But they are only beginnings. To the question of spontaneous generation they should have given the final answer. But the issue was to be fought out all over again a century later in a more spectacular setting and with consequences of the profoundest importance.

We have now reached a stage in the history of biology where the problem of what is the nature of life is found to be insoluble in terms of the naïve dilemma of a belief in a *vis formativa* engendering living creatures out of matter, *or*, on the other hand, of organisms being *simple* machines. By " simple " I mean mere mechanisms as conceived for instance by Borelli, however complex these mechanisms may be. During the eighteenth century a new approach to the nature of things had been cleared, free from the weeds of *pseudo*-Aristotelianism, alchemy and phlogistonism: this way was the way of chemistry. Moreover a chemistry, which in the hands of Scheele was beginning to attack the very stuff of living things. As the century drew to a close another powerful force—electricity—was revealed as acting upon both inert and living matter. And lastly, the problems of navigation which had beset the seventeenth century having been largely solved, new lands were yielding up an almost inexhaustible wealth of hitherto unknown forms of life. In the concluding chapters of this book therefore we shall see how four leading ideas grew and took shape —the ideas of form, of function, of historical tracks and, once again in a new setting, the nature of life itself. It is not for a moment suggested that these neatly rounded categories have at all times been the stimulus and the pattern for individual workers. Other factors have played a part and the dividing lines between the categories have never been sharply drawn. But to the author they seem the least artificial divisions into which the subject may be divided to avoid its degeneration into a disorderly mass of personalities and results.

One other idea however has played a very important part in biology and in all sciences at the purely descriptive level, that is the problem of classification and nomenclature. This is a kind of hybrid category; since although it starts at the purely descriptive level, it is impossible for it to remain there. Some ideal scheme, however arbitrary, must be at the back of every classification. And as the problem widens and rival theories as to the most effective system come into opposition, theory and speculation must keep on breaking in. To this problem, that of the systematic arrangement of individual kinds of living organisms, or as it is commonly called, taxonomy, we shall first address ourselves.

### SOURCES FOR CHAPTER XXVII

W. Harvey, *De Generatione Animalium* (Engl. trans., London, 1653).

Francesco Redi, *Experiments on the Generation of Insects* (Engl. trans., Chicago, 1909). The section on the maggots of flies is an excellent example of early experimental method in biology.

L. Spallanzani, *Tracts on the Nature of Animals and Vegetables* (Engl. trans., Edinburgh, 1799).

Clifford Dobell, *Antony van Leeuwenhoek and his " Little Animals "* (London, 1932). A beautiful volume giving an admirable idea of pioneer work in microscopy, which should be sufficient to correct the fiction that European thought was wholly given over to the mechanical interpretation of nature during the seventeenth century. In this connexion see also C. E. Raven's *Synthetic Philosophy in the Seventeenth Century* (Oxford, 1945).

G. W. Leibniz, *Monadology*, trans. of R. Latta (rev. ed., Oxford). Primarily a philosophical work, but important in that it displays one of the greatest mathematicians convinced of the *organic* view of the cosmos—to be developed in its full grandeur by A. N. Whitehead in our own time.

## CHAPTER XXVIII
## WHAT'S IN A NAME?

LIVING things—or at least most of the living things which are likely to have attracted the close attention of the pioneers—are individuals. But the book of Genesis bears testimony to the very early recognition of the fact that individuals belong to many different *kinds*. It is highly probable indeed that the first problem Man had to solve *vis-a-vis* the other living things was to give them names. Such a problem would have arisen in the age of the "food gatherers" before any interest was aroused as to *how* life worked in individuals. Names are based on similarity of form; so those primeval naturalists, immortalised as Adam and Eve in the Garden, who set about naming the kindly fruits of the earth were in a sense the founders of taxonomy—the department of biology devoted to classification. But between the naming of specific kinds and the recognition of *classification* yawns an intellectual gulf. It is as usual to the Greeks that we owe the *idea* of classification.

In the dialogues of Plato known as the *Statesman* and *Sophist* the idea is worked up in considerable detail, but rather for its logical implications than for use. It was Aristotle who took the decisive step. In his *Parts of Animals* he recognises that "we must make up our minds about the method of our investigation and decide whether we will consider, first, what the whole group has in common, and afterwards the individual peculiarities; or begin straightway with the particular instances." The latter procedure would involve a great deal of repetition—would in fact defeat the whole purpose of classification which in the first instance is to achieve economy of thought. Since classification there must be, it remains to decide what form it shall take. For *logical* purposes the only safe method—the one used by Plato in the *Sophist*—is the method of dichotomy, in which any group—

logical " genus "—under consideration is divided into two and only two sub-groups—logical " species "—one of which is and the other is not characterised by a single feature—logical " differentia ". It is Aristotle's chief claim to our regard in the matter of taxonomy that he rejected this method as wholly inappropriate to the kinds of living things. For no matter how carefully the *differentiae* are chosen, there is bound to be overlapping; " flying " would include some mammals, and " not-flying " some birds— and duplication—some *individual* ants fly and others do not. It is also to his credit that he recognised the importance of characters such as the possession of blood, which are overlooked in everyday classification. In the upshot he divides animals into: (*a*) Blooded (Man, viviparous quadrupeds, oviparous quadrupeds, and the footless serpents, birds, and fishes; (*b*) Bloodless (insects, shell-skinned, soft-skinned—the modern cephalopods—and crustacea). He recognised as intermediate forms, whales, seals, bats, ostriches, and the plant-like sponges, etc.

For Aristotle's views on the classification of plants we have no documentary evidence; but fortunately among the works of his co-worker, and successor in the direction of the Lyceum, Theophrastos, is a long book, *Enquiries into Plants*, which gives a very clear idea of the point of view of the school which, we may assume, would not differ essentially from that of its founder. There is however a marked individuality about Theophrastos' work[1]: it is clearly the work of a gardener rather than of a logician or medical man. How else would he have had such an exact knowledge of germination as to give us a clear distinction between seed and fruit (the word $Καρπός$ occurs distinct from $σπέρμα$), between monocotyledons and dicotyledons? And how, except for an expert knowledge of grafting and propagation of cuttings, could he have been led to his most famous saying that " a plant has the power of growth in all its parts, inasmuch as it has life in all its parts "? In regard to a working classification, the mark of his cautious genius is seen in the still useful division into trees, shrubs, under-shrubs, and herbs—and he warns us that these

[1] The *codices* on which our modern text is based are all more or less corrupt copies of the original, long since lost; but there are enough of them to provide internal checks.

PLATE VI

*Sonchus sp.* (Sowthistle) as depicted in the *Codex Aniciae Julianae*

divisions are by no means hard and fast, since a herb like a cabbage may in certain conditions of growth become a small tree. He was evidently puzzled by the fungi which possessed none of those parts upon which the above division is based.

Theophrastos, if not the first, was certainly the last botanist to attempt a natural classification—one, that is, based on essential similarity of form, habit, and habitat—until the Renaissance. But during the Alexandrian epoch of Hellenistic culture the rise of scientific medicine created a demand for a uniform system of *nomenclature*. Our two chief authorities for this activity are the Roman civil servant, the elder Pliny, whose *Natural History* contains an enormous mass of mainly second- or third-hand information, much of it fabulous, as well as some acute observations of his own; and the Asiatic Greek, Dioskurides, who was probably a surgeon in the army of Nero.

Our knowledge of the work of Dioskurides is based on one of the oldest of extant manuscripts—the so-called *Codex Aniciae Julianae* actually written about A.D. 512 and still to be seen at Vienna. This wonderful relic contains the descriptions of about 500 plants of medical importance, presumably copied from the original work of Dioskurides. It is not an exact copy, as it contains a number of beautiful tinted drawings of plants, almost certainly copied from originals executed by an earlier botanist, Krateuas, in the first century B.C. There is a tradition that the original MS. of Dioskurides was unillustrated. The special interest of this is that it perhaps marks the first occasion when an artist and a writer " collaborated " in the production of a text book—for such was *De Materia Medica libri quinque*, and as such it remained for the whole of Western Europe down to the end of the fifteenth century. According to Dr. Agnes Arber, whose book *Herbals* is the definitive modern work on this subject, it was still found to be in use among the monks of Athos when the Director of Kew visited the peninsula in 1934. Incidentally the names of many of the genera—*Aristolochia, Anagallis*, etc.—still apply to plants similar to those described by Dioskurides nearly two thousand years ago.

After Dioskurides even purely descriptive botany declined. Albertus Magnus certainly advanced our knowledge of plant structure in a manner which greatly helped later taxonomists, but taxonomy itself had no interest for him, and no further progress was made until near the end of the sixteenth century.

Printing from movable type dates from about 1450; within twenty years of this date there began that trickle of illustrated books on plants (the name " herbal " is probably a contraction of the Latin *Liber Herbalis*) which rapidly became a flood. Dr. Arber lists about a hundred published between 1470 and 1670, and regards the list as far from exhaustive. The round dozen which appeared previous to 1530 were mostly the first *printed* editions of " books " which had existed for some time in manuscript.[1] They are for the most part of no botanical interest: their illustrations are much inferior to those of Krateuas, and the information they supply is founded as much on folklore and superstition as on the exacter descriptions of Dioskurides. They are however valuable—and charming—historical documents for the light they shed not only on the naïve outlook of their compilers but also on the wholesale confusion of titles and texts made possible by the slow process of production and the frequent piracy before the laws of copyright had been established.

In 1530 appeared the *Herbarum vivae eicones*, commonly known as Brunfels' herbal. The importance of this work is not due to Brunfels' text which shows little appreciation of ecology (Theophrastos knew better) but for the startling improvement in the illustrations (the work of Hans Weiditz who had at last " gone back to nature " to produce images of real plants instead of the conventional forms with which endless copying had filled the earlier herbals. Henceforth this remained the chief line of progress). The great herbals of the sixteenth and seventeenth centuries are triumphs of the illustrator's art and evidence of the widening interest in nature. Doubtless the chief spur to their production was the demand for more accurate and more compendious lists of plants capable of yielding the only drugs which the majority of medical men were willing to employ; but " cheerfulness breaks in " in the form of delight in the display of native plants (cf. Turner " that I might in my herbal declare to the great

---

[1] A small book, first printed about 1480, known as the *Herbarium Apuleii Platonic* is believed to have been *written* in the fifth century.

honour of our countre what numbre of sovereine and strang herbes were in Englande that were not in other nations . . .") and in the knowledge gained in the planting of gardens. To this was added the novelty of the flavours and beauties brought from distant lands by the increasing number of voyagers of discovery. But in all this diligence there is little attempt at system. Enthusiasm overbore discretion; for such arrangement as there was not infrequently reflects a confusion of aims: plants are arranged partly by habitat and locality, partly by form, partly by their virtues, real and imaginary. The only advance is the gradual fixation of the custom of referring to a species by means of a generic term coupled with a specific *differentia*. When this started it is hard to say. The earliest recognition of generic status is probably contained in a letter of Konrad Gesner where he writes: " We may hold this for certain that there are hardly any plants that constitute a genus which may not be divided into two or more species. The ancients describe one species of gentian; I know of ten or more." Here is a clear recognition of *natural* affinity; but it is doubtful whether Gesner's views had any influence.

We have at this stage to distinguish between (a) classification as such, (b) systematic nomenclature, (c) descriptive terminology. As regards nomenclature Kaspar Bauhin (*Pinax theatri botanici*, 1623) was the first to attempt to name each plant by a generic term followed by a specific *differentia*—often a single word, but by no means always so. The first attempts at rigid classification were those of de l'Obel (*Stirpium adversaria nova*, 1570) using leaf form as a basis, and the better known Caesalpinus, who was the first to use floral structure for this purpose. We now know that both these systems are entirely artificial since they cut across groups which may be presumed to be closely related on much more extensive evidence. Their importance in the history of taxonomy is their tacit recognition that the plant itself must be the arbiter in the matter of its classification. Some realisation of this was revealed in the work of the Czech botanist, Zaluzian, who in 1592 made an attempt to sort plants out according to their grades of complexity of organisation. Had Caesalpinus applied this method to floral structure as a whole instead of restricting his attention to an arbitrarily selected part, he might have anticipated Linnaeus by nearly two centuries. He seems to have been influenced by the worst forms of Aristotelianism. For a consistent and concise terminology, by which the *homologous* parts of plants could be referred to in a description of a species, it was inevitable that botanists must wait until comparative studies had clarified the difficult problem of deciding what parts *are* homologous. With almost uncanny intuition (for he was completely ignorant of the sexual functions of flowers) Joachim Jung seized upon the importance of flower structure as a basis of classification, and sufficiently analysed these parts to be able to give names to the perianth, stamen, and style, as also to distinguish the ray from the disc florets of the capitulum of a composite inflorescence such as a daisy's. He also laid the basis of the extremely important classification of leaf form, distinguishing simple from compound, pinnate from digitate, opposite from alternate, leaves. Suspected of heresy, he published nothing, so his influence was very belated. The elaboration of technical terms, each with a signification, clearly defined and narrowly extended, must proceed *pari passu* with the advance of knowledge. But its absence retarded this progress, especially in sciences which are essentially descriptive, as were botany and zoology at this period. Without such terms each discoverer is driven back for similes on the furniture of his own mind, a repository variously stocked according to the nature of his interests and domestic circle. Confusion, not to speak of unwieldiness, is inevitable.

To find a way out of the morass of words, superstitions and confused aims, cramping systematisation on the one hand, and the looseness born of uncritical specimen-hunting on the other, demanded a man of uncommon powers. Such a one, John Ray, was born in 1627 in a blacksmith's cottage in the Essex village of Black Notley. Regarded with the greatest respect by some of the most famous biologists of all lands, Ray has lacked until recent years a fitting memorial—due largely to his own extraordinary reticence in those materials which provide the biographer with his means. This has now been rectified by the splendid work of Dr. C. E. Raven, to which I am indebted for the few remarks possible in this place.

To form anything like an adequate appreciation of Ray's importance in the growth of scientific ideas it is necessary to take note of the circumstances of his life. Proceeding to Cambridge in 1644 he found a community divided against itself. The scholastic tradition which had degenerated into a discipline " trivial " in fact as it had been, for a quite different reason, in name, was dying by inches. The old branches were being drastically pruned by the Earl of Manchester, mainly on political grounds, and vigorous young shoots (Milton, 1625-32; Henry More and the Cambridge Platonists, twenty years later) were beginning to push. To the old *régime* Ray owed his mastery of Latin and the meticulous use of words as symbols for exact ideas, which was one of the pillars over which the arch of his great achievement was built. To such young men as John Nidd, who probably taught him dissection, and Walter Needham, he owed the means to make good his conviction that in Dr. Raven's words, " the study of nature is essentially a religious duty, appropriate to mankind as a means to God's greater glory and a deeper understanding of His ways."

His apprenticeship was served in the compiling of the *Cambridge Catalogue* (1680)—the first, so to say, of the " County Floras "— in which he listed all the flowering plants to be found in the neighbourhood of Cambridge. His classification is mainly that of the earlier authorities, especially the Bauhins, but unlike many of the English herbalists he *studied every plant in its own habitat*. Dissatisfied with the chaos of nomenclature which, he saw, led to a needless multiplication of species, he modified many names, giving his own annotations. In 1666 he was further disciplined in this task by writing tables of British plants for Bishop Wilkins' *Essay towards a Real Character and Philosophical Language*, one of the earliest attempts at a universal language which should supersede Latin. But his main contribution to classification came in a series of works on all the chief divisions of living things. The first to be published was *Ornithologia* (1676) which, like the later *Historia Piscium* (1686), purported to be a pious labour of editing the material of his friend and patron, Francis Willughby. Dr. Raven has given good grounds for the view that Ray owed little more to Willughby than financial assistance, enthusiasm, and the ability to organise long journeys, and the stimulating commentary of a very great *amateur* naturalist. The former work had great influence on the course of natural studies in England in that it was the only one of Ray's works to be translated completely into English—and it is well known that ornithology numbers among its devotees in Great Britain far more than any other branch of natural history. It is however in his *Methodus Plantarum* (1682, amended 1703) that his greatest contributions to taxonomy are best seen.

In this book Ray retained the classical division into herbs and sub-shrubs, shrubs, and trees. The whole range of the former is divided into twenty-five " genera ". " Sub-marine plants ", " Fungi ", " Mosses " are comprised in non-flowering plants, and the remainder in the " flowering " plants. In the latter he makes the cardinal distinction between monocotyledons and dicotyledons (1703). Among these two groups may be recognised several " genera " which correspond roughly to the Linnaean " orders " (modern " families ") e.g. " Leguminosae ", " Tetrapetalae siliquosae aut siliculosae " (most Cruciferae); but these are subdivided in a manner quite different from the modern genera: for instance, two different species of *Brassica* are separated. Having cleared his mind of the confusion introduced by the existing floras, Ray set about the immense task of reviewing the whole array of living plants. The resulting *Historia Plantarum* in three volumes, published between 1686 and 1704, containing over 6,000 carefully described definitive species, gleaned, after severe pruning of repetitions, from all the well-known authorities, and checked wherever possible by personal inspection of the growing plant—an ideal, the pursuit of which led him and his collaborators over the greater part of Western Europe—forms his greatest memorial. Moreover it is prefaced by a long and critical account of the structure and habits of plants in general—a feature which marks Ray out as a modern, in that he was never content to distinguish species by any single character in isolation, knowing too well how size, colour, leafage, and habit may vary in different circumstances. Many of the plants he grew experimentally in his own garden. The *Historia* could never be a popular book for, apart from its huge bulk, it lies under the disadvantage of using descriptive phrases instead of single words for many, perhaps most, of the species. But it is generally recognised that if there had been no *Historia*, there would have been no *Species Plantarum* written by Linnaeus nearly seventy years later.

This was not the end of Ray's work on plants, but we must pass on to his studies of animals. Of these, the method of classification of " quadrupeds " and the studies he made for the uncompleted *Historia Insectorum* are perhaps the most significant for our purpose. The former is marked by the first effort to classify what we call mammals, according to their *nature*; to exorcise legendary beasts; and to seize upon characters distinctive of the natural kinds. With extraordinary sagacity he recognised the feet and the teeth as having the greatest significance. Thus his division into *Ungulates* with hoofs and *Unguiculates* with nails leads naturally to our present mammal orders. Moreover his use of dentition as the basis of subdivision carries them right down to such modern orders as *Carnivora* and *Rodentia*.

In the study of insects Ray had the relative advantage of a paucity of existing books. The only complete treatise on the subject was Mouffet's *Theatrum Insectorum*, in which a gallant attempt had been made to survey the already great number of described species; and it must be remembered that, as with the modern schoolboy, the term " insect " covered almost all animals not characterised as bird, fish, or quadruped. Ray had therefore to depend almost entirely on his own observations, not to mention those of his wife and daughters, and even of some of the villagers. It was in this department of natural history that he made the greatest progress in correlating field observations with planned experiments. With his breeding cages he was able to sweep away a huge number of species of " worms " which he showed to be the larval stages of lepidoptera, beetles, and diptera. The presence or absence of metamorphosis was recognised by him as a first criterion of classification, subdivision being effected in the first stage by the number of " feet " in the larvae, in the second by the nature of the wings—whether covered with scales (lepidoptera) or membranous (bees and " flies "). In 1705 the 16-page *Methodus Insectorum* was published; on 17th January in the New Year he died, worn out with pain and disease but vigorous to the end. When we consider that the books here reviewed form only a part of Ray's strictly *scientific* work; that his most influential book, *The Wisdom of God* (to be referred to in a later chapter) showed him to be standing on a promontory between the old age and the new; that it was as a teacher of *Greek* that he held his Fellowship for eleven years at Trinity; that he was a fellow-pupil with Isaac Barrow, later first Lucasian Professor of Mathematics; that he was esteemed one of the finest Latinists of the century, we may wonder why the average Englishman has never even heard of this modern Aristotle. So much for the scholar. The man can best be described in the words of a contemporary, John Worthington: " He is a person of great worth; and yet humble, and far from conceitedness and self-admiring . . . a conscientious Christian; and that's much said in little." A conscientious Christian, who resigned his Fellowship rather than compromise his conscience when the Church turned the tables on the reformers; a very gallant gentleman who, through years of unremitting pain, weakness, and maddening irritation due to ulcers of the legs, never complained, but worked on with his beloved plants, birds, and insects, and his colossal mass of printed books, whose weight he never allowed to crush the spirit of enquiry or that reverence for nature in which, everywhere, he saw the finger of God.

In the three desiderata for a sound taxonomy (see p. 369) Ray had made great progress in two; he had progressed hardly at all in the third, namely the attainment of a nomenclature at once uniquely descriptive and concise. A further stage was reached in this respect by Tournefort, Professor at the Jardin des Plantes, the cradle of so many botanical researches of the eighteenth century. But the man who uncompromisingly insisted on the binomial system in respect of every living organism was Karl Linnaeus, later von Linné. He shared with Ray the accident of humble birth, a youthful ardour for gardening, and an astonishing perseverance in the collection and cataloguing of species; in every other respect his character and scientific attitude were in direct contrast. Possessed of an encyclopaedic memory and a surpassing genius for ordering an immense mass of detail he devoted his whole life, on his own admission, to the ideal of being able to name every organism put before him. At a time when the scientific market was being flooded by the living wares of almost the whole globe, and the age of specialism was dawning, the execution of this task was a necessity which could not have been long deferred. That it was bought at a great cost, few would now deny; but the price had to be paid. The following phrases from his *Genera Plantarum* (1737) indicate the *simplifying assumptions* (by no means recognised as such by Linnaeus) which made his ideal possible of fulfilment: " There are as many species as the Infinite Being created as diverse forms from the beginning . . . just as there are now no more species than have been from the beginning. There are as many genera however as there are proximate common attributes of the distinct species which were created in the first instance. All genera and species are natural." Ray had doubts about hybrids, local varieties, and the possibility of gradual modification in some cases: such doubt must have been somewhat inhibiting towards the ascription of a simple *name* to each and every *kind*. Free from doubt, Linnaeus did just this; and without such dogmatic certainty a binomial nomenclature is impossible—it demands the " reality " of the genus as much as that of the species. It is perhaps significant that with our growing scepticism towards the existence of rigid " species ", let alone genera, we are accepting more and more *trinomials*—perhaps the first retrograde step to the descriptive phrase from which Linnaeus rescued us. Having settled this matter (first enunciated in *Species Plantarum*, 1753) to the lasting benefit of science, he proceeded to an expedient of more doubtful value—the sorting of genera into " orders ". The key to this he saw in the famous *Letter on the Sex of Plants* (1674) in which Camerarius had given some experimental backing to the belief which Grew had embraced as a result of his own microscopic observations of pollen, and Ray had accepted but tardily. Number is so much easier to classify than form: the temptation must have been great; and Linnaeus succumbed to it. His " classes " are determined *solely* by the number of stamens in the flower, the sub-division into " orders " by the number of styles or free female elements; thus all genera having five stamens and one style are packed into the box labelled " class pentandria, order monogynia ". In the absence of any characters common (and also differentiable) to all animals, Linnaeus had to cook the evidence by pretending that the invertebrates (a term not then invented) all have hearts with one ventricle and no auricle. However it was the definition of *genus* and *species* that was all important; and here he did posterity the signal service of framing every description on a definite plan in which all essential parts were passed in review and described in a terminology admirable for its terseness and clarity—he would have no truck with pictures, which were " for those with heads with nothing in them "; were not twenty-six letters enough for all possible descriptions? The summit of his labours was the *Systema Naturae* which in its 10th edition (1758) remained, and in some ways remains to-day, the Bible of taxonomy. The whole world is catalogued, including the four elements; and at the head, the summit and crown of glory, stands Man, proud Man, accompanied in the same genus, and much to the disgust of many of his contemporaries, by the Orang, *Homo troglodytes*.

The influence of Linnaeus was immense. He was an inspiring teacher, a successful publicist, and an effective organiser of expeditions to every corner of the world. His work formed the indispensable basis of the future, his method and outlook—static and formal—was the epitome of the past. Forces were gathering —to which in some measure he yielded later in life—which were to break through his iron-clad system, whose very completeness carried the implication of its own destruction.

The rigid artificiality of the Linnaean classification of plants was somewhat loosened by Bernard de Jussieu whose system was extended and published by his nephew, Antoine Laurent de Jussieu. Instead of the single diagnostic character provided by the stamens, de Jussieu adopted the form of the perianth leaves— sepals and petals. The fifteen " classes " of de Jussieu were characterised by the presence or absence of a " flower ", the union or separateness of the petals, and the insertion of the whole in relation to the gynaecium—above, below, or at the same level. These characters produce a system somewhat nearer to the modern more " natural " arrangement, but are still too artificially restricted. A further move towards giving to *all* characters their due weight was made by de Candolle. Throughout the middle of the nineteenth century his system was being gradually perfected, and by its attention to reproductive processes and vegetative anatomy it forms the basis of our present views on the plant kingdom.

Meanwhile Linnaeus's work on animals was being completed by a band of zoologists, mainly in France, of whom Cuvier was the outstanding leader and in fact dictator. Aided by Brogniart he brought out, in 1808, a study of the *fossil* animals of the geological region known as the Paris basin which shewed that a new note had been struck in the harmony of organic forms. It was this constant reference to extinct forms, coupled with the application of the principle of correlation of parts—whose explicit use as a *method* we owe largely to Goethe and the *Naturphilosophen* (see p. 387) which enabled Cuvier to transform animal taxonomy at one stroke into a *natural* system, despite his uncompromising acceptance of Linnaeus's dogma of the fixity of species. An Aristotelian in the best sense of the term he sought a meaning for every part of an animal in its probable function. By comparison of parts of different animals subserving the same function, he was able to build up a *generalised form*. Thence when faced with a single part—a jaw, a tooth, or a limb bone—won from the rocks, he proceeded to reconstruct the probable nature of the whole animal, not only its form, but by analogy its manner of life. With the huge extension of knowledge obtained in this way he set about the review of the whole animal kingdom, dividing it into four groups—Vertebrata, Mollusca, Articulata (including Annulata), Radiata—which were in turn divided into classes, whose names in many cases correspond roughly (in some cases exactly) with those in use to-day. In all this there is comparison between *all* the organ-systems—respiratory, skeletal, circulatory—making of it a truly *natural* system. The group Radiata alone shows lack of judgment: here the geometric form gives a misleading unification and the measuring calipers found no scope. It is a measure of Cuvier's greatness that by using the evidence of fossils at a time when geologists had still but a hazy notion of their significance, he created the first encyclopaedia, *Le Règne Animal* (1815), of the natural forms of animals. The measure of his weakness was his dogmatic silencing of the prophetic voice of his colleague, Lamarck, who saw, as he did not, the inescapable conclusion to which the fast accumulating evidence was pointing. Cuvier talked evolution while denying it.

## SOURCES FOR CHAPTER XXVIII

Book I of Aristotle's *Parts of Animals*, contains a discussion of the basis of classification.

Agnes Arber, *Herbals* (Cambridge, 1938). The greater part of this book is concerned with the so-called " Herbals " printed between 1470 and 1670; but it contains an introduction to taxonomy based on the system of Theophrastos and also some account of the origins of modern taxonomy in the sixteenth century.

C. E. Raven's *John Ray* (Cambridge, 1942) provides a very full account of the transition period in the seventeenth century when the correlation of groups in different parts of Europe, and beyond, became an urgent necessity.

F. J. Cole's *History of Comparative Anatomy* (London, 1944) should be consulted, especially the earlier chapters, for the background of animal taxonomy.

C. Singer's *Short History of Biology* gives a fairly detailed account of the whole development of the subject from the earliest times down to the late nineteenth century. More detailed notices of individual workers are to be found in Nordenskiöld, *op. cit.*, Gen. Bibl.

CHAPTER XXIX

# THE SEARCH FOR THE STRUCTURAL UNITS: THE STATIC THEORY OF FORMS

To give an adequate description of the architecture of a building we must be able to describe the materials which have been used, the units of structure—bricks, freestones, and the like—into which the materials have been fashioned, and the arrangement of the units in the finished structure. But although a great cathedral does in some sense " grow ", the course of its growth is very different from that of an organism. It is quite likely that habitual knowledge of the growth of cathedrals was one of the causes why the early biologists do not seem to have been concerned very much with the mode of growth of organisms. A cathedral starts more or less full size, and its " growth " is a matter of raising the walls and spanning them with a roof. Once it is " finished " it stops growing, until a returning crusader adds a retrochoir or increases the length of the nave. But a living and working puppy continues to " swell ". It is true that Galen took note of this " assimilation " of food into the fabric of the animal, but he was too fogged with metaphysical *words* to regard it as a problem which had to be solved. Nor was it merely a matter of scale; for the invention of the microscope, by revealing a new world of *complete* organisms, if anything retarded interest in the growth of more familiar ones. In fact the discovery of the spermatozoon was possibly a disaster; for those who were dogmatically in favour of male dominance at once saw in them prefabricated " homunculi ". Thus the problem ceased to exist. The homunculus must have been conceived on the pattern of those once familiar pig balloons, which had merely to be blown up to grow. The ovists, with rather more excuse, saw similar objects in the egg.

The problem was first posited by Caspar Friedrich Wolff in his *Theoria Generationis* (1759). This very influential book is devoted to a miscroscopic study of the structure and development of both plants and animals. Though the cellular structure of plants had been noted by Hooke in his *Micrographia*, Malpighi and Grew first distinguished different types of cells. Malpighi's observations, as for instance of the spirally thickened vessels, were very accurate; but he was too ready to follow such false analogies as the comparison of these spiral vessels with the breathing tubes (tracheae) which he had discovered in insects. Grew, though Aristotelian in outlook, made the important observation that the cells, or " bladders " as he often called them, were, in the *growing* parts of plants, thin walled and juicy. He did not however follow up this clue, but was content to compare the texture of a plant with that of the lace woven by women on upright pins. Wolff however made the cardinal advance of recognising that " leaves grow for the most part by new vesicles ("*Bläschen*") pushing themselves between the old ones, to some extent also by the expansion of the vesicles ". This is an excellent start; unfortunately almost in the next breath he asserts that " since the vesicle in a young leaf is nothing but a pore or cell, the vessel nothing but a canal filled and enlarged with fluid, it necessarily follows that the delicate solid substance that fills the spaces between the vesicles is enlarged by its contained fluid into a vesicle ". From which it follows with equal necessity that Wolff was as far from understanding the true nature of the vesicles as were his predecessors. They are mere bubbles in the delicate, solid (*zarte, feste*) substance in which evidently the *vis essentialis*, which he later described, must reside.

But although Wolff had misconceived the nature of the architectural units, he made two further contributions towards our understanding of the process of building.

" In every living bud . . . and at every point where growth is taking place, one meets younger and smaller leaves wrapped up in older and larger ones. If one presses forwards from the outside inwards, one meets at last the inner substance of the plant, which is moist and full of sap, and which no longer shows the protuberances (*Vorragungen*) of leaves." In this substance (the growing apex of the stem) vessels become more prominent as one traces it backwards. The leaf rudiments have their edges more and

## THE SEARCH FOR THE STRUCTURAL UNITS

more crinkled the further they are from the growing point. Here for plants at any rate is the complete refutation of the doctrine of preformation; for the leaves are clearly growing out of the "inner substance" which is fed by the vessels. In the case of animals no such simple "epigenesis" was to be seen before the days of serial sections of an embryo embedded in paraffin. But with great sagacity Wolff seized on the fact that the blood islands appear outside the embryo proper before the heart; that under the influence of heat the white substance in which they are embedded turns first yellow then reddish, and at last traverses channels leading to the heart, which is thus stimulated to contract. The dissolution of the yolk as the embryo grows, he regards as no isolated phenomenon peculiar to embryonic growth, but "it takes place in every growth and in every case of the nourishment of animals". Thus the contents of the stomach are known to be dissolved as a preliminary to absorption into the blood.

Wolff's *dis*proof of crude preformation is based on unambiguous facts. His rival theory of "epigenesis" was hardly more than a name for these facts; and in so far as it was explicit, it was retrograde in calling in another "name"—the *vis essentialis*—as an "explanation" of what was still a mystery. In his elaboration of the relation of matter and life he puts forward views which must be reckoned with in any *philosophy* of nature.

In the course of his observations on animal embryos he thought he had detected an arrangement of rudiments similar to the leaf rudiments of plants. This could only have been a guess, but it happened to be right. In the hands of his successors, Pander and, particularly, von Baer, it became established as one of the most fruitful principles in animal embryology. But it also influenced Goethe and Oken, the founders of the, for a time, very influential school of *Naturphilosophie*.

In his examination of plant rudiments Wolff was struck by the fact that at a sufficiently early stage of development it was impossible to detect the difference between the primordia of leaves and those of flowers, from which he concluded that plants are nothing but stems and leaves from which the other parts are differentiated. Linnaeus, drawing attention to those monstrous forms, familiar to gardeners, in which leaves appear in the centre of flowers, considered the flower to be a modified and compressed shoot. But it was Goethe who, apparently in ignorance of Linnaeus's work, extended the idea of metamorphosis into every realm of biology. Believing that organisms in all their wonderful variety are the result of the development of some simple ideal unit, he sought for instances to illustrate this. He found such an example in the human upper jaw which was well known to consist of only one pair of components, whereas in other mammals there are two pairs. With the eye of faith he saw where the line of division ought to come between the two hypothetical components on each side and foretold its discovery in the human foetus: this was confirmed nearly forty years later. Another more familiar example is the functional adaptation of a common vertebral plan in various parts of the spinal column. Unfortunately he went further and saw in the cranium a mere extension of the vertebral column. Although the ground plan is still conveniently referred to as occipital, parietal, and frontal " rings ", the view that these represent specialised vetebrae is quite untenable.

The value of Goethe's principle of metamorphosis is in viewing structure as subservient to the functions of the *living* organism: to him we owe the more suggestive word " morphology " to replace " anatomy ", whose association with the dismembered corpse is only too apt to be reflected in the outlook of the investigator. In his own words, " Now whether the plant puts forth a shoot, comes into bloom, or bears fruit, the organs are still always the same, which in multifarious vocations, and under forms often varied, perform the precept of Nature." It would however be a mistake to attribute to Goethe the dynamic interpretation made possible, as we shall see, by the advance of embryology. Though he also emphasised the importance of this aspect of morphology it is clear that his interpretation was purely Platonic—the object of morphology was for him the discovery of " ideal " unities of pattern, nowhere existing in any actual animal, but to which the forms of the various functional organs more or less approximated. This viewpoint, adopted consciously as a *guide* and not as an *end*, still plays a helpful part in our modern " generalised " structures. It played an important part in the development of comparative anatomy in the nineteenth century (see below, p. 416).

During the period (1790-1807) when Goethe was exploring the wider implications of organic form, there was working in France a young man who made the first definite advance on the main road since Wolff's refutation of preformation. This was Bichat. Within the two years prior to his death he had displayed the evidence which led inevitably to the foundation of a special branch of biology, namely histology. " The general doctrine of this work ", he writes, " consists in analysing with precision the properties of living bodies, in showing that every physiological phenomenon is ultimately referable to these properties considered in their natural state; that every pathological phenomenon derives from their augmentation, diminution, or alteration; that every therapeutic phenomenon has for its principle the restoration of that part of the natural type . . ." It is not at first clear what Bichat meant by " properties ". He admits that his doctrine " differs from that of Stahl and those authors who, like him, refer everything in the living economy to a single principle, purely speculative, ideal, and imaginary, whether designated by the name of soul, vital principle, or Archeus ". It is clear then that Bichat is trying to systematise the *mechanical* attitude to living things. " It is a theory like that which shows in the physical sciences, gravity, elasticity, affinity, etc., as the primitive principles of the facts observed in these sciences." But it is a mechanism with a difference, for " the properties, whose influence we have just analysed, are not absolutely inherent in the particles of matter that are the seat of them. They *disappear when these scattered particles have lost their organic arrangement*" (italics mine). The " machines " (organs) then of the living organism, are not ultimate but " these separate machines are themselves formed by many textures of a very different nature, and which really compose the elements of these organs. Chemistry has its simple bodies which form, by the combination of which they are susceptible, the compound bodies; such are caloric, light, hydrogen, oxygen . . . In the same way anatomy has its simple textures which, by their combinations . . . make the organs." His systematisation therefore seems to follow two lines, of which the fundamental is conceived on an analogy with the newly-founded science of chemistry (the *Traité sur les Membranes* appeared in 1800, Lavoisier's *Traité* in 1789).

He was in fact searching for the *elements* of living organisms, and found them as he thought in the " textures " whose various combinations formed the familiar organs of the body. The second line is the recognition that these " elements " departed from the chemical ideal in being themselves *organised*—their " properties " disappear when the material particles have lost their *organic* arrangement. He does not further consider in what this special arrangement consists; nor could he, being ignorant of the " atoms " (cells) of which the " elements " consist.

What are these " elements "? By calling them " textures " and enumerating their kind, i.e. nervous, arterial, venous, osseous, fibrous, etc., he clearly was groping towards the modern assumption of the existence of more or less stable sheaths of specialised cells, namely *tissues*. That he was little more than groping is shown by his regarding " venous " and " arterial " textures as distinct units, whereas we know them as still " compounded " of several elements—nerve, muscle, collagen fibres, elastic fibres, in different proportions as befits the different functions of the two kinds of vessels.

The evidence for this rather sweeping but fruitful theory was twofold, namely the different reactions of different textures when submitted to physical and chemical tests, and, perhaps more cogent, the fact that disease tends to be limited to contiguous *tissues* rather than to organs. Of the latter Bichat gives many examples of which one must suffice. " I have examined a great number of bodies in which the peritoneum was inflamed either upon the intestine, the stomach, the pelvis, or universally; now very often when this affection is chronic, and almost always when it is acute, the subjacent organs remain sound. I have never seen this membrane exclusively diseased upon one organ, while that of neighbouring ones remain untouched; its affection is propagated more or less remotely." It is easy to see under how great a debt biology is placed to the practice of medicine.

After Bichat, the search for the elements of organic structure becomes exceedingly confused. It is dangerous to dogmatise, but one might hazard a guess that the confusion was in large measure a result of speculation running far ahead of the facts of observation. The facts in this case are difficult enough to ascertain with all modern resources, but it must be remembered that in 1800 the microscope was little improved beyond the state to which Leeuwenhoek's contemporaries had brought it. Although the achromatic combination had been made by Dollond as early as 1758, the quality of the glass (see Ch. XXII) was such that it had little application till much later. By 1823 however greatly improved instruments were being turned out by Amici, and within a few years all the classical work on the basis of cytology (the science of cell structure) had been done. But it is difficult to pick out any one of the half-dozen contributors as one whose views were free from error. It seems at least likely that their eyes were clouded by the welter of uncritical speculation which emanated chiefly from Germany during the preceding half-century.

During the eighteenth century much ink was spilt in the attempts to decide whether organisms were pure machines or were directed by an entelechy, soul, *vis essentialis*, or what you will. The philosopher, Kant, attempted a reconciliation of these antagonistic views in his *Critique of Teleological Judgment*, the last part of his *Critique* of the human mind. In the *Critique of Pure Reason* Kant established the position which has been the base of most modern philosophy, that any individual act of knowledge is a unified whole from which the " subject " and " object " may be only ideally abstracted. And that this " synthetic unity " comes about only because human understanding (not to be confused with *reason*) " prescribes to nature her laws ", that is, provides a framework into which knowledge of " things in themselves " may be fitted and thereby distorted. The " objects " of the understanding are the " phenomena " of science; of things-in-themselves the understanding gives us no knowledge. But in the business of life we have to behave in the light of our intuitions, e.g. of the " categorical imperative " ( = " thou shalt " or " shalt not ") and to make judgments on a similar basis. It is this last point which was, and is, of such importance to biology. For although we have no " knowledge " of the purposiveness or harmony of nature, we are compelled by our own nature to interpret the natural world in such terms: to refuse to do so would be needlessly to cramp and limit the possibility of extending our knowledge. But teleology must be ascribed to nature only as " a regulative (i.e. interpretative) not as a constitutive judgment". Bearing this caution in mind we may readily admit that to call an organism a *mere* machine is to misuse terms—no machine in the accepted sense of the term is self-*generative*. The most fertile and, at the same time, well-based attitude is to regard every " organised product of nature as one in which every part is reciprocally purpose and means ".

Now although Kant's discovery of the " transcendental unity of apperception " was perhaps the most fruitful discovery in the realm of pure thought, its very fruitfulness provided the possibility of a tangled growth of weeds. It is clear that the dichotomy of phenomena and noumena (things-in-themselves) was in itself a source of dissatisfaction. It is not our business here to consider the possible modes of reconciliation, but only to note the one taken by Kant's successor, Fichte. Fichte was no biologist—even less so than the mathematically-minded Kant—but his resolution of the Kantian dilemma had the gravest consequences for science. Without, I hope, doing violence to Fichte's long and ingenious argument, the upshot may be summarised as follows: If every act of cognition involves an ego and " object ", and if we agree that one must be reduced to the other, then since the ego is *immediately* given, but the " noumenon " only under certain forms of cognition (e.g. time, space) dictated by the ego, then the noumenon is forever unknowable, a mere fiction. Consequently the ego is the sole *reality*—not, of course, any individual ego which, so far as we know, comes into being and may pass away, but the world-ego or universal reason. Such a philosophy has little place for natural science, and Fichte made little attempt to find any; but his pupil, Schelling, while accepting the idea of a universal mind, made it his life's work to show that a universal nature exists side by side with it, but in such a way that each was, as it were, a reflection of the other. Into the bog of words which he generated we need not follow him; but although most of his *Naturphilosophie* is now no more than one of the larger exhibits in the Museum of Monstrosities of the Human Mind, the school of thought which owed allegiance to him was far from negligible. Goethe's fanciful, but by no means fruitless, analogies owed if not their origin at least their popularisation to him.

## THE SEARCH FOR THE STRUCTURAL UNITS 389

encapsulated egg-embryos, but instead a remarkable " guess " at the process of gametogenesis, as it is studied in every testis, ovary, anther, or ovule.

It is perhaps unnecessary to warn the reader that I am making no claim for Oken as the *discoverer* of the cell theory— " Everything of importance has already been said by someone who did not discover it ", as Whitehead reminds us. But may it not be the case that such a prodigal generator of ideas, however confused, as Oken, may be necessary when some great *new* concept is struggling to be born, upon which more critical and cautious, but perhaps less resourceful, minds may even unconsciously draw as stimulants? Oken was no mere recluse fermenting in the wort of his own imagination, but, as Rudolph Virchow (p. 394) was to remind German men of science, the man who " gathered the first assembly of German naturalists and physicians ". This was in 1822 " when the few persons who composed the first Conference of German naturalists met at Leipsic "; moreover " it still appeared so dangerous to hold such an assembly that it actually met in the darkness of a secret session." As a champion of free enquiry he was persecuted and died in exile. Nine years after the German meeting, the similar British Association was founded, not in danger indeed, but to be the object of the ridicule of Charles Dickens and of many University professors, who ought to have known better.

Somewhat under the influence of *Naturphilosophie* was Schleiden who, having failed as a lawyer, decided to shoot himself. Even less successful with firearms than with briefs, he merely wounded himself and lived to found cytological botany. This he based on the discovery by Robert Brown, in 1831, of the cell *nucleus*. Examining in 1831 the cells of the dividing ovule he realised the cardinal importance of the nucleus, which he called the " cytoblast ", but wrongly concluded that this body grows by accretion and thereby generates the cell. In his reference to the plant body as a *Polypstock*, he reminds us of Oken's theory of the genesis of the higher organisms by concretions of " infusoria ".

Parallel to the work of Schleiden on plants was that of Schwann. Influenced by the former's work on the nucleus, he published a book a year later in which he sought to determine "whether the cause of organic phenomena lies in the whole organism, or in its separate elementary parts"; and also whether this power is mechanically or purposefully determined, and if the former, what are the physico-chemical characters of its make-up. In answer to the first question he replies:

> We have seen that all organised bodies are composed of essentially similar parts, namely cells, that these cells are formed and grow in accordance with essentially similar laws . . . Now if we find that some of these elementary parts not differing from others, are capable of separating themselves from the organism and pursuing an independent growth, we may conclude that each of the other elementary parts, each cell is already possessed of power to take up fresh molecules and grow; and that therefore each elementary part possesses a power of its own, an independent life . . . if the relations which it bore to external parts were but similar to those in which it stands in the organism. The ova of mammals afford us an example of such independent cells growing apart from the organism.

In the concluding sentence Schwann refers to the identification, in 1827, by von Baer of the mammalian ovum as a minute cell inside the follicle which de Graaf had falsely supposed to be the ovum itself. Moreover von Baer had come to this discovery by comparing embryos at various stages of growth, from those easily visible to the naked eye to those at the earliest stage of cleavage. In 1837 Siebold and Sars had actually observed the cleavage of invertebrate ova; and Schwann himself saw it in the hen's egg, thereby proving that the " eggs " of different animals may vary enormously in size, not in accordance with the size of the animal but with the amount of food reserve stored as " yolk ". The argument thus set forth established beyond any doubt that the growth of animals takes place by the formation of cells. But though Schwann gave reasons to show that the same may be said of plants, he failed to rid himself of the common error that the cells are formed by the increase of size of the nucleus. He even went so far as to develop a highly ingenious physico-chemical theory in which growth was regarded as something akin to the growth of crystals but out of materials known to be capable of imbibition of water. He was in fact, without realising it, foreshadowing the later discovery of growth of plant cells by vacuolation.

There was at this time, then, a clear understanding of the

cellular structure alike of animals and plants, combined with a very confused idea as to how and out of what these cells arose. Although Purkinje, the founder of the Czech national university, had used the word " protoplasm " in 1839 and had watched ciliary movement in various situations, it was Dujardin who first recognised the *activities* of the essential living stuff in animalcules, though he, not without reason, called it " sarcode ". During the years 1842-6 Nägeli applied to plant cells chemical tests recalling those of Boyle in the first identification of the products of distillation (p. 174), as a consequence of which he was able to differentiate sharply between the nitrogenous *contents* and the non-nitrogenous wall and inclusions. He saw reason also to reject the view that the cell is formed by the budding of the nucleus. Von Mohl however must be given the credit for recognising the importance of these discoveries and of pointing out that even in cells apparently filled with sap, there is a " primordial utricle " of protoplasm often to be seen in streaming movement within its own cell. Thus by 1858 the situation was sufficiently clear for the first great triumph of the cell doctrine, the *Cellular Pathology* of Virchow. Just as Bichat had demonstrated that disease is more commonly due to disordered *tissues*, so Virchow developed with far more skill and in greater detail, based on the finely mounted preparations then possible,[1] the view that disease is in the last resort the outward sign of damaged *cells*. In a famous passage he sums up the greatest generalisation in biology:

> No development of any kind begins *de novo* and consequently we reject the theory of spontaneous generation just as much in the history of the development of individual parts as we do in the development of entire organisms. Just as little as we now admit that a *Taenia* can arise out of a suburral mucus, or that out of the residue of decomposition of vegetable or animal matter an infusorial animalcule, a fungus, or an alga, can be formed, equally little are we disposed to concede either in physiological or pathological histology, that a new cell can build itself up out of any non-cellular substance. Where a cell arises, there a cell must have previously existed (*omnis cellula e cellula*), just as an animal can spring only from an animal, a plant only from a plant.

Virchow's reference to Taenia (an intestinal worm of mammals) shows how a new branch of biology—parasitology—had

---

[1] Fixatives were introduced soon after 1830, carmine stain in 1849. The microtome was not introduced till 1870.

been founded, and helped to clarify the general problem. Pallas was the first to combat the universal view that parasites were spontaneously generated within the body of the host. Rudolphi, although he gave far more accurate descriptions and discovered many new species, unfortunately fogged the issue by returning to the old view in a modified form: while denying complete *abiogenesis* he nevertheless accepted the possibility that the tissues of the host might in some way change into living parasites. It was Siebold, one of the founders of comparative anatomy, who drew attention to the masses of eggs inside mature parasitic worms, and who solved the problem of the transfer of the egg after the whole question of life histories had been put in a clearer light—chiefly the work of Steenstrup (1842; see later p. 422).

It is difficult for the modern student of medicine to realise the revolution in thought which Virchow's *Cellular Pathology* started. At a single blow the whole theory and practice of medicine in respect of the understanding of the *mechanism* of disease was illuminated as never before. The living organism appears as a society of co-operating individuals; some, the amoebocytes, relatively free throughout their lives; others, the cells of growing tissue, rapidly losing their individuality, while retaining in greater or less measure their power of living; yet others, the red cells of the blood, doomed in the mammals to lose their nuclei and so become virtually " dead " and specialised to a narrow range of exchanges; and finally, the germ cells, budded off from the parent stock, sometimes voyaging forth into unknown seas as in the coelenterates, others retained within the parents up to a relatively advanced state of development. Disease of the organism is proximately the fault of the cells composing it; though shock, damage, or attack by some external agency may in some cases be the occasion of such degeneration. The microbian theory of disease was yet to come, but in relation to purrulent and cancerous growth Virchow correctly pointed out that " there is a stage when it is impossible to decide with certainty whether we have in a part to deal with a simple process of growth, or with the development of a heteroplastic, destructive form ". It is perhaps not surprising to know that this great teacher, though originating little, but realising as no one else the far-reaching implications of the new theory, was also a pioneer of public health services. With physical courage in the face of typhus and the moral courage to lay the cause of the scourge where it belonged—in the disgraceful social conditions of an industrialised region—he was constantly in trouble with the Prussian bureaucracy. But despite his liberal views, in which he actually opposed Bismarck, he ended his days as one of the most respected leaders of thought in Berlin.

Virchow had however left one aspect of cell structure unclarified; the word " protoplasm " does not occur in the full index to the *Cellular Pathology*. This last great synthesis was made by Max Schultze. Once again it may be said that there is nothing completely new about Schultze's work; but in seizing on the essentials of the existing knowledge, in ranging far and wide, and in particular in coining the definitive description of a cell as a " lump of nucleated protoplasm ", he left the problem of structure in a very different position from that in which he found it. Gone is any dalliance with " internuclear substance " and " membranes ". Underlying all is the self-developing, self-regulating, self-differentiating, plastic stuff, *protoplasm*. Associated with every *active living* mass, constituting the *unit*, is the nucleus. Two extensions alone remained to be achieved. Extending the view of Siebold that many of the " infusoria " were single-celled animals, Ernst Haeckel proposed a neutral kingdom of " protista " composed of organisms whose differentiation was so little marked that they must be regarded as neither animals nor plants. The division " protista " no longer appears in our text books, but it served a purpose in directing attention to the fact that certain organisms show slight differentiation along *both* lines and so are found in text books both of plant and animal biology.

The remaining advance in the cell theory was not so much an extension as the fashioning of a " missing link ". Cells had been observed to divide, but what was the *mechanism* of this division whereby a single fertilised cell may multiply to the compass of a whale or an elephant? In 1842 Nägeli had observed what he called the " transitional cytoblast " or disappearance of the nucleus at cell division. Progress waited on the introduction of adequate technique. By 1875 the microtome was in regular use; and the synthesis of the first aniline dye by Perkin in 1856, provided a means of discriminating detail within the cell and even the nucleus. This year was also marked by the accurate description of the phenomena of nuclear division in plants by the great botanical teacher, Strasburger. The more difficult demonstration in the case of animals was carried out by Walther Flemming, whose pioneer work in animal cytology appeared in 1872. The details of this process, which have since been corrected in many particulars, are well known to biological students and may be found in any text book of moderate compass. It may be as well to remind modern students however that the complete picture was arrived at even in its primitive and somewhat misleading form, only after several years of work to which Waldeyer, Boveri, and Golgi added essential contributions. We shall have occasion to refer to the significance of this process in the concluding chapter.

## SOURCES FOR CHAPTER XXIX

C. F. Wolff, *Theoria Generationis*. Excerpts in the above chapter are my translations of the *German* translation in Ostwald's *Klassiker*.

Lorenz Oken, *Elements of Physiophilosophy* (London, 1847). This is only a historical curiosity, but is important as an indication of what was actually taken seriously by men of high scientific standing, also as a psychological document of an aberrant genius.

R. Virchow, *Cellular Pathology*. Introduction (Berlin, 1858; trans. London, 1860).

R. Virchow, *The Freedom of Science in the Modern State* (trans., London, 1878). An important document for the history of the freedom of thought and the social relations of science.

C. Singer, *Short History of Biology*.

E. Nordenskiöld, *op. cit.*, more detailed account.

Two long extracts of Bichat's and Schwann's writings are given in Knickerbocker, *Classics of Modern Science* (New York, 1927).

C. Sherrington, *Goethe on Nature and on Science* (Camb., 1949). As authoritative an estimate of Goethe's place in the history of scientific thought as we are ever likely to get.

CHAPTER XXX

# THE HISTORY OF HISTORY:
# THE DYNAMIC THEORY OF FORMS

THE Big Idea of biology is evolution. It stands in relation to biology as the atomic theory to chemistry, and the theory of gravitation to physics. Once firmly established it transformed every branch of the science of living things—some rapidly, like taxonomy and anatomy, others like physiology, only in recent years. Its actual establishment was the work of one man, Charles Darwin. But the *Origin of Species* (1859) was the culmination not only of forty years of meditation by Darwin himself, but of more than two thousand years of groping: two thousand years marked by meteoric flashes of prophetic insight, long periods of tacit acceptance of the opposite view, punctuated with periods of downright rejection. The doctrine of evolution as such, with which alone we are at present concerned, is merely the belief in more or less orderly *change*, the assertion that the present is only the *product* and not the replica of the past. The history of the idea of evolution demands, and has often in the past called forth, a volume to itself. All we can do here is to scrutinise with some care the pronouncements of the accepted leaders in the movement, and to place the subject in a wider background than has often been the case.

The doctrine was first unambiguously asserted by Herakleitos (p. 16) " you do not step twice into the same rivers "; first systematically, but still speculatively, developed by Demokritos, and sung in " immortal verse " by Lucretius. While not expressly denied by Plato as regards the world of *phaenomena*, its relegation to that world in contrast to the " real " world of changeless intelligible forms influenced the greatest minds down to Linnaeus and Cuvier with their insistence on a fixed number of God-given immutable species. Reference to what we have said about

Aristotle (p. 327) shows as usual the admirable spirit of compromise to which this great man was prone. There is a " scale " of living creatures, each in some measure an imperfect realisation of the one above; process is not ruled out—at any rate at certain levels.

The Middle Ages, ruled by Aristotle's logic, and in complete ignorance until St. Albert and St. Thomas of his natural history, had perforce to deal only with determinate " substances ". And when later the *pseudo*-Aristotelianism of this period gave place to a more liberal culture, the latter was tinged rather by Platonism than by the whole Aristotelianism which, as we have seen, reached full stature only as it was ripe for supersession—in Fabricius and Harvey. An explicit disavowal of the changelessness of the form of the universe was liable to lead to martyrdom, as was the case with Giordano Bruno (1600)—perhaps the first modern pioneer of evolution.

How difficult it was, even in the comparatively liberal atmosphere of seventeenth-century England, to grasp the nettle of evolution is well illustrated in the case of Ray. Good naturalist as he was, he was struck by the marked variation in appearance of what seemed undoubtedly the same species growing in different conditions. Even more remarkable was his interpretation of fossils as evidence of dramatically changed conditions of the land surface and of the organisms it had borne when submerged beneath the waters. At that time there were three current views as to the nature of fossils, or " formed stones " as they were naïvely called. The *pseudo*-Aristotelians believed that they represented the abortive efforts of a *vis plastica* or *formativa* to induce the " form " of organisms upon the " matter " of the earth. In an age dominated by attempts to rationalise theology this was largely replaced by the view that fossils represented the detritus of living organisms stirred up by Noah's deluge and allowed to sink back into the mud. Even Woodward, a judicious collector and founder of the famous museum at Cambridge, adopted it. Lastly there were a few bolder spirits who inclined to the belief hinted at by Herodotus, revived by Leonardo and Fracastorio, that fossils were indeed the actual remains of organisms preserved in the earth. Ray's good sense inclined him to this view, but his critical scientific method suggested two objections: First, that no living organisms were identical with fossils; second, how came the latter to the tops of mountains or the middle of land masses? The first—involving a wide-spread extinction and replacement at least of genera, he could find no answer to. The second is more plausible but " since the most ancient times recorded in history the face of the earth hath suffered little change ", and therefore if the mountains were not there from the beginning, either the world is a great deal older than is imagined, there being an incredible space of time required to work such changes ... or in the primitive times the creation of the earth suffered far more concussions and mutations in its superficial parts than afterwards. But the temper of the age was too strong for this unassuming, cautious man. The date of the creation had been calculated to a nicety as only 5,600 years previous to his observation. In such a short time how could so much have happened as in his disciplined imagination he recreated? So the chance was lost.

It is impossible to put a date to the renewal of the claim for the fact of evolution, but it is generally agreed that it was in Buffon's *Histoire Naturelle*, whose publication occupied fifty-five years, that the idea was expressed as interpretative of more than one series of natural objects. At this time, despite Linnaeus, evolution was in the air, or rather in the earth, for it was from the rapidly increasing study of the rocks, in the light of knowledge recently gained from physics and chemistry, as also from systematic botany and geology, that the evidence came which permitted *of no alternative theory*. The doctrine of evolution had not at this time, nor has it essentially, any necessary connexion with the " origin of species " as set forth by Darwin or anyone else. Let us digress a little to consider the " testimony of the rocks ".

Though " theories of the earth " had not been wanting from the time of the writers of *Genesis*, the first to be based on a systematic, though as it turned out, a one-sided, observation of nature was the so-called " Neptunist " theory of Werner. Owing to the increased demand for ores consequent on the improvement of smelting technique, there had grown up at Freiberg a school of mines, to which Werner, himself a member of a family connected with mining for three centuries, was appointed Inspector of Collections and Teacher in 1775. Writing hardly anything, but reading almost everything from the pioneer work of Steno (who, owing to his involvement in a bitter religious controversy, had far less influence on his contemporaries than the quality of his work demanded), to those of his own contemporaries, of whom the chief was Bergman, Werner explained the origin of the rock formation as being due to a series of depositions beneath a primordial ocean which covered the surface of the earth. His theory forms the basis of the first systematic study of the earth's surface and paved the way for more accurate and less ambitious views. Fossils were treated with scant respect, and were used not at all in the placing of the rocks in a temporal sequence. The system is therefore of interest to us only as the spur to an amiable Scottish gentleman, whose amateur studies of the rocks of Berwickshire and the glacial débris of East Anglia had prompted him to different views.

James Hutton's *Theory of the Earth* was printed in the Transactions of the Royal Society of Edinburgh for 1788. There had been many—too many—high-flown speculations printed under a similar title, so it is not altogether surprising that it attracted hardly any attention. Fortunately for the progress of geology, the most completely novel, though not in the end most valuable, part called forth a scathing attack from the celebrated chemist and mineralogist, Kirwan. Hutton's pride was roused and he responded by rewriting the whole paper in a book of three large volumes published in 1795. It is appallingly repetitive and leisurely in style, but the thought inspiring it is so vital as to break through in many memorable passages. His purpose is " to examine the construction of the present earth, in order to understand the natural operations of time past; to acquire principles by which we may conclude with regard to the future course of things, or judge of those operations by which a world so wisely ordered, goes into decay; and to learn by what means such a decayed world may be renovated, or the waste of habitable land be repaired ". The stimulus to this attempt came perhaps from the farmer in him; for near the beginning of the book he observes that:

A solid body of land could not have answered the purpose of a habitable world; for a soil is necessary for the growth of plants; and a soil is nothing but

## THE DYNAMIC THEORY OF FORMS 399

the materials collected from the destruction of the solid land. Therefore the surface of this land ... is made by nature to decay, in dissolving from that hard and compact state in which it is found below the soil; and this soil is necessarily washed away, by the continual circulation of the water, running from the summits of the mountains towards the general receptacle of that fluid.

The heights of our land are thus level with our shores; our fertile planes are formed from the ruins of the mountains; and those travelling materials are still pursued by the moving water, and propelled along the inclined surface of the earth.

Thus we are to explain the formation of valleys and plains—not so obvious to his generation, seeing that de Luc, one of the leading geologists, flatly denied that "gravel and sand are carried from our continents to the sea". The Aristotelian approach to the problem is interesting in one of Hutton's generation; but in him the relation of means to ends is not used as a *proof*. We must at all times examine " the natural operation of the globe in order to see if *there actually exist such operations* as, from the nature of the solid bodies, *appear to have been necessary to their formation* " (italics mine). This is the key to Hutton's greatness. Gone are the cataclysms demanded by the speculators; gone on the other hand the neat and orderly Wernerian ocean from which all the rocks crystallise in due season; and in their place the inexorable processes *that the imaginative eye can see acting all around us every day*. The world is evolving (in one place he actually likens it to an organism, harking back perhaps to the Greek hylozoists) and we are set in the midst thereof.

In considering the cause of the consolidation and upraising of the sea bed (the *fact* of which is attested by de Saussure's report of molluscs embedded in the rocks of the high Alps), he is less happy. Chemistry was too little advanced to teach him the " abnormal " effects of pressure on the solubility of silica. So he had to look for evidences of the alternative—the melting power of *heat*. And he found indisputable vestiges of its action in the fact that " the regular form and horizontal direction of strata throughout this country of coal ... has been broken and disordered by the irruption and interjection of those masses of basaltic stone or subterraneous lava." From this he " would ask those naturalists who adhere to the theory of infiltration and the operation of water alone, how they are to conceive those strata formed and consolidated." Of the molten origin of basalt he was further assured by finding *coke* on the edge of an Ayrshire coalfield, where it was in contact with a basalt dyke.

Hutton's preaching of "uniformitarianism", as the doctrine is usually called, might have gone unchallenged; but its corollary, the igneous origin[1] (so called because of his belief in subterranean *fire*) of many great rock masses undermined the reputation of Werner and his many followers. It is doubtful whether the painstaking effort which ended the old man's life would have established his theory; but fortunately there were, as there are to-day, glassworks in the neighbourhood of Edinburgh, and there his devoted friends Hall and Playfair demonstrated by actual experiment that a *mixture* of substances having the total composition of basalt, on being suitably cooled from a state of fusion, produced material indistinguishable from natural basalt. The record of their *Illustrations of the Huttonian Theory of the Earth* appeared in 1802; but the result was slow of acceptance.

Ten years before Hutton was telling the Royal Society of Edinburgh that the earth was continually evolving, Kant published similar views about the cosmos. From the distribution of stars in the milky way he concluded that it is in fact a mature member of the select band of island "universes" or spiral nebulae.[2] The reason for its saucer-like appearance is the aspect from which we view it—somewhere near the middle. The spirality of these systems is due, he thought, to the supposition that attenuated masses of gas gravitating on themselves would start to spin. When the centrifugal force of rotation exceeded the gravitational force, pieces would fly off, thus giving rise to other spinning systems in which the process is repeated until planetary systems are produced about a central star or sun. A magnificent sweep of imaginative genius, based unfortunately on an incorrect premise, namely that a contracting mass can acquire *rotation*— a violation of the law of conservation of *angular* momentum. But

---

[1] Desmarest had expressed a similar opinion in 1774, but he was in error in regarding granite as a *volcanic* rock.
[2] The task of cataloguing these cosmical objects was being carried out at this time by Messier, Lacaille, and Herschel. Their correct interpretation was almost entirely the work of Herschel, who with uncanny insight distinguished between nebulae within the galaxy and those extra-galactic "universes" referred to above.

though the details are incorrect, astronomers have little doubt that the nebulae and stars are evolving systems, and the earth an end-product of such cosmic evolution, produced more probably by tidal action between two " colliding " stars—an exceedingly unlikely event, which thus restores to man the pride of uniqueness to compensate him for his spectacular fall from the centre of the universe to an inhabitant of a third-rate planet circling about a third-rate sun in one of a vast number of nebulae.[1]

Neither of the works we have just reviewed had any marked influence on their contemporaries; the one we are about to consider lies in a different category. It was certainly with no thought of the light it might cast on the problem of evolution that the practical-minded William Smith set about his hobby of collecting fossils and *correlating them with the nature of the rocks in which he found them.* His employment as civil engineer and surveyor in the construction of roads and of one of the canals which began to supplement the inadequate and badly constructed roads of those days, gave him unrivalled opportunity to follow his bent. Possessed of a phenomenal memory, he committed nothing to writing, but in 1799 dictated a list of fossils in the order in which they could be gathered from a succession of rocks ranging, as we now know, from the Carboniferous to the Cretaceous systems. The enthusiastic reception of this aid by his fellow-geologists encouraged him to prepare a series of maps which culminated in 1815 in the coloured geological map of England and Wales—the first of its kind in the world, and a model for all subsequent ones. Smith's work, undertaken with no thought of pecuniary reward (which he never got), or of application to social or economic use, provided geologists for all time with a *language* and a *scale of reference* for the correlation of rocks all over the world. The rapid spread of such methods must have been of inestimable influence in the subsequent spread of the idea of geological and biological evolution: in particular, as we shall see, it was from this vantage point, and not from that of comparative anatomy in which he was deficient, that Darwin first glimpsed the Promised Land.

[1] For an alternative view see the article " The Origin of the Solar System " in *Nature*, Feb. 12, 1949.

While all this was going on, the first definitive step towards a theory of biological evolution was taken by Lamarck. Lamarck's own " evolution " was of the catastrophic kind. Fleeing from his Jesuit college at the age of 16 he caught up the French army on the eve of the battle of Fissinghausen. In the following days' fighting Lamarck had greatness thrust upon him by the death of all the officers of his company. His courage and resource in command earned him an immediate commission. After five years in the army he was pensioned as the result of an accident and lived in poverty as a bank clerk, hack journalist, medical student, musician, and botanist. In the last category he rose to the highest position of trust—keeper of the herbarium in the *Jardin du Roi*. Then came the Revolution. After many delays—the Chair of Botany in the new *Museum d'Histoire Naturelle* being filled—he was appointed to the Chair of Zoology, devoted to the " inferior animals ". Thus at the age of fifty, married for the second time, with six children, and a miserable pittance to live on, he was faced with the task of professing a subject of which he was largely ignorant.[1] Yet within ten years he began the publication of the *Histoire Naturelle des Animaux sans Vertèbres*, which was the culmination of a series of classifications and in which order was at last brought into the Sub-Kingdom *Invertebrata*—a term first introduced by Lamarck—comprising far more species than the group *Vertebrata*. The hotch-potch of forms called *Vermes* by Cuvier was broken up, as was the group *Radiata*: these were rearranged to give *Radiaria* (comprising echinoderms and many of what are now called *Coelenterata*) *Polypes*, and soft worms. The *Articulata* of Cuvier were broken down into *Crustacea* (which Lamarck wrongly grouped with *Annelida*), *Insecta*, and *Arachnida*. The final system was far in advance, as regards natural relationships, of anything which had gone before.

Our chief concern however is with his *Philosophie Zoologique*, published in 1809, and in the preface of which he tells us that " in order to fix the principles and establish rules for guidance in study [and teaching] I found myself compelled to consider the organisation of the various known animals, to pay attention to the singular differences which it presents in those of each family,

[1] Except for a, at that time, rather sketchy knowledge of " shelled " animals.

each order, and especially each class; to compare the faculties which these animals derive according to its degree of complexity in each race . . ." The stimulus was therefore that of classification. As a result of these studies he asks himself " How could I avoid the conclusion that nature had *successively* produced the different bodies endowed with life, from the simplest worm upwards? " (italics mine). The acceptance of this belief in successive development of forms of higher and higher complexity of organisation forced him to deny the existence of any taxonomic divisions in nature, in which only an ascending ladder of individuals exists; though owing to various circumstances, particularly the approximation to a specific kind, convenience dictates the setting up of such artificial divisions. But how does this ladder of forms come into being? Not by any special creation, nor by spontaneous generation (except of the *lowest* forms of life), but by the modification of structure by use and disuse, and its subsequent inheritance. Since this is the crux of the whole Lamarckian theory of evolution it is imperative to quote:

> In every animal which has not passed the limit of its development, a more frequent and continuous use of an organ gradually strengthens, develops, and enlarges that organ, and gives it a power proportional to the length of time it has been so used; while the permanent disuse of any organ imperceptibly weakens and deteriorates it, and progressively diminishes its functional capacity, until it finally disappears.
>
> All the acquisitions or losses wrought by nature or individuals, through the influence of the environment in which their race has long been placed, and hence through the influence of the predominant use or permanent disuse of any organ; all these are preserved by reproduction to the new individuals which arise, provided that the acquired modifications are common to both sexes, or at least to the individuals which produce the young.

These are the famous, or ill-famed, according to taste, laws of the inheritance of acquired characters. Probably few biologists would question the truth of the first law, which is almost an obvious fact of observation, but whose importance was bound to be overlooked by those to whom fixity of species and perfection of means to end were dogmas. The second is in an entirely different category. So far it has never been possible to give unambiguous experimental proof of its accuracy; but such failure has not been a justification for the scorn poured upon the principle, nor for the many peculiarly irrelevant experiments made in connexion with it. I refer to the absurd attempts to breed rats with short tails by amputation through a long series of generations. Lamarck's theory is evidently not concerned with *mutilations* or *accidents*; his emphasis is laid on *use*. The only relevant process would therefore be to dissect out the tail muscles so that the use of the tail became impossible; but such drastic procedure would hardly be evidence even were it to succeed. Though Lamarck often pushed the hypothesis to limits which to us seem absurd, in drawing attention to " race-horses shaped like those in England, . . . draught horses so heavy and different from the former, . . . basset hounds with crooked legs, . . . greyhounds so fleet of foot, . . . fowls without tails, fantails, etc." he was stating a perfectly reasonable hypothesis (not of course, as he called it, a " law "). The weakness of the Lamarckian hypothesis is not its absurdity, but its one-sidedness. In the examples quoted above he ignores the other factors involved, namely the *selection* exercised by man —incidentally, why should the basset hounds' legs become crooked through either use or disuse? But it is as unjust to brand Lamarck with the mark of dogmatic stupidity; on the contrary he deserves a place in the history of biology no less honourable than that of Copernicus in astronomy. In a biological atmosphere of static forms he created a revolution; he also created the science of *biology* when in 1802 he first used the term. (The botanist Treviranus used it in the same year.) We do not laugh at Copernicus because, ignorant of the inverse square law of gravitation, he maintained a system of epicycles for the planets; nor should we belittle the achievement of Lamarck because he failed to take account of selection.

One other feature of Lamarck's teaching must be mentioned in this place, namely his firm belief in an unbroken ladder to which has been added rung after rung until man appeared as the highest. For complete consistency it would be necessary to show that every change in every organ should be an improvement; but this is manifestly not the case, for surely no one would go so far as to suggest that the sucking lamprey is a more " perfect " animal than the incredibly swift and beautifully controlled dragonfly? Lamarck has to admit that several of the divisions which in their main characteristics are allied to the more "perfect" animals must be regarded as off-shoots which have fallen away from grace by " degeneration ", and consequently are not to be regarded as being on the main sequence from which all " higher " types have evolved. Unfortunately he lays himself open to misrepresentation by describing his *belief* in an innate urge to complexity and consequent perfection in terms of a *degradation* from the highest to the lowest—an inversion in exposition presumably out of deference to the lingering Aristotelianism implied in all contemporary natural history.

Lamarck's theory had comparatively little influence, even in a general way. During the years that he held the Chair of the Inferior Animals, Cuvier also was in Paris. The latter's genius was of a very different sort from Lamarck's. A mastery of detail, a reverence for fact, the use of the imagination only within the limits prescribed by observations: these were his strength; and at a time when natural history was struggling to become an exact science, it was perhaps well that his influence should have prevailed. It may seem strange to us that the man who, more than any other, forged the palaeontological weapons which are the evolutionist's most potent aids, should have refused to budge an inch from a belief in the fixity of species; should have called in the aid of hypothetical world catastrophes to *remove* from the earth those dominant species whose forms he had reconstructed with such skill; but history often reveals such paradoxes. He died in 1832; two years earlier appeared the first volume of Lyell's *Principles of Geology*, which perhaps more than any other scientific work provided a climate of opinion in which Darwin's work could ultimately flourish.

The sub-title of Lyell's great work is *An Attempt to Explain the Former Changes of the Earth's Surface by References to Causes now in Operation*. The method by which he proposes to carry out this task is the *historical* method; his own words are worth quoting:

> We often discover with surprise, on looking back into the chronicles of nations, how the fortune of some battle has influenced the fate of millions of our contemporaries, when it has long been forgotten by the mass of the population. With this remote event we may find inseparably connected the geographical boundaries of a great state, the language now spoken by the in-

habitants, their peculiar manners, laws, and religious opinions. But far more astonishing and unexpected are the connexions brought to light when we carry back our researches into the history of nature. The form of a coast, the configuration of the interior of a country, the existence and extent of lakes, valleys and mountains, can often be traced to the former prevalence of earthquakes and volcanoes in regions which have long been undisturbed.

The hypothesis to be maintained is no less than that the whole *earth* is in a process of evolution, that the present is the product of past conditions, and is in process of change to a future which may be as different as is the present from the past. For the *causes* whereby the present has been fashioned out of the past *are still in operation*. There was nothing fundamentally new about this hypothesis; although Lyell made original contributions to certain departments of knowledge, it is rather as a synthetic thinker of unusual power that he promoted the progress of science. Moreover he had the quality, unusual in such a thinker, of allowing his views to be *guided* only by the writings of others, but to be *based* only on observation. It was in fact only to reassure himself about certain aspects of the tertiary formations that he postponed the completion of his book and toured France and Italy with Murchison. It thus came about that the Third Volume, which he found himself compelled to add to the two originals contemplated, was not published till 1833. It is impossible even to summarise this great work; it amounted to a re-writing of the whole of the existing science of physical and historical geology, as contained in the works of its founders, from the special point of view of the author, the key points being driven home by detailed descriptions and interpretations of his own observations. What was chiefly new in the result was the recognition of the mutual relations between the solid and aqueous shell of the earth and of the living creatures which have inhabited it. The greater part of the Second Volume is therefore taken up with the question as to the constancy or otherwise of species, the finding of an answer to which demanded a thorough investigation of the conditions of fossilisation and consequent preservation of the " documents " upon which any conclusion must be based. His great service in this volume is in making the question of the mutability of species " respectable ". A glance at the full table of contents reveals an open-mindedness at the outset of the quest such as never characterised the earlier writers. Lamarck's *Zoologie Philosophique* reads like an oration designed to convince an audience of the truth of a proposition, which only their ignorance or prejudice could prevent of a ready acceptance. Lyell's work consists of a series of questions which are to be considered in the light of the evidence. It must however be added that his conclusion—namely that although a considerable degree of variation of each species must be accepted, there was no evidence for any actual transformation—was arrived at a little too easily in view of the paucity of the geological remains. It is perhaps significant that Lyell was trained for the legal profession, but through an early weakness of the eyes was compelled to renounce all close work—his subsequent studies being made possible only by the whole-hearted devotion of his wife to the labours which were so dear to him. If on the one hand he cleared the air of the wilder fantasies of Lamarck, on the other he removed the cramping influence of Cuvier's catastrophism. The latter's explanation of the complete disappearance of whole classes of living forms by the postulation of recurring disasters on a planetary scale, was in fact a refusal to admit that any problem existed. Of such catastrophes no evidence existed: their postulation was simply the shift of a mind wholly absorbed in the delineation of " static forms "—a simple means of getting rid of an awkward question. Here however once again Lyell's caution in *method* deserted him in the final statement of his opinion, namely that although changes in the earth's surface were continuous, they had never differed in kind from those which may be witnessed to-day; he even went so far at one time as to deny the *possibility* of the earth's solar origin—despite the cogent arguments for such an " evolutionary " stage by Laplace and Kant. It must be remembered however that in attempting to reconstruct the growth of the ideas of evolution we are at present confining our attention to the *Principles of Geology* (1830-3) (re-arranged and written in the form of a text-book as *The Elements of Geology* first published in 1838). But if Darwin was greatly influenced by Lyell, the debt was repaid in some measure, so that in *The Geological Evidences for the Antiquity of Man (with Remarks on Theories of the*

*Origin of Species by Variation*), published in 1863, Lyell modified his views; we shall refer to this later.

When Charles Darwin sailed from Plymouth on 27th December 1831 in *H.M.S. Beagle*, he took with him the first volume of the *Principles*; the second reached him at Monte Video a year later. He could hardly have had a better guide to the study of the geology of the American continent, which was his main purpose in joining the *Beagle* as (unpaid) naturalist. His knowledge so far had been gained almost entirely by casual contacts with fishermen, game-keepers and fellow-students of medicine. His academic training had been entirely negative: at Edinburgh the anatomy lectures of Monro were in his estimation equalled in dullness only by those of Jameson on geology. At Cambridge he did not even risk going to hear Sedgwick; but he became sufficiently acquainted with him to accompany him on a geological tour of Wales—doubtless a most valuable experience. It was to Henslow, Professor of Botany at Cambridge, that he owed the development of his capacity for patient collection of data and that enthusiasm for the study of nature which became an obsession almost to the exclusion of every other interest. It was through Henslow that he received the " generous " offer to collect data for H.M. Government unpaid and at his own expense. Well might his father jib at such a proposal; but being at last persuaded that natural history was very suitable for a clergyman (for which calling Charles was preparing himself at Cambridge), he yielded to his son's entreaties. Had he known the sort of " natural history " his son's genius was to give birth to, he might have been even more reluctant to yield!

It was thus as a geologist that he began his life's work. " There is nothing like geology "; he wrote home from the Falkland Islands " the pleasure of the first day's partridge shooting or first day's hunting cannot be compared to finding a fine group of fossil bones, which tell their story of former times with almost a living tongue." The influence of Lyell is clear; but as yet there is no questioning of the nature of species. *The Origin of Species*, written twenty years after, opens with the words: " When on board *H.M.S. Beagle* as naturalist I was much struck with certain facts in the distribution of the inhabitants of South America, and in the geological relations of the present to the past inhabitants of that continent." What were these "facts"? The first recorded hint is a note (in his diary for 22nd September 1832) of the finding of bones of extinct species of armadilloes which he compared with those of living species. This is no more than a hint; but the crucial observations were made in the Galapagos Archipelago, about five hundred miles from the mainland, in September-October 1835. The only written evidence that it was at this time and place that his doubts on the immutability of species were stirred, is an obscure note discovered posthumously, and printed in *Nature*, 7th September 1935. It is unnecessary to quote it as his reactions were set out much more fully in the *Origin*:

Here almost every product of the land and water bears the unmistakeable stamp of the American continent. There are twenty-six land birds, and twenty-five of these are ranked by Mr. Gould as distinct species, supposed to have been created here; yet the close affinity of most of these birds to American species in every character, in their habits, gestures, and tones of voice, was manifest. So it is with the other animals, and with nearly all the plants, as shown by Dr. Hooker in his admirable memoir on the Flora of this archipelago. The naturalist, looking at the inhabitants of these volcanic islands in the Pacific, distant several hundred miles from the continent, yet feels that he is standing on American land. Why should this be so? Why should the species which are supposed to have been created in the Galapagos Archipelago, and nowhere else, bear so plain a stamp of affinity to those created in America? There is nothing in the conditions of life, in the geological nature of the islands, in their height or climate, or in the proportions in which the several classes are associated together, which resembles closely the conditions of the South American coast; in fact there is a considerable dissimilarity in all these respects. On the other hand, there is a considerable degree of resemblance in the volcanic nature of the soil, in climate, height, and size of the islands between the Galapagos and Cape de Verde Archipelagos; but what an entire and absolute difference in their inhabitants! The inhabitants of the Cape de Verde islands are related to those of Africa, like those of the Galapagos to America. I believe this grand fact can receive no sort of explanation on the ordinary view of independent creation.

Now it is a strange fact that this passage occurs near the end of his book, if it was indeed this and similar facts which compelled him to renounce the orthodox belief in separate creation and fixity of species. Nor was this a matter of chance, for exactly the same kind of observation at the other side of the world prompted Alfred Russell Wallace to send home to Darwin a paper on what he believed to be an entirely novel view on the origin of species. Wallace had in the course of his survey of the Malay Archipelago been dumbfounded by the complete difference between the fauna and flora of the islands of Bali and Lombok, separated by only a narrow but very deep channel; whereas the former is markedly Indian in character, the latter is just as definitely Australian. The probable reason for placing these cardinal facts at a late stage in the book was that the conclusions to be drawn from them were of a negative character: granted the existence of endemic species on geographically *isolated* islands, it was difficult to believe that Providence was so prodigal of designs as to provide every tiny island with a special creation. But if such special creation was denied, how were such peculiar species to be accounted for? To this Wallace had no answer. It is not too much to say that it was the elaboration of the *answer* to this question, rather than the recognition of the need for an answer, which gave Darwin his unique position in popular as well as scientific esteem. Before we turn to consider the answer, we may note that the other piece of circumstantial evidence for the non-fixity of species, namely the sequence of living forms preserved in the rocks, which we know to have been the first to awaken Darwin's doubts, is also reserved to the latter part of the book. In Darwin's day the record, as he himself strongly emphasised, was such that " of this history we possess the last volume alone relating to only two or three countries. Of this volume, only here and there a short chapter has been preserved; and of each page, only here and there a few lines." From such evidence mutually opposite conclusions could be, and were, drawn according to the temperament and outlook of the individual concerned.

Convinced in his heart of the origin of species by descent, like varieties from other species, he sought a mechanism and found it, as he thought, by the study of domesticated animals and cultivated plants, the results of such studies being taken in connexion with the teaching of Malthus (by whose book, *Essay on Population*, 1817, he was much influenced) that in consequence of the power of reproduction tending to outrun the means of subsistence, there must ensue a constant struggle for existence. Malthus had taught this only in relation to *human* populations. Darwin applied it to

## THE DYNAMIC THEORY OF FORMS

the whole animal and vegetable kingdoms. "As many more of each species are born than can possibly survive, and as, consequently there is a frequently recurring struggle for existence, it follows that any being, if it vary however slightly in a manner profitable to itself under the complex and sometimes varying conditions of life, will have a better chance of surviving, and thus be *naturally selected*. From the strong principle of inheritance, any selected variety will tend to propagate its new and modified form." The hypothesis is ingenious, plausible, and simple. But the possibility of its truth depends on two assumptions, namely that organisms do continually vary, and that those which survive as a result of their advantageous variations *will propagate these variations*. The first assumption Darwin was able to make good by a wealth of incontestable facts. The second was an entirely different matter. What Darwin meant by " the strong principle of inheritance " it is difficult to say; it is probably an example of that looseness of expression and use of question-begging phrases due to a lack of formal academic training in the sciences—the debit side to that marvellous freshness of outlook and innate virtuosity apparent even from the days when the pebbles of his father's drive awakened his childish curiosity. There was in any case no *science* of genetics when the *Origin* was written (" the laws governing inheritance are quite unknown ")—all unbeknown to Darwin it was being founded by Mendel at the time the *Origin* was published, but the latter's work attracted no notice until about 1900. To discover the nature of inheritance, Darwin had to rely on the reports and stud-books of animal breeders. As Prof. J. B. S. Haldane has pointed out, he was at a disadvantage here owing to a recent political event—the repeal of the Corn Laws—which had brought about an almost complete lack of interest in plant breeding. Had Darwin been living in France he might have learnt from de Vilmorin—what was later proved with statistical thoroughness by Johannsen—that variations in wheat were of two kinds—some which remained " fixed " and were therefore of use to the breeder, and others which were unreliable.

Darwin's supposed " proof " of the origin of species by natural selection depended on an analogy alone. It was impossible to detect the production of a new species in nature; he therefore directed attention to what he called " artificial selection ". In a chapter which has become famous he describes the many types of domestic pigeons: " at least a score of pigeons might be chosen, which if shown to an ornithologist, and he were told that they were wild birds, would certainly, I think, be ranked by him as well-defined species ". Some forms indeed he ranked as belonging to different genera; yet " I am fully convinced that the common opinion of naturalists is correct, namely that all have descended from the Rock Pigeon (*Columba livia*), including under this term several geographical races or sub-species which differ from each other in the most trifling respects." If it be objected that from " varieties " only new " varieties " can be derived, Darwin asks how it is to be decided when a " variety " is sufficiently distinct to be ranked as a species, and quotes several authorities as being at variance over many groups of both plants and animals. He concludes that " to discuss whether they are rightly called species or varieties, before any definition of these terms has been generally accepted, is vainly to beat the air." The point is that by artificial selection the fancier has produced divergences of the most extreme kind. What the fancier can do in a century surely the haphazard methods of nature can achieve in millenia. To Darwin the cogency of the analogy seemed overwhelming. But with the return of interest in plant-breeding it was generally demonstrated that so far as the evidence goes, it is only the marked and abrupt changes (mutations) which are inherited; and such changes are commonly of such a nature as to upset the harmonious development or adaptation of the organism. We are consequently compelled to regard " the origin of species by natural selection or the preservation of favoured races in the struggle for life " as largely illusory. That it is *a* means of producing new species few would now deny; but our increased knowledge of the criteria of specificity through deeper understanding of the heritable material of the cell has driven us further along the sceptical path than Darwin ventured. We not only are unable to produce a " generally accepted definition of species " but are compelled to admit that in so far as species can be recognised as such they must be regarded as of different kinds—that is as acquiring their specific character through different causes.

If then one of the fundamental assumptions underlying Darwin's hypothesis was false, why was he acclaimed throughout the civilised world as being, in Kant's prophetic phrase, the " Newton of the blade of grass "? Nordenskiöld gives an answer, which I think has been too little emphasised in the older histories of science, namely, that Darwin was born into a world in which the liberal ideology of Rousseau, Thomas Paine, Bentham and James Mill was reaching the climax of its powers. Everywhere was being heard the call to cast off the shackles of authority so that the free spirit of man might develop its full stature. Only let nature take its course and progress was inevitable. The doctrine was given systematic scientific expression by Comte (*Cours de Philosophie Positive*, 1830-42) and was employed as a political instrument in the *Communist Manifesto* of Marx and Engels in 1848. It is true that the dictatorship of the proletariat called for in the *Manifesto* was to be brought about only by revolutionary seizure of power, but the whole tenor of Marx's teaching is that this act of violence is but the culmination of an inevitable process—an *evolutionary* process in the character of human society. This new emphasis on history as, so to say, a *cause* and not merely a record of events, finds an echo in Darwin himself. In a memorable passage towards the end of the *Origin* he writes: " When we no longer look at an organic being as a savage looks at a ship, as at something wholly beyond his comprehension; when we regard every production of nature *as one which has had a history*; when we contemplate every complex structure and instinct as the summing up of many contrivances, each useful to the possessor, nearly in the same way as we look at any great mechanical invention as the summing up of the labour, the experience, the reason, and even the blunders of numerous workmen; when we thus view each organic being, how far more interesting, I speak from experience, will the study of natural history become."

Darwin had the good fortune to be born into a world that was waiting for him. Even the materials of his theory—variation, the fossil record, embryological correlation, unity of morphological plan, the struggle for existence, artificial selection—lay ready at hand. He was fortunate also in combining the aid of comfortable worldly circumstances with a certain independent intellectual obstinacy. To these he had to add (or selection had added for him) an outstanding intellectual honesty and immense patience in the face of physical debility and the need for handling a vast range of facts. He showed evolution—however caused—to be a *vera causa* in the sense that no biological fact is isolated either in space or time, but it speaks of whence it came while hinting at whither it is going, balanced precariously within the web of relationships of its environment, seen and unseen. The " origin of species " was a comparatively unimportant and rather faulty by-product of a transvaluation of all biological values. In the next chapter we shall have to glance at some of the changes in outlook which were thus brought about.

## SOURCES FOR CHAPTER XXX

Nordenskiöld, *op. cit.*, Chs. X-XII. An excellent account of the historical background and climate of opinion of the age of Da win.

Charles Darwin, *The Origin of Species* (London) and *The Voyage of the Beagle*, ed. Nora Barlow (London, 1945). The latter contains much interesting autobiographical and biographical matter relating to Darwin's early life and crucial voyage.

James Hutton, *Theory of the Earth* (Edinburgh, 1795-9).

Karl von Zittel, *History of Geology and Palaeontology* (London, 1901).

Sir Charles Lyell, *Principles of Geology* (London, 1830-3; 12th ed., 1875).

J. B. Lamarck, *Zoological Philosophy*, trans. Elliot (London, 1914). A long book, largely obsolete, but giving an invaluable insight into the ferment of ideas on the problem of evolution and taxonomy at the beginning of the nineteenth century.

C. E. Raven, *John Ray*, Chs. XVI and XVII

Julian Huxley, *Evolution—The Modern Synthesis* (London, 1942). The historical introduction provides a valuable critique of " Darwinism " and a re-assessment in the light of modern knowledge.

David Lack, *Darwin's Finches* (Cambridge, 1947). The Galapagos finches to-day. It would be difficult to find a better introduction to the "species problem" than this beautifully illustrated work.

## CHAPTER XXXI

## THE DARWINIAN REVOLUTION

LOOKING back over the history of science we may discern three revolutions. The Copernican Revolution reversed the accepted relation between the earth and the universe; the Newtonian Revolution reversed for the time being the relation between what may be roughly called the material and spiritual powers in the ordering of events in inorganic nature. The Darwinian Revolution extended the idea of ceaseless, but in principle predictable, change to the world of organised beings *including Man himself.* Compared with these world revolutions the *Révolution Chimique* was for all its importance almost a domestic affair.

To get the " feel " of living in the early days of the Darwinian Revolution one can hardly do better than glance through the closing chapters of Lyell's *Antiquity of Man*, published in 1863. Lyell occupied an almost unique vantage point to survey the scene of action. He had himself, as we have seen, prepared the way for the revolution by detailed elaboration of Hutton's demonstration of the evolution of the earth's crust. He had been the recipient, as early as 1844, of Darwin's confidence; and had with Hooker " repeatedly urged him to publish without delay, but in vain as he was always unwilling to interrupt the course of his investigations ". Finally it was to him that Darwin sent Wallace's paper at the latter's request, which brought about the publication of both men's theories. In the *Antiquity of Man* Lyell points out that there were at this time two schools of thought, namely those who believed in " progression ", and those, at that time almost restricted to Darwin, Wallace, and Hooker, who believed in " transmutation ". He quotes Adam Sidgwick as having defined the former view in 1860 as follows: " This historical development of the forms and functions of organic life during successive epochs seems to mark a gradual evolution of creative power manifested by a gradual ascent towards a higher type of being. But the elevation of the fauna of successive periods was not made by transmutation but by creative additions." Owen, Agassiz, and Hugh Miller are also quoted as speaking in the same vein as regards animals; Brogniart is stated to have come to the same opinion as the result of his great work on fossil plants. It is clear that this view is a development of Cuvier's teaching. The facts—an observed succession in time of more and more highly developed dominant types—are correctly, or almost correctly, stated; though Lyell was able to show that this abrupt "stratification" of types was an over-simplification based on the presumed *absence* of overlapping—an assumption which wider investigation was continually negating. From the facts moreover a rich store of knowledge of *functional* correspondence had been derived. Owen, that implacable opponent of Darwin, had indeed introduced the terms "analogous" and "homologous" in relation to every comparison of similar organs. Thus the wing of a bat is *analogous* to the wing of a bird, but dissection reveals its *homology* (or unity of morphological plan) with the *hand* of other mammals. What was denied by the "progressionists" was not "progress"—the general climate of opinion was too powerful an influence for that—but progressive *change*. It is the old story of the Platonic forms still haunting the stage. The wing of the bat, the flipper of the seal, the hand of the violinist, have not developed out of a primordial five-membered limb by any historical succession, but only in the mind of God, out of which they have become actualised by separate acts of creation.

How deeply ingrained was this reverence for "static forms" may be inferred from the almost wilful misinterpretation of the results of that branch of comparative anatomy which dealt with the embryonic stages of animals. As with comparative anatomy in the broad sense, it is difficult to put a date and a name to the beginning of comparative embryology. It may be fairly stated that up to and including C. F. Wolff, all the emphasis in studies of the embryo was laid upon the problem of its *origin*; embryology was in fact concerned with the nature of life itself. In the work of Meckel however we find a complete shift of emphasis to the comparison of corresponding stages of development. Meckel's greatest service to science would probably be reckoned his awakening of his native Germany to the zest for exact anatomical description which he learnt from Cuvier. But for our present concern more significant was his pre-occupation with the idea of a primitive life-form from which all others had been developed. To account for this development he allowed himself a latitude of interpretation such as enabled him to see in a colonial polyp the sign of a power which in Man expressed itself in the form of Siamese twins, or in the possibility of new species being the offspring of a cat and a hare! It was characteristic of his age that such a farrago of nonsense should proceed from the same mind which recognised the swim bladder of fish as the forerunner of the lung of higher types, and which laid a sound basis for the homologising of the bones of the cranium and jaw—perpetuated in the names *Meckel's cartilage* and *Mentomeckelian* bone. On similar evidence, some but not much of which has stood the test of time, he announced his belief that every organism in its development passes through stages in which it resembles in the closest degree the structure of existing lower organisms. This was almost pure guess work; but five years before Meckel's death in 1833 there appeared a book which may fairly claim to be the foundation of modern comparative embryology: this was von Baer's *Ueber Entwickelungsgeschichte der Tiere*. This followed close on his epoch-making discovery of the mammalian ovum; and though in the absence of the cell theory his detailed descriptions are expressed in Aristotelian terms, his comparisons unlike those of Meckel are firmly rooted in observation. Thus he was able to reformulate Meckel's *theory* in a form which approximates closely to the actual facts, namely that the embryos of higher animals pass through developmental stages closely resembling the *embryos* of lower forms. Using this principle he was able to put Rathke's discoveries in a true light. These discoveries, and certain inferences he drew from them, place Rathke on a level little inferior to von Baer himself, as one of the most sagacious of pre-Darwinian anatomists. Studying the development of the respiratory organs of birds and mammals, he was surprised to find that at a certain stage the arterial system was in the form of *loops* similar to the branchial loops of adult fishes. Moreover with these loops were for a time associated clear indications of actual gill clefts. His other most famous work was on the development of the urinogenital organs. He noted that in the early stages of the development of vertebrates excretory organs were present in the anterior regions of the organism, and to these he gave the name *pronephros*. Later these degenerated, or were metamorphosed into sexual organs, the excretory function being taken over by newly developed ducts placed more posteriorly. These phenomena, and many more of a similar kind, suggested to him a type of " evolution " whose importance was not fully recognised until much later: to it he gave the name " *rückschreitende Metamorphose* " (lit. " backward-stepping metamorphosis ").

It might have been thought that von Baer's recapitulation theory—almost a " law ", though not of universal application, since the recapitulation of any particular organism is not of the embryonic characters of *all* lower types—would have given the impetus to a more critical and thorough examination of " evolution by transformation " as suggested by Lamarck, but not necessarily by such a crudely-conceived mechanism. But it was not so; and when the " recapitulation theory " was ultimately pressed into use by enthusiastic Darwinians, it was at first in the form enunciated by Meckel.

Darwin himself relegated the embryological evidence to the last pages of the *Origin*; but what he has to say is of great interest. He is aware of the views described above, and having drawn attention to the evidence for the belief that " at whatever age any variation at first appears in the parent, it tends to reappear at a corresponding age in the offspring ", he concludes that " the leading facts in embryology, which are second in importance to none in natural history, are explained on the principle of slight modifications not appearing, in the many descendants from some one ancient progenitor, at a very early period in the life of each, though perhaps caused at the earliest and, being inherited, at a corresponding not early period. *Embryology rises greatly in interest when we thus look at the embryo as a picture, more or less obscured of the common parent form of each great class of animals* " (italics mine). The earlier part of this citation is an example of Darwin's style at its worst; but presumably he means that the similarity of the *embryos* of related classes is due to the " delayed action " of the forces which are the cause of their ultimate divergence.

The idea of the italicised sentence was quickly seized upon by his followers and applied with an uncritical extravagance quite foreign to Darwin. Of these followers Ernst Haeckel was by far the most fertile in extension of Darwin's ideas beyond the realm to which they were relevant, and for a time his influence was world-wide. Space does not admit of even a gloss on the naïve materialistic philosophy which was apparently adequately suited to the intellectual appetite of the products of popular " education" throughout the world.[1] What concerns us here is his so-called " biogenetic law "—a perverted form of von Baer's " recapitulation theory " decked out with all the detail brought to light in the ensuing half-century. It will be necessary to summarise these discoveries, and we may suitably start with the work of Fritz Müller on the life histories of the Crustacea—the more suitably since it was Müller's account of this work in the strangely named book, *Für Darwin* (1864), which seized Haeckel's imagination. Müller emigrated to Brazil at an early age and there found a great variety of Crustacea, both freshwater and marine, on which to test Darwin's theory in its embryological aspects. The Crustacea display a complete range of life histories from that of the Crayfish, which emerges from the egg in a form little different from that of its parents, to the Prawn, which passes through three stages, each of which not only differs markedly from the adult but closely resembles a distinct lower order of Crustacea. To Müller then the evolution of the individual appeared as an historical document, sometimes partially " effaced ", sometimes " counterfeited ", as when the struggle for existence caused the various free-living larval forms to diverge somewhat from the supposed ancestral type.

The second strand which was woven by Haeckel into his all-embracing theory was the discovery by T. H. Huxley that one large and distinct group of animals—the Coelenterata—no matter how varied in form, are constructed by the multiplication of two

---

[1] The reader interested in the matter may refer to my *Science and Monism* (Allen Unwin, 1933).
2E

known to involve a great over-simplification in detail; both acted as a prodigious stimulus—both in support and opposition—at the time. Darwin's being at the same time closer to the facts, more fundamental, and relating to a wider field—the whole field of biology—has weathered better. For half-a-century after its promulgation however, Haeckel's biogenetic theory—it clearly was " law " only unto itself—dominated comparative anatomy, and through that taxonomy. The development of the mesodermal structures after the formation of the gastrula was studied in detail by Oscar and Richard Hertwig in Germany and by Ray Lankester in England: their findings gave considerable evidence for regarding all the higher animals—Vertebrata, Arthropoda, Mollusca, Annelida, Echinodermata—as being bound together in a major group by virtue of their possession of a body cavity or coelome, completely lined with mesoderm, though only vestiges of a true coelome remain in the mature forms of some of the groups. The recognised continuity of the coelome with its associated fluids has certainly been a fruitful aid in medicine and physiology generally. In another sphere also Haeckel's theory found startling confirmation, namely in the discovery by Kowalewsky of the free swimming larval stage of the colonial organisms known as Ascidians. The fact that such a sedentary, typical " invertebrate " organism passes through a stage superficially like a tadpole is remarkable enough, but what was of more theoretical significance was the possession by these " tadpoles " of a notochord—the strip of tissue supporting the nerve cord throughout life in *Amphioxus* and *recognisable as at least a transitory structure in all vertebrates*. This discovery placed the vertebrates in the natural division henceforth to be called the *Chordata*, and incidentally provided further confirmation of Haeckel's biogenetic theory and his " corollary ", that where higher forms lack any of the characters typical of their class, this may be due to the special conditions of their environment, the notochord of the mature sedentary ascidian having shrunk to a vestige while being retained in the closely related but more active *Amphioxus*.

Over the remainder of Haeckel's manifold activities—his dogmatic insistence on a " mechanical " explanation of the universe, coupled with an equally vehement and even abusive rejection of the embryologist His's (vide inf. p. 467) alternative *mechanical* explanation of the successive phases of embryonic growth; and his apparently deliberate falsification of the diagrams in one of his works, it is perhaps better to draw a veil. Nordenskiöld holds the view that he allowed political fervour—a doctrinaire radicalism, which may well have sprung from the sympathy with the oppressed, natural to a man of highly emotional temperament—to dictate his biological policy. It is too big a question to discuss here; but it is a good example of the powerful use to which a *pseudo*-scientific dogma may be put for political ends. It is said that his book *The Riddles of the Universe*, in which, he held, the riddles of the universe were in principle solved, sold in hundreds of thousands, from Greenland to Japan. The war of 1914, in which the countrymen of Darwin fought against his own, stunned him; he never really recovered.

It remains for us to consider briefly the effect of all this on taxonomy. We left this subject, as far as zoology is concerned, with the system of Lamarck (p. 402), in which the Animal Kingdom comprised Vertebrata, Crustacea, Insecta, Arachnida, Mollusca, Annelida, Radiaria, Polyps, Soft Worms and Infusoria. To follow in detail the successive modification of this scheme would take us too long; suffice it to indicate the kind of advances in general biology whereby improvements in classification were made possible. We have already seen how the researches of Kowalewsky brought the Vertebrata and Tunicata (which Gegenbaur still included in the *Vermes* as late as 1874) into one major group, Chordata. We have also mentioned the regrouping of "polyps", "medusae", and "sea anemones" into one phylum Coelenterata, characterised by their development from only two germinal layers. It is necessary to point out here that such a complete re-grouping would not have been possible without a discovery of major importance—that of the so-called "alternation of generations". Though isolated cases had been observed of a sessile animal giving rise by a kind of "budding" to free swimming individuals, it was Steenstrup who recognised this mode of reproduction and dispersal as characteristic of otherwise markedly different organisms. Thus he recognised that certain types of free-swimming "medusae" are not distinct genera, but individuals liberated from the " hydroid polyps " of the order *Campanulariae*. These medusae alone bear sexual organs whose gametes, when released into the water, unite to form an embryo which grows up into a hydranth polyp. With great sagacity, though giving rather a fanciful account of the phenomenon, he extended it to the case of the parasitic flatworms of the class Trematoda, whose sexual individuals pass their life in a different host and were formerly regarded as completely different organisms. We have already seen (p. 392) that it was Siebold who by experiment placed the whole question of parasite distribution on a sure foundation. To Leuckhart fell the task of completing Steenstrup's observations on the " polyps "; this he did by showing that in certain types of " polyps " the apparent confusion of forms is a remarkable adaptation of the colonial habit. The famous Portuguese Man o' War for instance is in a sense not one animal but a society of individuals of various forms, each distinct group of which may display a unanimity of purpose (as in the grave injury they may inflict on the unfortunate bather, against whom the colony may happen to be drifted) reminiscent of a " colony " of social insects. Leuckhart was also the first to demonstrate that the internal anatomy of the starfish, sea urchins, and their allies which resemble the Coelenterata externally in being the only large animals to display radial instead of bilateral symmetry, is entirely different from that of the latter group. He therefore split Cuvier's Radiata into Coelenterata and Echinodermata. It is important to note that this was before the publication of the *Origin*; Leuckhart was in fact opposed to any evolutionary theory.

A few years after Leuckhart's new division of the invertebrates, appeared Milne-Edwards' classification of the whole animal kingdom. The change of special interest is the recognition of the fundamental relationship between reptiles, birds, and mammals on the one hand and fishes and amphibia on the other; this was in the possession by the former of the foetal membranes—chorion, amnion, and allantois—which are absent from the development of the latter. Thus comparative embryology, even when no question of evolution was raised, was providing a valuable basis for classification. By 1880 the debt was being repaid; then for the first time appeared a work with the title of *Comparative Embryo-*

*logy*. The author was F. M. Balfour who at an early age had virtually created a new department of biology. The *Introduction* to this work, unfortunately too long to quote, lays down the lines on which the subject has since developed. Evolutionary in outlook, though cautious in the acceptance of the " biogenetic law ", Balfour expresses the view that, among other things, the purposes of embryology are to reconstruct the history of the race from observations on the life history of the individual, and to differentiate between what is ancestral and what is specialised for the adaptation of the particular individual species. It is in this spirit that the study of morphology is most fruitfully entered on to-day. The classification corresponds closely to that now accepted, the Chordata being recognised as a definite phylum.

We have as yet said nothing about the status of the microscopic organisms in the classificatory scheme. Though Ehrenberg prepared the way for an understanding of their true nature, yet, led on perhaps by prejudice, perhaps by faulty optical devices—most likely the latter playing into the hands of the former—he saw in these creatures " complete animals ", with minute organs comparable to those of the more familiar animals; this was perhaps the dying gasp of the " preformation theory ". Its ghost was successfully laid by Dujardin (see above p. 391). It was however Siebold who established their claim to be regarded as a phylum, which he named Protozoa, and characterised with admirable conciseness as follows: "Animals in which the different systems of organs are not sharply distinguished, and whose irregular form and simple organisation may be reduced (*auf eine Zelle sich reduzieren lassen*) to one cell." On this basis he removed the (multicellular) Rotifera to a separate phylum. Those organisms which, though bearing flagella or cilia like the majority of the true Protozoa, contained chlorophyll, he placed in the vegetable kingdom. Later Haeckel reformed all unicellular (better, non-cellular) organisms into the group Protista, regarding them as a primitive life-form not definitely animals or plants. While there is much to be said for this view in a few cases (e.g. *Euglena*) an effort is always made, usually with a fair sense of conviction, to establish every organism as either animal or plant.

We have now carried the classification of animals up to a

point at which it falls little short, as regards the major groups, of what is accepted to-day. As to plants, progress during the nineteenth century was much less marked. One discovery however eclipses all others in its far-reaching implications. It was made by a music-seller, more or less in his spare time, and ignorant of the true nature of cellular organisation. Yet it was the sort of discovery which would be described as " academic " in the highest degree—not at all the sort of correlation such as might reasonably be expected of the " born naturalist ", working on his own like Leeuwenhoek. What made it possible, we may suppose, was the fact that it required no elaborate apparatus (other than a microscope) and few experiments; only endless patience, and an obsession for one idea—the right idea.

The life of Wilhelm Hofmeister holds out for us the same sort of lessons as that of Darwin. But as is symptomatic of much of the " history of science ", it is usually dismissed in a few lines; in one well-known recent " History of Science " his name does not even appear in the index. Like Darwin, he was born into a cultured home and had the advantage of stimulating experiences both in conversation and in seeing his father's herbarium and miniature botanic garden. The direction of his ultimate life work was not at first apparent; he spent most of his leisure in playing and studying music, and in collecting insects; he showed no inclination for botany. His school seems to have been an admirable instrument of education, but it provided no recognised " certificate " such as was usually regarded as essential in Germany. He attended no University. Yet at the age of nineteen, almost self-taught, he began the studies which revolutionised the contemporary views of the development of plants.

To understand the significance of what he did, it is necessary to see the problem as he must at first have seen it—through the spectacles of others. In 1823, armed with the compound microscope which he himself had done so much to improve, Amici had seen the germination of pollen grains. About seven years later he was able to make preparations which showed the growth of the tube down the stigma and its penetration into the ovule. It is necessary to recall here that the cell theory was at this time as embryonic as the bodies to which attention was being drawn.

In 1842 appeared Schleiden's *Grundzüge der Wissenschaftlichen Botanik*, which did so much to stimulate research and discussion, but at the same time gave the authority of its author's great name to the fantastic view that the pollen tube was in fact the embryo whose subsequent development was initiated by its contact with the ovule. This was only one of several competing and equally erroneous hypotheses. The error was exposed by Amici in 1842 when he made known the existence of a *cell* in the apparently empty region of the ovule, which had come to be known as the embryo sac; this cell underwent division only *after* the pollen tube reached the sac. These observations were confirmed in 1846 by von Mohl and also by Hofmeister; but Schleiden and his followers were quite unconvinced. However Hofmeister had by this time found his *grande passion* in more ways than one: He was happily married, and his work in his father's music business left him sufficient energy to examine (mainly between four and six o'clock every summer morning) nineteen families of angiosperms. In the ovules of all of these he was able to show that there exist, before the pollen tube arrives, two *groups* of cells, from only one of which an embryo normally develops. Of the three cells at one end he says: " I hold them as originally equivalent, and all of them qualified to develop themselves into embryos even though they [i.e. two of them] fail to do so in the great majority of species." The other group, at the chalazal end, he believed " to have no other importance than the preparation of nutritive material for the growing embryo ". All this was published in 1849—an astonishing achievement at a time when every preparation was made by hand cutting from fresh material. The only advantage which Hofmeister had was, paradoxically, exceptionally short sight, which not only compelled him to specialise in close work but also enabled him to pick out with a needle an object literally invisible to those of normal sight.

Great and important as was this achievement, establishing for ever the correct course of fertilisation (though the details of the *nuclear* division were worked out much later by Strasburger), it was only a platform from which a much more adventurous leap into the unknown was to be taken. A hint of this occurs at the end of the short work in which the above results were made known:

## THE DARWINIAN REVOLUTION 427

this was a note on the similarity of the reproductive process in the conifers and heterosporous Pteridophyta. Possibly this revolutionary hypothesis had provided the spur to all his work. At any rate the outcome stands in almost the same relation to botany as Ampère's *Electrodynamics* does to current electricity—it generalised a number of apparently unrelated phenomena and thereby created a new department of the subject—a comparative morphology of the whole plant kingdom. The title gives a summary of the whole work: *Vergleichende Untersuchungen der Keimung, Entfaltung und Fruchtbildung höherer Kryptogamen (Moose, etc.) und der Samenbildung der Coniferen* (Comparative Researches of the Germination, Development, and Fructification of the Higher Cryptogams (Mosses, etc.) and the Seed Formation of Conifers). When this book appeared in 1851 botanists recognised the two major groups of plants—flowering and non-flowering—between which functional relationships had gradually been discovered. Thus in many of the Cryptogams—so called because their sexual parts were inconspicuous as compared with the showy flowers of the Phanerogams—antheridia had been seen releasing spermatozoa, and other organs called pistillidia (later archegonia) from which, for instance, the " capsule " of mosses developed. In the case of the ferns however the formation of the prothallus from the " seeds " on the leaf, which had been known from the time of Malpighi, was regarded as a normal germination, and this despite the observation of " spiral filaments " (spermatozoa) in the neighbourhood of the prothallus. Put rather crudely it was generally agreed that the Cryptogams are reproduced by a process of fertilisation—the functional correlation with the Phanerogams was thus accepted—but as to what organs in the former corresponded to anther, pollen, ovary, and ovule in the latter, there was complete confusion, even among those who ventured to suggest any correlation. The events in the fertilised ovule of the Gymnosperms (those Phanerogams whose seeds occur " naked " on leaf-like bracts instead of in a closed ovary) were better known; the embryo was seen to be growing on a mass of nutritive tissue (endosperm). The cardinal demonstrations whereby Hofmeister unravelled this tangle (far worse than the above bare outline would suggest) were stated by himself to be " the genesis of the germinal vesicle in the central cell of the archegonium, also of the dependence of the germinal vesicle's development upon the incident of its fertilisation, also the proof that the course of formation of the leafy moss plants corresponds with that of the prothallus of the vascular Cryptogam, and with that of the endosperm in the Conifers, and that the course of development of the Moss fruit corresponds with that of the embryo in the Vascular Cryptogams, the Conifers, and the Phanerogams." We may amplify this as follows: The act of fertilisation occurs between elements produced from the prothallus of the fern or its equivalent (e.g. the "leafy Moss plant"), the prothallus itself being produced by the germination of a spore or detached cell of a plant *which does not itself produce sex cells*. There is thus an "alternating of generations" superficially similar to that shown by Steenstrup (p. 422) for certain animals, though actually, as later research was to show, of a far more fundamental form, since the spore is the result of the halving of the chromosome number—a process which does not precede medusa formation. In plants higher than the ferns two kinds of spores are formed, giving rise to two kinds of prothalli —one relatively evanescent, the other forming the endosperm of Conifers and the embryo sac apparatus of Angiosperms. With remarkable insight Hofmeister surmised that in the Gymnosperms the function of the pollen tube might be to transmit actual "spiral filaments" to the ovum; this was confirmed later in the case of the two orders, Cycadinae and Ginkgoinae.

The author of this superb result had to pay the penalty for its very clarity and completeness. Although elected to a Chair, first at Heidelberg, then at Tübingen, he is no longer named in the elementary text books of botany. His generalisation seems not to be regarded so much as a product of individual genius and labour, but rather as an Institution, which is taken for granted in the exposition of the subject. By many of his contemporaries it was taken up as the foundation of botanical phylogeny akin to Haeckel's biogenetic "law" for animals. Hofmeister himself never viewed it in this light; nor should it be so viewed, as it is not to be supposed that the modern Phanerogams are descended from *existing* heterosporous Cryptogams, even though a remarkable gradation of modifications in the process of "alternation" is apparent. And we are certainly not justified in *assuming* that the ancestral forms revealed any such gradation. How Hofmeister would have viewed the impact of the Darwinian Revolution on botany had he outlived the first years of *ultra*-enthusiasm which swept over Germany, it is difficult to say. The years between 1851 and 1867 were spent on the development of his studies of *causal* morphology. In 1877 he died, worn out by overwork and the effects of a shattering series of domestic tragedies.

## SOURCES FOR CHAPTER XXXI

C. Lyell, *Geological Evidences for the Antiquity of Man* (London, 1863).

C. Darwin, *The Origin of Species*, Chs. XIII and XIV.

F. M. Balfour, *Comparative Embryology* (London, 1880).

C. Gegenbaur, *Grundriss der vergleichenden Anatomie* (Leipzic, 1874).

V. Goebel, *Wilhelm Hofmeister*, trans. ed. F. O. Bower (Ray Society, London, 1926). A sympathetic and critical account of Hofmeister's work by one of his pupils. It contains also a short biography by Hofmeister's daughter.

E. Nordenskiöld, *op. cit.*, Gen. Bibl.

## CHAPTER XXXII

## THE INSIDE STORY OF LIFE

THE fundamental problem of biology, viewed as the science of *living* things, is undoubtedly that of assimilation and growth. To that end all else is subservient. For reproduction is merely a device for maintaining continued growth; and movement and co-ordination for more effective growth—neither of them essential, since they are little developed in the plants. The importance of *form*, with which we have been concerned during the last chapters, is twofold: it is the most obvious though not the only means to classification—a convenience to Man, though clearly of no interest to the organisms themselves—and more important, it is, as has been shown, a plastic record of the *history* of organisms. The importance of this is its philosophical implications, of which a word may be said in the conclusion. From the organism's point of view, *form* is the embodiment of *function*, which in turn is the mode of realisation of *end*. If this sounds Aristotelian, let it be so; if the reader objects, let him find an alternative *meaning* to life. This is not the place to dispute his possible objection that life has no meaning; or that if it has, it is none of the business of science to determine it.

In recent chapters we have been concerned entirely with form and its implications. It must not be thought however that all the great morphologists were entirely blind to function. The old dissension—superficially between Plato and Aristotle—was revealed in these men as in everything else. While Gegenbaur for instance saw in every organism and in all its organs the embodied development of an archetypal *ensemble*, Cuvier was so influenced by function that in his *Règne d'Animal*, after dividing animals into four major *formal* groups, he subdivided the Vertebrates for instance according to their solution of the functional problem of respiration. Moreover Mammals were separated from Birds on account of their superior *sensitiveness*; the mammary glands and viviparity are mentioned almost casually. Both these attitudes have enabled us to see further into the heart of things; both have produced valuable disciplines, the marks of which are still to be seen in our text books. But while it is true that many of the great anatomists have had a functional rather than a formal bias, it remains the case that the *inner workings* of the living organism—what in the last resort *drives* these self-producing, self-guiding, self-reproducing machines—has been the concern of a different group of workers. To these we must now return. We shall consider first the idea of assimilation and growth—though the earlier workers did not envisage the problem in those terms—and second, that of co-ordination and direction.

In our view of the growth of the idea of combustion we found that closely linked with it was that of respiration on the one hand, and the assimilation of carbon into the green plant from the air on the other. In the first gropings towards the idea of chemistry we found van Helmont (p. 168) unconsciously positing the second problem, and also in his efforts to improve the contemporary practice of medicine (his principal work was called *Ortus Medicinae*) putting forward sound views on the *chemical* nature of the process of digestion, recognising that acid promotes digestion in the stomach, alkali in the intestine. In postulating further "digestions" (including one in the heart!) he shows himself to be still befogged by the Galenic humours. His chief successor, Sylvius, was less original as a chemist, but better versed in physiological anatomy. He does not seem to have done much original work, but was an influential teacher at Leyden, where the youthful de Graaf carried out in 1664 what is believed to have been the first collection of secretion from the living pancreas. This he did by fixing one end of a duck's quill into the orifice of the pancreatic duct and the other into a flask which hung below the dog's body. We may shudder at the callousness implied by such an experiment before the days of anaesthetics, but admire the superb dexterity with which such common materials as quills, paste and paper, were made to serve before the days of glass canulas, rubber tubing and spring clips. What strikes us as so odd was the absence of adequate *chemical* test of the collected fluid; the whole difficult and ingenious operation was vitiated by the dogmatic insistence of de Graaf (perhaps reflecting his master's prejudice) that the pancreatic secretion was *acid*. But this, it must be remembered, was before Boyle's *Sceptical Chymist* had had time to spread the technique of analytical test. Sylvius is of importance in the history of physiology for his recognition of the purely *chemical* nature of the transformations of food under the influence of the saliva, gastric juice, pancreatic secretion and bile. There is no mention of any *Archeus*; his only serious error apart from the one already mentioned, is his confusion of true fermentation with simple effervescence, as when acid acts on marble—natural enough to a man intent on a naturalistic interpretation.

The chemical idea of digestion made little progress for a century—Stahl's work, though not without an underlying wisdom, was retrograde in that it re-enthroned the *Archeus* in a form more suited to the contemporary climate of opinion. For a time a mechanical theory such as we should expect in the time of Newton held the day. This started from the exact experiments of Borelli in which glass bubbles, pieces of wood and lead were introduced into a turkey and later recovered in a pulverised state. By comparing the crushing force with his calculated value for the human jaw (by which bubbles were only just broken) he estimated the force of the stomach contraction as 1350 lbs. He admitted however that digestive juices exert a " corrosive " action in the case of animals whose stomachs are not so well-adapted for crushing. Later exaggeration of these views is of importance only in prejudicing Boerhaave, the most influential teacher of the early eighteenth century, who extended the life of the false hypothesis that *nutriment* is extracted from food largely by mere trituration. A more balanced view was adopted by Haller in his great book, *Elementa Physiologiae*, which, as the late Sir Michael Foster says (in his *Lectures on the History of Physiology*, to which I owe most of the information in this section) indicated " the dividing line between modern physiology and all that went before ". The account contained nothing new except a clear appreciation of the work of the bile in emulsifying fat—a fact proved by ligaturing the duodenum just above the orifice of the bile duct, with the result that food above the ligature remains in the form of grey lumps, that below being a white homogeneous mass. In this account the actual mechanism of digestion proper was not explained, but while the *Elementa* was appearing—in 1752—the basis was being laid for the solution of this problem by a new experimental technique. Réaumur, who was the author of this paper, was one of those men of science who by virtue of their versatility seem to be almost entirely overlooked in elementary text books, except in respect of some trivial by-product of their work. Just as Gibbs is the shadowy author of the " phase rule " —in itself a not very remarkable generalisation—so Réaumur is the " author " of a now almost obsolete and never very important temperature scale. Actually this same man may be said to have played a leading part in the foundation of three such widely different enterprises as the steel industry, the general biology of insects, and the physiology of digestion. If Fahrenheit had not done the job rather more thoroughly, Réaumur would also have been regarded as the founder of exact thermometry. All this by no means exhausts his contributions to science.

Réaumur had a Kite, a bird well known (like the Owl) to regurgitate what it cannot digest. Since it could hardly be expected to digest a sponge Réaumur fed the creature on small sponges. These it delivered saturated with digestive juice. The sponges were contained in small perforated tubes, which were also used for the inclusion of pieces of meat, bone, etc. The upshot of these experiments was the first unambiguous indication of the nature of gastric digestion. The tubes when returned revealed that the meat had been partially dissolved, fragments of bone much less so, starch not at all; and there was no sign of putrefaction. " What then is this liquid which acts on meat and on bone in some such way as *l'eau régale* acts on gold, but has not the same power over starch that *l'eau régale* has over silver? To which of the solvents that chemistry offers us can this liquid be compared? " (quoted from Michael Foster *op. cit.*) He thus recognised that it is a *chemical*, not a mechanical question; but the *specificity* of the attack puzzled him. So he set about further enquiry, as a modern biochemist would do. He found the liquid brought up in the sponge, free from food, to be acid. He tried to make this liquid digest *in vitro*; and he used a control experiment consisting of meat placed in plain water. After 24 hours at 34°R he found the meat to have been digested in the former and to have putrified in the latter. Digestion was therefore not the same as putrefaction. Then the Kite died—one of the many unsung martyrs in the cause of natural knowledge. He continued experiments with dogs and sheep, but learnt little more. The world was hardly ripe for such exact methods; even Haller referred to them only casually. Twenty-five years later however the ingenious Spallanzani (p. 362) repeated and extended Réaumur's experiments, confirmed their results in the main, and made the additional observation that by constantly renewing the juice by allowing it to drop on and drain off the meat, the process of digestion was greatly accelerated—due as we now know to the *reversibility* of the action. John Hunter, while scoffing at Spallanzani's methods—he was a convinced adherent to Stahl's school—confirmed his results, and noted that the stomach of a dead man begins to digest *itself*, which seemed to confirm the Stahlian doctrine that the living body possesses an animal spirit which prevents its attack by substances capable of dissolving it after death—an interesting observation on which the last word has probably not yet been said.

No further progress was likely nor was in fact achieved,[1] until the science of organic chemistry had been established as a separate discipline. Scheele was already at work on the isolation of pure natural products such as citric acid and other acids, casein, and glycerol, between 1770 and 1785. At the turn of the century came Berzelius's *Lectures on Animal Chemistry*, where he showed that such " vital " products are subject to the ordinary laws and tests of inorganic chemicals. From 1810 onwards, Chevreul was discovering the nature of fats, oils, and soaps, and isolating the higher fatty acids. In 1815 Gay-Lussac published a description of the behaviour of the first recognised compound radical (cyanide) destined to play so important a part in the under-

[1] An outstanding synthesis was, in fact, achieved in 1822 by the American surgeon Beaumont, who, taking advantage of a severe stomach injury to a French Canadian, Alexis St. Martin, observed over a period of several years—through the orifice guarded by the prolapse of the membrane, the operation of the three muscle coats, the antiseptic character of the secretion which prevents putrefaction, and the course of digestion in a wide range of circumstances (quoted by Ryle in *The Background of Modern Science*, ed. Needham and Pagel).

## THE INSIDE STORY OF LIFE

standing of organic compounds. Finally, in 1828, Wöhler prepared the distinctive end-product of animal metabolism, urea.

This last-named preparation gave rise to one of those legends which are of such great interest to the modern historian of science. In almost every text book of organic chemistry known to recent generations the statement has been made to the effect that " Wöhler's synthesis of urea from its elements (or at least from purely inorganic materials) killed at last the erroneous view that the productions of living organisms are possible only by virtue of a ' vital force ' and consequently could never be synthesised by the chemist." Growing doubts of the historical truth of this statement have culminated in a thorough investigation of the sources by Dr. McKie, whose article (*Nature*, Vol. 153, p. 608) is commended for study. In the paper in which the preparation of urea was announced, Wöhler himself made no claim that it was either a " synthesis " or that the materials were inorganic (it has never been denied that the ferrocyanide which is the key substance was made from horns and hair), or that it had any particular bearing on the problem of the existence of a vital force as such. He did not deny its importance however; and a communication to Berzelius called forth from that great man a ribald reply on the possible connexion between the production of urine and the problem of immortality. But Berzelius remained an uncompromising vitalist to the day of his death. Nor were the younger leaders inclined to give up allegiance to a vital force. Fourteen years after Wöhler's synthesis Gerhardt was developing his view that life is a distinctive force striving to build up the complex materials characteristic of living bodies from the simpler substances of the inorganic world; with the corollary that the " death " of the organism is its failure to maintain this equilibrium, which rapidly shifts in the direction of disruption or decomposition. Even Liebig confined his mechanistic views on the nature of growth and decay within the strictest limits, the nature of which will appear more clearly in the sequel.

The mechanistic revolution when it came was heralded in by no sensational and crucial experiment. Such revolutions in thought seem to grow like tropical fungi underground or in the deeper layers of the host, suddenly to appear in a riot of fructification on all sides. This discontinuity, or occult process, seems to cause historians such mental *malaise* that they are driven to invent a crucial instance in which the scientific Elijah may bring discomfiture upon the priests of Baal. For seventeenth-century dynamics it was the entirely unauthenticated dropping of the weights from the tower of Pisa; for the laying of the ghost of vitalism it was Wöhler's " synthesis " of urea. But while the origin of the former fable has so far remained undiscovered, Dr. McKie claims Hofmann as the creator of the Wöhler myth in his encomium of 1882. That the myth was not at once established in the museum of intellectual antiquities is evidenced by the absence of any reference to this " death knell of vitalism " (though the word " synthesis " of urea occurs in the index) in E. von Meyer's celebrated *History of Chemistry* whose publication took place in 1888.

Actually, as Dr. McKie reminds us, the first " synthesis " of an " organic " substance from its elements was that of acetic acid by Kolbe. The substance obtained in Kolbe's synthesis was trichloracetic acid, but this could easily be converted into acetic acid by potassium amalgam (Melsens). The significance of this far more rigorous demonstration of the possibility of non-vital synthesis of organic compounds appears to have passed unnoticed. But major revolutions of thought are no more matters for nice logical distinctions than are revolutions in social relations. The rapid accumulation of new derivatives of " organic " substances and the manifold interconversion of the latter had really made the question meaningless. By what criteria for instance was it to be decided whether the carbon used in Kolbe's synthesis was free from " organic " taint? Chemical vitalism had not been killed, it had not even died; but, like the old soldiers in the song, had simply faded away.

It was far otherwise with the broader question of the relation between the " vital force " and the chemical materials which it was supposed to operate on. Here we can do no better than take a look at the views of Justus von Liebig, Wöhler's lifelong friend and collaborator. Liebig with all his faults was one of the great men of the nineteenth century. In his laboratory at Giessen, by personal supervision, guidance and inspiration, he brought into the teaching of chemistry the technique of the great teachers of medicine, the art of the Masters of painting and sculpture. As Davy's greatest discovery has been said to have been Faraday, so Liebig's greatest " synthesis " was in the minds of Hofmann, Playfair, Gerhardt, Williamson, Würtz, Frankland and scores of others. In rivalry with J. B. A. Dumas of Paris, he was the co-founder of the whole *idea* of organic chemistry—his share being the recognition of the compound radical as the unit more important than the individual atom, Dumas adding the complementary idea of the possibility of an almost limitless substitution of new atoms within the framework of the radical. Of the long, highly involved, and, often irritating story of the fashioning of the immensely powerful idea of molecular structure, we cannot here treat. Fundamental as it is for the understanding of science, its actual genesis, owing in part to the contemporary confusion in regard to the relation of atoms to molecules and the consequent uncertainty as to the atomic weight of carbon and oxygen, involved so much fruitless controversy carried on in terms now rendered obsolete, that its study is inevitably the affair of specialists: its elucidation forms no part of the main cultural advance.

Liebig's third contribution is however very much our concern. Under the stress of population increase, Europe was becoming anxious about the need for increased food production, one result of which was the recognition that chemistry might be called in to improve agriculture. Thus when Liebig was asked by the newly-formed British Association for the Advancement of Science to draw up a Report on the State of Organic Chemistry, what he provided was in fact a short treatise which, together with those of Davy and de Saussure, formed the foundation of Agricultural, and to a less extent Physiological, Chemistry. Its importance for us lies in the statement of a theory of the relation between the living and the inorganic world which was for a long time highly influential.

Having insisted upon the fundamental status of the green plant for the living world, Liebig proceeds to enunciate the principle by which all research into plant growth has subsequently been guided, namely that " Vitality is the power which each organ possesses of constantly reproducing itself; for this it requires a supply of substances which contain the constituent elements of its own substance, and are capable of undergoing transformation. All the organs together cannot generate a single element, carbon, nitrogen, or a metallic oxide." To understand life it is first and foremost necessary to understand the power and subtlety of chemical action. " Is it true vitality which generates sugar in the germ for the nutrition of young plants, or which gives to the stomach the power to dissolve, and to prepare for assimilation, all the matter introduced into it? A decoction of malt possesses as little power to reproduce itself as the stomach of a dead calf; both are unquestionably destitute of life." The actions which these lifeless bodies bring about in the living body can as well be imitated in a test tube. " The power therefore to effect transformations does not belong to the vital principle: each transformation is owing to a disturbance in the attraction of the elements of a compound, and is consequently a purely chemical process." Liebig nowhere denies the possibility of, as it were, a vital *directiveness*: he merely insists on the exorcising of learned mumbo-jumbo. " The expression ' vital principle ' must in the meantime be considered as of equal value with the terms *specific* or *dynamic* in medicine: everything is specific which we cannot explain, and dynamic is the explanation of all which we do not understand; the terms having been invented merely for the purpose of concealing ignorance by the application of learned epithets."

How slowly do we learn! Of all humbug, scientific humbug is the most objectionable. The newly-coined concepts of one age become perverted into the catchwords of the next: instead of aiding Archeus against Tartarus we are told to " alkalinise the mucous membrane ", " cleanse the blood ", " soothe the nerve cells " or " energise " anything and everything. Fortunately the human body (vital principle?) appears to have almost unlimited reserves wherewith to buffer itself against the progress of medical science.

To aid the living plant then it is necessary to provide it with the chemical bricks in just quality and quantity from which it may build its body. As elementary bricks, Liebig correctly recognised carbon and nitrogen as of prime importance. Further he recognised that the elements as such cannot be assimilated by the plant, carbon becoming available only in the form of carbon dioxide; nitrogen in the form of ammonia. Unfortunately accepting, contrary to the opinion of many of his contemporaries, the demonstration by de Saussure that the atmosphere is the chief, if not the only, source of carbon dioxide utilised by plants, he dogmatically asserted that " In spring . . . the component substance of the seeds is exclusively employed in the formation of roots . . . they extract from the soil their proper nutriment, namely the carbonic acid generated by the humus." Blinded by this prejudice Liebig missed the whole point of the importance of humus in sound husbandry. " It is the greatest possible mistake to suppose that the temporary diminution of fertility in a soil [through continuous cropping] is owing to the loss of humus; it is the mere consequence of the exhaustion of the alkalis." It was this and the hypothesis that the whole of the nitrogen needed by plants is supplied in the form of ammonia—mostly from the air, partly by the decomposition of humus and dung—that gave rise to the agricultural practice of trying to boost fertility with sulphate of ammonia (then becoming a by-product of coal gas) and minerals. The later discovery that nitrate was the form actually absorbed most readily by plants led to the use of nitrate of soda, but failed to halt the disastrous progression to the dust bowl. The influence of humus on tilth was a sealed book to Liebig; it was only in 1849 that Graham even began to study the chemistry of the colloidal state.

In demanding that where a chemical action could explain a phenomenon accompanying life, no other " forces " were to be called in question, Liebig was following the ideal of science. But as must happen to every pioneer in such a complex field, he went far beyond the evidence and was too easily satisfied with mechanico-chemical " explanations " for which there was no evidence or which were just as empty forms as the " vital principles " he so justly scorned. His cavalier treatment of the function of humus is an example of the former; for an illustration of the latter we shall turn to his theory of organic disintegration—a topic all the more important as it touches much more closely than his chemical theory of nutrition the problem of the boundary between the living and the non-living.

He distinguishes three forms of what I have called " organic disintegration ", namely putrefaction and fermentation which occur (according to him) in absence of oxygen, and decay or " eremacausis " which, since it needs oxygen, is essentially a form of slow combustion. It is with the former that we are chiefly concerned. Yeast, for instance, according to Liebig, is formed from gluten and is " a body in the state of decomposition, the atoms of which consequently are in a state of motion or transposition. Yeast placed in contact with sugar communicates to the elements of that compound the same state, in consequence of which the constituents of the sugar arrange themselves into new and simpler forms, namely into alcohol and carbonic acid." He was led to this hypothesis by a number of chemical analogies, such as the reduction of hydrogen peroxide by silver oxide, the induced crystallisation of supersaturated solutions, etc. The production of yeast from gluten occurs " in consequence of the transposition of the elements of the sugar exciting a similar change in this gluten. After this explanation [!] the idea that yeast reproduces itself, as seeds reproduce seeds, cannot for a moment be entertained." At the conclusion of the book Liebig extended the same idea to the action of organic poisons. The septicaemia following on the contact of pus with a trivial wound is explained as being due to the imparting to the substances of the healthy tissue the state of violent degradation proceeding in the decomposing matters of the pus. He goes so far as to say that it is merely a special application of the principle enunciated by Laplace and Berthollet that " a molecule set in motion by any power can impart its own motion to another molecule with which it may be in contact." This purely *quantitative* theory of induced action bears the birth marks of an age when every phenomenon was " explained " by recourse to the attraction, repulsion, or impact of particles in motion. Its inadequacy in the present connexion was soon to be demonstrated by Liebig's opponent, Pasteur; but it is not on that account to be ignored. From our present vantage point may we not see in it the first crude groping towards that concept of *activity* which, combined with the notion of chain reaction, has already played so great a part in physical chemistry, and allied to the *qualitative* theory of molecular structure and polarity, bids fair to provide a rational account of the action of enzymes? In putting forward this suggestion, we are anticipating; but it is done in the hope of convincing the reader that had not Liebig put forward the " wrong " hypothesis, the " right " one later urged by Pasteur might have been too easily accepted and its deeper significance thereby overlooked. In conclusion it is necessary to recall that Liebig's mechanico-chemical hypothesis of fermentation and putrefaction was restricted to the mechanism of the supposed degradation. To account for the failure of living tissue to ferment or to decay he is content to assert that the " vital " principle holds the chemical action in check. Although superficially he is dogmatic and prejudiced, the following passage shows how near his genius brought him to a great discovery: " In smallpox, plague, and syphilis, substances of a peculiar nature are formed from the constituents of the blood. These matters are capable of inducing in the blood of a healthy individual a decomposition similar to that of which they themselves are the subjects; in other words they produce the same disease. The morbid virus *appears to reproduce itself just as seeds appear to reproduce seeds* " (italics mine). The inconsistency of this with the passage quoted above about yeast hardly needs to be pointed out. But such inconsistency is the privilege, if not the mark, of genius.

Though Liebig played a foremost part in proclaiming the principle that the working of the living body is to be sought in the chemical action going on within it, he could not establish the principle. For just as he castigated the botanists for their ignorance of chemistry, many of his contemporaries were able to show that his own glaring errors were the result of ignorance of structure. He was a chemist for whom chemistry acquired a new and exciting quality in its application to biology—especially, in his experience, plant biology, for of animal biology he was even more ignorant. To establish the principle it must be revealed in action by one who, having first-hand knowledge of biological form, could devise experiments in which the physico-chemical conditions were altered in a controlled manner, and who could interpret the response of the organism in physico-chemical terms—one who in a word could point the way to the physico-chemical determinants of vital action. Such a one must also possess so sure an analytical grasp of general ideas as to be able to restate the dilemma of vitalism and mechanism, which as we have seen was becoming confused to the point of meaninglessness. This man, who has been described as "not merely a physiologist, but physiology", was Claude Bernard.

Bernard arrived in Paris at the age of nineteen with few possessions but an unacted tragedy. Fortunately for science he was persuaded not to regard play-writing as a promising profession. Apparently bored by the pedestrian teaching in the hospital, he came to life only on entering the laboratory of the Collège de France. Here Magendie was already beginning the task which his young pupil was destined to complete. Critical, callous, and completely unprejudiced, Magendie was bringing into medicine the experimental approach demanded by Liebig for biology in general. His utter disregard for the sufferings of his experimental animals was matched only by his exposure of himself to the danger of cholera or the threats of the ignorant and desperate mob among whom it was raging. His insistence on the unprejudiced observation of facts did not stop short of the admission of complete ignorance, as testified by his cynical query from a pupil whose treatment was not proving successful, whether he had tried doing nothing—which in the existing state of medical knowledge was at least as likely to be of equal benefit to the patient. His contribution to biological thought was not entirely negative; he was quick to seize on the importance of the phenomena of osmosis lately observed by Dutrochet. He has considerable claim moreover to have been the first observer of the differential action of the roots of the spinal nerves, usually attributed to Charles Bell.

Schooled thus in the practice of ruthless criticism, in the belief of the absolute sovereignty of facts and in the emptiness of words, to which no precisely defined concept is known to correspond, Bernard created "General Physiology", that is physiology as a branch of science whose methods and concepts were no less, and no more, distinct than those of physics and chemistry. In nature, Bernard reminds us, there are no sciences but only phenomena;

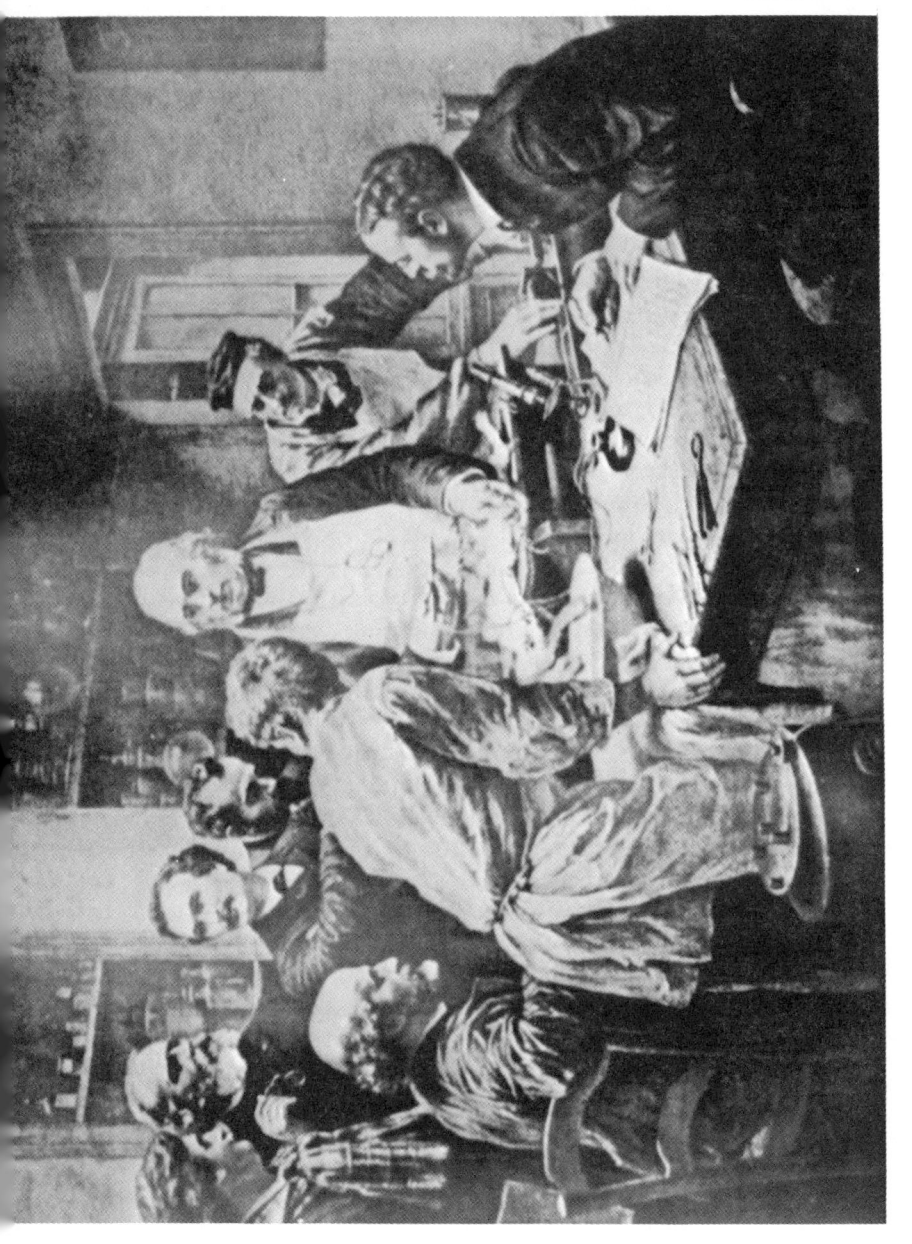

Claude Bernard demonstrating an experiment to his friends. The famous physiologist, Paul Bert, stands, with arms folded, at Bernard's right hand

PLATE VII

but in so far as our human frailty compels us to break down each concrete problem into ideally separate parts, so we must create distinct disciplines, armed with appropriate units of thought, to deal with them. Bernard was the first to demonstrate that physiology is such a discipline. Whereas Liebig had virtually claimed physiology as a branch of chemistry, Bernard writes: " Though we can succeed in separating living tissues into chemical elements or bodies, still these elementary chemical bodies are not elements for physiologists . . . Organic individual compounds, though well defined in their properties, are still not active elements in physiological phenomena; like mineral matter they are as it were only passive elements in the organism. For physiologists, the truly active elements are what we call anatomical or histological units. . . . Certain naturalists refuse to give them the names of elements and propose to call them elementary organisms. This appellation is in fact more appropriate; we can perfectly well picture to ourselves a complex organism made up of a quantity of distinct elementary organisms, uniting, joining, grouping together in various ways, to give birth first to the different tissues of the body, then to its various organs . . ." This was in 1865, only a few years after the publication of Virchow's *Cellular Pathology*—deep calling unto deep; but there were still many in the Medical Faculties to whom these ideas were worse than Greek, in which subject they were at that time likely to be well versed.

Bernard's biological philosophy was based on years of masterly practice in every branch of animal physiology; we have space here to deal with two which are of special importance in the understanding of metabolism—that aspect of life which I have claimed as the most fundamental.

One day Bernard noticed that the urine of rabbits just brought into the laboratory was clear and acid like that of carnivora. " I assumed that they had probably not eaten for a long time and that they had been transformed by fasting into veritable carnivorous animals, living on their own blood . . . I gave the rabbits grass to eat; and a few hours later, their urine became turbid and alkaline." He found he could repeat each phase at will, and also on horses as well as rabbits. Clearly we are all on the brink of literal dissolution. Bernard however demanded further evidence, which he obtained, first, by feeding normal rabbits on cold boiled meat, and, secondly, by carrying out an autopsy on one after it had been killed. He not only found all the signs of carnivorous digestion but by a miracle of exact observation was led to the discovery of the function of the pancreas in the emulsification and digestion (i.e. chemical decomposition into glycerol and fatty acids) of fat by a " specific ferment ".

The second example is concerned with what is perhaps his most famous piece of work. His object was to discover the *fate* of the sugar eaten by animals, it then being assumed that the animal life consists in the conversion by oxidation of carbohydrates into carbon dioxide, water, and energy. He was shocked to find that there is always sugar (glucose) in the blood of animals even when they are receiving no sugar, nor even any other carbohydrate, in their food. This disagreement of fact with theory was not a matter for regret, but a challenge. For some time he could make nothing of it. Then he " began to estimate the sugar in the livers of animals placed in various physiologically defined circumstances. I always made two determinations of carbohydrate for the same liver tissue. But pressed for time one day I could not make my analyses at the same moment; I quickly made one determination just after the animal's death and postponed the other analysis till next day. But then I found much larger amounts of sugar than those which I had got the night before with the same material. I noticed on the other hand that the proportion of sugar which I had found just after the animal's death the night before, was much smaller than I had found in the experiments which I had announced as giving the normal proportion of liver sugar." Here was a dilemma. Should he average the results? Was it an accident? No—" Nothing is accidental." " By forcibly injecting a current of cold water through the hepatic vessels, and passing it through a liver that was still warm, just after an animal's death, I showed that the tissue was completely freed from the sugar which it contained; but next day or a few hours later, if we keep the washed liver at a mild temperature we again find its tissue charged with a large amount of sugar produced after it was washed." Thus was discovered the formation of glycogen in the liver and its progressive conversion into glucose by another enzyme. From which it followed that not only is food broken down in the gut by a series of enzymes, but that the materials of metabolism themselves, or rather the chief of them, are controlled by enzymes. It showed further that the animal body is capable of synthesis of large molecules from small (though not perhaps as Bernard said " by a mechanism in every respect like the mechanism found in vegetables "). Lastly it was the first demonstration of a gland of internal secretion; the secretion of bile by the liver is carried down a duct into the lumen of the gut, but the glucose, which is as much a secretion as adrenalin, though much less drastic in its immediate effect, is, like the last-mentioned substance, passed directly into the blood.

Here we must leave Bernard for the present and turn our attention to his great colleague, Pasteur, who, cultivating just that attitude to the problem of life which Bernard tended to neglect, brought about a revolution in biological thought which, as Professor Lawrence Henderson has pointed out, delayed the realisation of the " physiological revolution ".

If Bernard achieved immortality as the personification of physiology, Pasteur's name will live as long as it is necessary to destroy the latent danger in mass-produced milk, collected in unhygienic conditions from unhealthy herds. In France he was (I believe) the first man of science to compete with dissolute monarchs and opportunist dictators for the privilege of adorning a postage stamp. And with justice. For Bernard only brought scientific method into the hospital; Pasteur brought it into the brewery, the dairy, the vineyard, the silk industry, childbed, and the operating theatre. All these, except the last, where if he penetrate it at all, it is mercifully in a state of unconsciousness, are more familiar to the common man and woman than the laboratories of the hospital. So while Bernard remains a great thinker and experimentalist, the development of whose work will stretch far into the future, Pasteur almost at once became a Public Figure, lovable, ingenious, painstaking, courageous, yet wrapt round with a legend greater than himself. If despite the efforts of Hollywood and the B.B.C., there are still any who are unfamiliar with Pasteur's achievements in making biology pay (not for himself) these are all set out in the affectionate but sober account of his life by René Vallery-Radot. Here we are concerned with his final erasure from scientific speculation of the myth of spontaneous generation so far as the visible world is concerned. This long disputed question has been a matter of contemporary technique and instrumental precision. Each stage in the controversy has been reached and passed by an advance in experimental procedure. The last stage was the result of preoccupation with canned meat for the citizen armies of the nineteenth century, made possible by the extended use of steam; the resolution of the problem was effected by the greater precision of the compound microscopes then in use.

Although the final battle in the long struggle commenced in 1859, as a result of the dogmatic assertions of the biologist Pouchet, Pasteur was unconsciously preparing himself for the contest by his demonstrations that milk may be rapidly soured by inoculation with the " ferment " which can be obtained from sour milk itself, and by his systematic investigation of the fermentative processes in brewing. Although Cagniard de Latour had observed the self-reproductive nature of yeast obtained from the brewing vats, and Schwann had at about the same time shown that air drawn through a hot pipe no longer causes putrefaction in infusions, no one seemed inclined to doubt the views of Liebig and of Berzelius, who regarded Latour's " cells " as mere " copies " of the simplest life forms, but not themselves alive. No one except Pasteur, whose doubts were awakened by noticing that " globules " of different *shapes* were associated with different types of fermentation in the beer vats; but even to him the cause of fermentation remained a mystery.

In January 1860 Pasteur was honoured by the Academy for his superb work on the connexion between the crystallographic forms of the tartrates and the rotation of the plane of polarisation of light by their solutions. The following month he sent in the first instalment of the account of his investigations on spontaneous generation, which he by then suspected to be connected with the problem of fermentation. He at once recognised that the whole question turned on the presence or absence of " germs " in the *air*. If putrescible material were to remain unchanged *in free communication with the air,* the point at issue would be settled. Suspecting that these " germs " were present in all ordinary air but absent from mountain air, and could not "fly", he constructed a large number of glass vessels with necks drawn out and bent. These vessels being filled with infusion and then boiled, remained sterile even though the ends were left open to the air; the germs could not get round the corner. Similar vessels whose necks were broken at once became cloudy and teeming with microscopic objects; but if the fracture was carried out in the clear air of the Montanvert above Chamonix and the tubes subsequently closed, no alteration occurred. The germs of life are therefore omnipresent—or almost so. Kill them by heat and they will not reappear. Admit air in which they have been similarly killed by heat (Schwann) or from which they have been filtered by cotton wool (Pasteur), and they will not reappear. Admit air in which they have never been (why mountain air is free from them will be seen in a moment), and they will not reappear. Hence *omne vivum ex vivo.* Unfortunately some of them can feign death for years, and such spores are highly resistant to heat. Their existence (unrecognised at that time) enabled the dead dogma of spontaneous generation to give a post-mortem twitch.

According to Tyndall, Bastian re-opened the question in 1870, announcing that even after prolonged boiling, infusions which he had prepared came to life. The obsequies of the dogma had therefore to be repeated by Tyndall, who with experiments rivalling in elegance, clarity, and simplicity, those of Pasteur, showed (1) that only air which does not scatter light (as does dusty air) is completely sterile, hence microbes live on the motes floating in the sunbeam; (2) infusions which are boiled *twice* or, better, three times, on successive days, remain sterile; from which he inferred that the first action of heat may not kill the resistant spores, which become active as the infusion cools, and are thus caught on the hop, so to say, by the second heating. Incidentally these researches, described in Tyndall's book, *The Floating Matter of the Air,* led to the invention of a most potent weapon, the ultramicroscope, for the investigation of the " world of neglected dimensions ", of particles smaller than the smallest microbe.

Had Pasteur's researches stopped at this point they would have effected no more than to demonstrate with uncompromising exactness what was already settled in principle by Spallanzani's experiments (p. 362). But in his memoir to the Academy in 1860 there occurs the sentence: " Ce qu'il y aurait de plus désirable serait de conduire assez loin ces études pour préparer la voie a une recherche sérieuse de l'origine de diverses maladies."

The supposed connexion between microbes and disease was by no means novel. A hint of the communication of human disease by " atoms " was advanced by Fracastor[1] in the sixteenth century: the theory was revived in a systematic form by Henle in 1840. In the case of *plant* diseases Tillet had proved as early as 1752, by experiments admirably conceived and executed, that the Bunt disease of cereals was spread by minute black particles from the " bunted " heads; in 1807 Bénédict Prévost had actually observed the germination of these " seeds ". But again it was only Pasteur who regarded the hypothesis as anything but heretical. There is little doubt that his preoccupation with the problems of industry was an immense benefit; as was also his theoretical bias owing to the fact that his academic training had been in the physical sciences. In these latter days of gargles, nasal sprays, and antiseptic soaps it is difficult to get the history of this great discovery in proper perspective. Its very success has blinded us to the difficulty of its birth and has at the same time given it perhaps almost the sanctity of a dogma. It is therefore necessary to state the facts as they were known in about 1875 when matters came to a head, to compare the rival views as to the genesis of contagious and epidemic diseases, and to see what exactly was established by Pasteur and those who were of his mind. Such a study is essential to an understanding of the growth of the idea of life in action, for the result clarified the concept of life while enlarging the boundaries of the living. No detailed history will be attempted such as would be necessary in a history of medicine; but our purpose may, I believe, be achieved by confining our attention to the disease whose in-

[1] The significance of this guess has probably been over-rated. Fracastor certainly mentioned "air borne germs" (in the inappropriate case of syphilis) but he combined the theory with the customary renaissance abracadabra of astrological signs and portents.

PLATE VIII

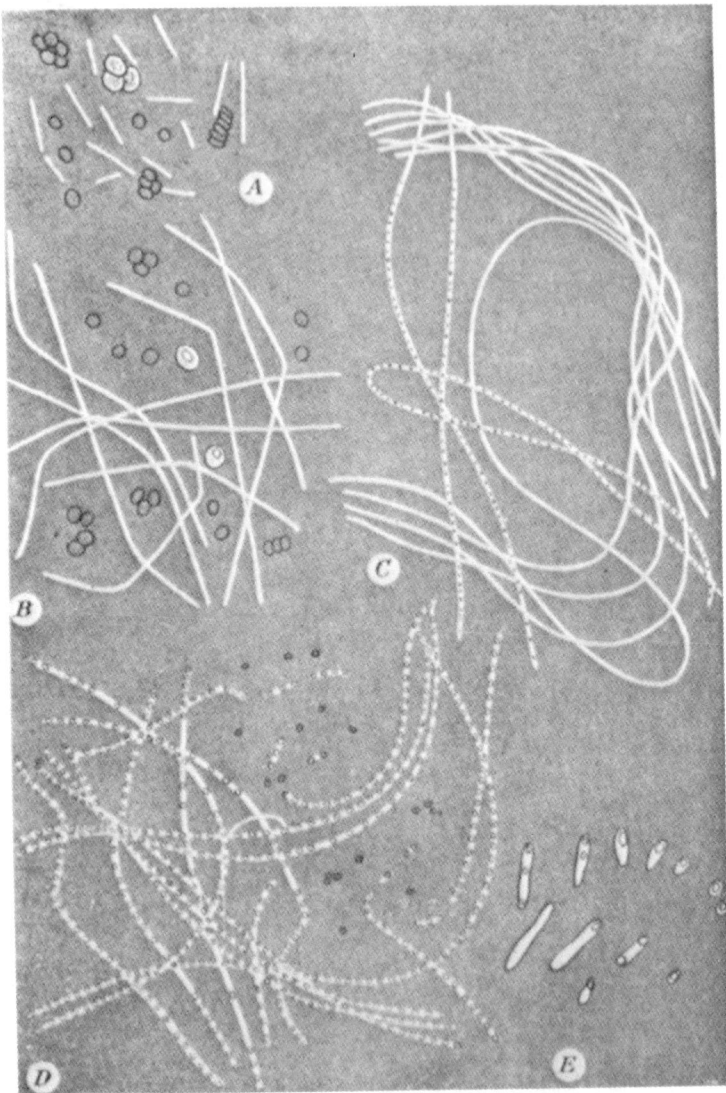

Drawing made by Koch in 1876 to illustrate development of Anthrax Spores

vestigation was on the whole the most crucial; this was anthrax, or " charbon ", as it was known of French veterinarians.

As early as 1838 Delafond had shown to his pupils the minute rod-shaped bodies present in the blood taken from animals which had died of anthrax; Davaine also recorded them in 1850. But neither regarded them as of any importance except perhaps as singular *effects* of the disease. Eleven years later, however, Davaine's curiosity was aroused by Pasteur's discovery of the butyric " ferment "—one of special interest in that it apparently worked in absence of air. Might not the rods be the *cause* of anthrax? Davaine confirmed his suspicion by inoculating some rabbits with blood from a dead sheep: the rabbits died of anthrax, and their blood contained the " bactéridies ", as he called them. His triumph was short-lived: two other investigators, Jaillard and Leplat, inoculated rabbits with blood from a cow which had died of anthrax. The rabbits died alright, *but their blood contained no " bactéridies "*. Davaine, rightly as it turned out, denied that anthrax was the cause of the rabbits' death; but a repetition of the experiments by Jaillard and Leplat gave the same result. Meanwhile, a young German country doctor, having at his disposal a little leisure time, much enthusiasm, the kitchen of his lodgings, and the insight of genius, attempted successfully to cultivate the " bactéridies " in a medium different from blood— actually the secretion of the eyes of rabbits or cows. By this means Robert Koch was able to follow by eye the growth and multiplication of the *Bacillus anthracis* as it was now possible to call the " bactéridies ". He was able to see after a sufficient lapse of time small spherical bodies appear at intervals along the rods; both fresh cultures and those with the spherical spores were as fatal to animals as the blood of a diseased animal, *and the complicating factor of the blood had been thereby eliminated.*[1]

Yet there was one loophole by which the opponents of the microbian theory of anthrax and similar diseases tried to escape. Admittedly the rod-shaped bodies were capable of reproduction, but that was no proof that they were the cause of the disease; for there was no guarantee that some matter from the original blood

---

[1] The absence of this technique had been the cause of the death of Jaillard and Leplat's rabbits by *septicaemia* from the *stale* blood of the animal which had *died* of anthrax.

might not have been passed with them into the experimental animal. No; it was still possible—in fact more than possible—that the cause was a *virus*, whatever that may have meant. Pasteur closed this last loophole by successive inoculation of sterile liquid medium to the fortieth generation—one drop of which, the " virus " having thereby been attenuated to an unimaginable degree, was as fatal as the original drop of blood. Whatever was the *cause* of the disease must increase spontaneously, and this could only be something alive, namely the *Bacillus anthracis*.

The battle was won for the sheep, for the defence of women in confinement, and for the victims of contemporary surgery, only in the teeth of bigoted opposition from the Medical Faculty. Serum therapy for rabies followed. The science of bacteriology had been created by a " mere chemist ". The invention by Koch of solid media, whereby pure cultures could be attained, and the technique of staining, established the new science as one of the corner stones of practical medicine.

Once the existence of a whole world of *specific* organisms beyond the limits of normal vision had been established the whole picture of the living world was changed. Fermentation, putrefaction, the cycle of nitrogen in the soil were progressively revealed as the work of specific microbes including the yeasts and other minute fungi. The flux of Herakleitos, the vision of the balance of nature by Lucretius, became a living reality for harmony or destruction. The " digestion " of humus by the plant, of cellulose by the herbiverous animal, became clear. Seventy years after the first unambiguous proof of the power of self-reproducing microbes the possibilities seem almost limitless.

But how do they " work "? It is in connexion with this deeper problem that the universal success of the microbian theory for a time proved a stumbling block. Liebig's hypothesis of " activation by impact " was inadequate in that it assumed units of molecular complexity no greater than contemporary chemists were accustomed to conceive. Also it left out half the truth by denying the existence of any organism as agent. Pasteur's theory, adequate at the one level, left out of account the *mechanism* whereby the yeasts and putrefactive bacteria were able to bring about their observed actions—a strange omission for a chemist, but one not so surprising perhaps in an investigator working mainly at the level of concrete problems—beer, silkworms, puerperitis, and so on. Bernard seems to have sensed the one-sidedness of the new views: " I have sometimes heard it said that the property of decomposing sugar was due to the life inherent in a globule of yeast. This vitalistic explanation means nothing and explains nothing about the action of yeast. We do not know the nature of this property, but it must necessarily belong to the physico-chemical order and be as precisely defined as, for instance, the property of platinum sponge . . ." Bernard here shows his philosophical insight in action. When he asserts that the property of fermentation belongs to the " physico-chemical order " he presumably means that a chemical action—the conversion of sugar into alcohol—must be brought about by a *chemical* or system of chemicals. It is the business of the physiologist to go on from where the mycologist left off and discover, if he can, what this system is which the living organism uses, for Bernard is not denying the possibility of a living body being more than " merely " a physico-chemical system. After his death, some ambiguous notes were found which might have been statements of results or merely a programme for investigation. Pasteur, who worked in the same building, and was on terms of great cordiality with Bernard, was naturally taken aback to find an apparent denial of the validity of his own results. Bernard's notes certainly went beyond the facts as subsequent research has shown; but he saw, what Pasteur did not, the possibility of fermentation in the absence of any *living cell*. This prophetic insight received its fulfilment, in 1897, when Buchner and Hahn showed that by grinding yeast with clean sand a cell-free filtrate could be obtained capable of bringing about normal fermentation of sugar. The study of these *enzymes*, as they were named by Kühne, has shown them to be of the nature of chemical catalysts, whose action depends not at all on life as such but is conditioned in a manner rather different from that of inorganic catalysts, in that they are more highly specific, seem always to contain a protein basis, and are to some extent " used up " during the course of the action. We shall have occasion to refer to them again in the conclusion.

Although we have seen (p. 430) reasons to regard metabolism as the most fundamental characteristic of the inner working of life, we cannot conclude this section without at least a glance at the growth of the idea of the control of its activities exercised by the organism. We have already seen how the mystery of nerve action was " explained " by whatever *physico-chemical concept* happened to be in vogue at the time. Erasistratos' vague passages for the πνεῦμα gave way to Borelli's hydraulic *machinery*. But with the rise of chemistry Willis materialised the " animal spirit " into spirit of wine or hartshorn. This crudity was hardly likely to be refined by the first great systematic physiologist, Haller, whose chemical insight rose no higher than to regard muscle as compounded of lime and earth. He did, however, introduce system into the study of the relation of the nervous system and the organs of the body by distinguishing every organ, and many parts of organs (the concept of tissue had not as yet been thought of) as *sensible* or *irritable*. Thus the muscles, *as such*, are not sensible, but irritable, in that they themselves respond to stimuli; the tendons are neither. The " sensation " transmitted from the " sensible " organs by the nerves is carried by one of those " subtle fluids " which was all the eighteenth century could conceive of. In some ways Haller reminds us of Galen. In his 650 printed works, many of them large, he summarised all that was known, or thought to be known, in the greater part of the field of biology. Like Galen also, he stands at the close of an epoch: a humanist, poet, lover of nature, and conscientious in the performance of civic duties. Posterity has appraised his contribution to science as variously as it did Galen's.

The first recognition of the futility of pretending to explain nerve action as due to something else—be it " soul ", spirit of wine, fire, or uncharacterised fluid—seems to have come from Vicq d'Azyr. He himself admitted the strong influence of the teaching of Condillac, which may be summarised as the view that consciousness is the sum of the impressions communicated by the senses, and which in turn derives from the sceptical side of the philosophy of Hume.

Complementary to this view—both in respect of contrast and fulfilment—is that of Lamarck. "A subtle substance, remarkable for the rapidity of its movements and receiving little attention because it cannot be directly observed, collected nor experimentally examined; a substance of this character is the very strange and wonderful agent that nature employs for producing the muscular movement, feeling, inner emotions, ideas, and acts of intelligence, which many animals are able to carry out." The natural response of a modern critic to this passage from the *Philosophie Zoologique* would be that *no* attention—scientific attention—can be paid to a " substance " which can neither be directly observed, collected, nor experimentally examined. Second thoughts however may recognise the operative word " directly ". That Lamarck had this in mind is suggested by the following words: " We can only know this substance through its effects." The same is true of atoms, which are no more susceptible of *direct* observation than the " nervous fluid ". The only question is whether a " nervous fluid " is as adequate a scientific entity for purposes of interpretation as are atoms. The elaboration of the concept of this fluid by Lamarck shows a far deeper insight into the *form* of the problem than his predecessors had possessed. The nerves are no longer hollow, no *observable* substance—blood, lymph, and the like—could possibly operate *fast* enough; " assuredly it can only be a fluid, moving almost with the swiftness of light . . .; now we have some knowledge of fluids which possess this faculty." We can see the way his thoughts are tending. With the experiments of Galvani fresh in the memory, and their interpretation by Volta imperfectly grasped, it is not surprising to read that he is " convinced even that it *is* electric fluid, which has been modified in the animal economy and to some extent animalised by its residence in the blood". Here Lamarck loses his grip of the form and pursues vague shadows of his own imagining. In this pursuit we shall not accompany him. His achievement was to rid earlier views of much of their crudity, and to recognise the *specificity* of nerve action. He was unlucky in being misled by the experiments of Galvani whose unrealised complexity has already (p. 219) been discussed. In trying to account for sensation, he asserts dogmatically that " no kind of matter whatever does or can enjoy any such faculty "; and he warns the reader against the belief that feeling is centred in *any* particular *parts* of an animal. These warnings went unheeded; they are still unheeded in some quarters.

Lamarck's ideas reflect alike his uncertain genius peering into the future, and the limitations imposed by the state of physics which had not then rid itself of the concept of the " imponderable fluid "—a concept of matter in motion, but matter deprived of the one characteristic—ponderability—which Newton had recognised as constituting its nature.[1] Further progress in the understanding of the nature of nerve action was impossible until physics had put its own house in order. Meanwhile, in 1811, a modest pamphlet was being privately circulated by a typical Scottish lad o' pairts, who, graduating at Edinburgh after working in great poverty, lived to become the Curator of the Hunterian Collection in London before returning to his native land as Professor of Anatomy, and where, in 1842, he died, full of honours, a pious and beloved man. Such was Sir Charles Bell, whose pamphlet, *Idea of a New Anatomy of the Brain*, gave a hint that the divisions or roots, into which the spinal nerves of vertebrates separate on entering the cord, are separate in function—the posterior (dorsal) roots alone being concerned with sensation, the anterior (ventral) with motion of the muscles. Disinclined to cause the great suffering entailed by the necessary operation, he never gave adequate experimental proof of this view. Magendie, unconcerned with such scruples, put it to much more complete test; but it was Johannes Müller who first established it without reservation and brought the physiology of nerve action to a state of perfection, from which progress on a wide front became possible in comparatively recent years.

Combining in one person the worst excesses of the *Naturphilosophie* school, in which words took the place of facts, with the power of exact and critical observation, Müller was perhaps the greatest physiologist before Bernard altered the whole basis of the subject. His doctorate thesis, *On the Relation of Numbers in Connection with the Movements of Animals*, contained so much extravagant nonsense that in later years he is said to have burnt every copy that

---

[1] " We must . . . universally allow that all bodies whatsoever are endowed with a principle of mutual gravitation " (*Rules of Reasoning in Philosophy III*). The property of gravitation was not however considered by Newton to be *innate* to matter (see above p. 101).

## THE INSIDE STORY OF LIFE 455

he could lay his hands on. He was saved by Rudolphi from sinking deeper into this bog. With one foot upon the solid ground of nature he set to work to develop Goethe's observations on the nature of sense perceptions. On the basis of many personal observations he succeeded in putting the whole subject on a new footing. " Sensation therefore consists in the communication to the sensorium, not of the quality or state of the external body, but of the condition of the nerves themselves excited by the external cause. We do not feel the knife which gives us pain, but the painful state of our nerves: the probably mechanical oscillation of light is itself not luminous: even if it could act on the sensorium, it would be perceived merely as an oscillation; it is only by affecting the optic nerve that it gives rise to the sensation of light. . . . The nerves themselves are not passive communicators of specific actions, but specific communicators of general stimulation. It is only too well known that a blow or an electric shock or an increase of blood pressure on the eye causes, if anything, a sensation of *light*; and similarly for the other nerves." This new analysis of the specificity of nervous response not only guided research in the physiology of sensation along fruitful lines, but gave an experimental basis to psychology. Opinions differ as to how far the latter is merely a branch of natural science: the Behaviourist school regard it as wholly so; but the view that an adequate psychology involves assumptions drawn from philosophical analysis seems to be gaining ground. Since the same is becoming daily more true of physics itself, the whole argument may ultimately prove to be merely verbal. However this may be, every psychological hypothesis must include Müller's observations as *minimum* assumptions.

Though the above citation is taken from Müller's *Handbook of Human Physiology* (1837), the work itself was inspired by the outlook of *Naturphilosophie*, to which he was a prey in his earlier years. The theoretical basis has proved to be thoroughly unsound; but his " other nature "—that of the critical observer—was apparently able to work within its own categories to the permanent enrichment of exact science: a strange, but by no means unique, example of a phenomenon about which much remains to be discovered. By the time he came to write the *Handbook* he had grown out of his early contempt for the *experimental* (as against the almost purely " observational ") method, and was thus able to complete and generalise the work of Bell and Magendie. Experiments on the mammals revolted his sensitive nature; moreover he recognised that an animal in pain provides most unreliable and contradictory evidence. He therefore determined to test the hypothesis by means of the frog. After several preliminary experiments he performed a crucial one which has become one of the classics of the science of physiology. " If in the same frog the three posterior roots of the nerves going to the hinder extremity be divided on the left side, and the three anterior roots on the right side, the left extremity will be deprived of sensation, the right of motion. If the foot of the right leg, which is still endowed with sensation, but not with the power of motion, be cut off, the frog gives evidence of feeling pain by movements of all parts of the body, except the right leg itself, in which he feels the pain. If on the contrary, the foot of the left leg is cut off, the frog does not feel it." A modern physiologist would choose his words more cautiously: we have no *certain* knowledge that the frog, or for that matter any animal without the power of verbal communication, *feels* anything; we merely observe an entire absence of any response, from which we infer that the stimulus failed to reach any centre from which movement could be initiated. Apart from this—psychological—assumption the experiment is ideal in being, as it were, internally controlled.

At the beginning of this chapter we set out to discover the growth of the idea of how the living organism *works*. In the preceding pages we have seen how the idea of the working of the nervous system grew, in a twisted shape, by the influence of the prevailing modes of thought—mechanical, hydromechanical, chemical, physical. In the early nineteenth century this action was identified with the latest novelty—the electric fluid, or rather fluids, since " galvanic " and " frictional " fluids were still distinguished. With clear insight into the true nature of scientific explanation, Müller who, in other respects was an Aristotelian vitalist, expressed the truth as nearly as it could be ascertained at a time when sensitive galvanometers were first coming into the hands of the physiologist, in these memorable words: " Of the nature of the nervous principle we are as ignorant as of the nature of light and electricity; but with its properties we are nearly as well acquainted as with those of light and other imponderable agents. However much these various principles differ from each other, the same question applies to all, namely, are their effects produced by currents of an imponderable matter travelling through space, or by the undulation of a fluid? Whichever theory may be correct, in the case of the nervous principle, is at present a matter not affecting the study of the laws of its action; just as the laws of optics must remain the same, whichever theory of the nature of light be adopted." The long shadow of " imponderable fluids " still casts an obscurity over this otherwise luminous passage, which points the way towards the critical physiology to be founded by Bernard. After completing his *Handbook*, however, Müller turned more and more from physiology to what may not be inaptly called comparative biology. In these studies, the value of which he had already shown in the introductory passages to the several sections of his physiology, he emphasised the outstanding importance of " boundary " forms (though denying the mutability of species) and devoted ten years to the study of the primitive cyclostomes (allies of the lamprey), whence came the masterly comparative interpretation of the urinogenital organs of vertebrates. Into this and his dogmatic Aristotelianism we cannot enter. He had the saving grace, as one of his pupils said, of teaching not dogma, but method. And when we take account of the fact that he taught Virchow, Remak, Henle, Schwann, Kölliker and Helmholtz, we may agree that his teaching was to good effect.

## SOURCES FOR CHAPTER XXXII

M. Foster, *Lectures on the History of Physiology* (Cambridge, 1901).

D. McKie, *Wöhler's ' Synthetic ' Urea—A Chemical Legend* (art. in *Nature*, vol. 153, p. 608).

J. von Liebig, *Chemistry in its Relation to Agriculture and Physiology*, ed. Lyon Playfair (2nd ed., London, 1842).

C. Bernard, *An Introduction to Experimental Medicine*, trans. H. C. Greene (New York, 1927). Almost unique in the history of science, in that it is a brilliant attempt of an outstanding experimentalist to portray and criticise his own method.

R. Valléry Radot, *La Vie de Pasteur* (Paris, 1900).

E. C. Large, *The Advance of the Fungi* (London, 1940). This work, which is fully documented in an appendix, seems to have been overlooked by recent historians of biology. It gives a very full account of the sporadic beginnings of mycology and provides an excellent illustration of President Conant's contention that—even in recent times—the advance of science is anything but uniform and orderly.

Johannes Müller, *Elements of Physiology*, 2 vols., trans. H. Baly (London, 1838-42).

## CHAPTER XXXIII

## CONTINUITY AND ORGANISATION

DARWIN made the individual organism not only the "heir of all the ages" but the determinant of the "shape of things to come". The facts of the geological succession and of the endemic species of oceanic islands leave us in little doubt that descent with modification is also a fact. But Darwin also put forward an hypothesis, that of natural selection, to account for the *mechanism* of this theory of descent. The hypothesis was exceedingly plausible, but it involved the massive assumption of the inheritance of those deviations from the norm which Darwin called *variations*. Although he does not seem ever to have realised that this was no more than an assumption he was so well aware that it was an essential part of the general theory that in his *Variation of Animals and Plants Under Domestication* (1868) he put forward a subsidiary hypothesis to account for the mechanism of the mechanism—that is as a material link between successive generations whereby the "variations" of the former could be impressed upon the latter. This was his hypothesis of *pangenesis*, which was to the effect that every cell of every organ puts forth " gemmules ", which being distributed to every region of the body, are necessarily included in the germ cells; here especially, as elsewhere, they are capable of development into cells with all the characteristics of those from which they came. Moreover " gemmules " from the remote ancestry are thus present in every body. It is interesting to note that the at one time very influential " philosopher of evolution ", Herbert Spencer, had suggested a superficially similar hypothesis a few years earlier; but whereas Darwin's " gemmules " are all as it were potential *cells*, Spencer's " physiological units " are the potential *components* of cells. What the theories have in common is a more or less tacit assumption of the inheritance of acquired characters—an assumption which in his later years Darwin explicitly stated in the words: " In my opinion, the greatest error which I have committed has been not allowing sufficient weight to the direct action of the environment, i.e. food, climate, etc., independently of natural selection " (1876). Some critics have held the view that the acceptance of this dogma by an otherwise cautious thinker was inevitable in an age which regarded " progress " as a certain consequence of suitable education and propitious environment. The other common feature of these two theories is that they are really little more than the formal re-statement of the problem of inheritance; they explicitly state that some material determinant must be handed on from one generation to another, but as to the structure and mode of operation of this determinant we are left entirely in the dark.

In trying to reconstruct the history of the attempts to give precision to these ideas one is struck by the extraordinary isolation of the various workers. The data for such an advance were available in the practice of the well-known firm of seedsmen, de Vilmorin, and in the published experiments of Mendel. It has been customary to account for the complete neglect of the latter as a result of publication having taken place in an obscure journal, but Professor Waddington has recently drawn attention to the fact that copies were sent to many leading biologists. The reason certainly seems to lie deeper: perhaps in the lack of stimulus to plant-breeding in Britain as a result of the repeal of the Corn Laws (Professor J. B. S. Haldane's view); perhaps in the fact that the ruling liberal dogma of " progress " prejudiced minds otherwise alert and critical in favour of the possibility of unlimited improvement by the effects of the environment guided by selection. With such a view, which was undoubtedly dominant at that time, Mendel's results and de Vilmorin's methods would have been in little accord. This may likewise have accounted for the lack of interest shown in Galton's papers of 1875 and later, which criticised the indiscriminate diffusion of " gemmules " throughout the body cells, and laid the basis of the accurate statistical treatment of inherited characters. Weismann, writing in 1892, admits that he had never even heard of these papers until ten years after the publication of the first series.

## CONTINUITY AND ORGANISATION

Of all the workers just mentioned none seems to have exercised much influence on his contemporaries with the possible exception of Weismann; we shall therefore consider his work first—a procedure justified on the ground that in his great book, *The Germ-Plasm*, published both in German and in English in the same year, 1892, he surveyed the whole field (except Mendel) and provided the first articulated theory of the cellular mechanism of heredity.

August Weismann dedicated his book " To the Memory of Charles Darwin ", a fact to be borne in mind when attempting to appreciate justly his relation to the evolution theory. Despite an impression at one time rather generally held, Weismann undoubtedly regarded his task as that of seeking out and correcting the errors in the Darwinian theory, and amplifying it in detail in respect of the mechanism of inheritance. The outline of this task was sketched as early as 1883 in an essay *On Heredity*. In Weismann's own words:

> I contested at first in general terms not only the existence but also the theoretical possibility of the transmission of acquired characters and tried to release the theory from the necessity of an explanation which deprived it of any further development. In this essay I further assumed the existence in the germ cell of a reproductive substance, the *germ-plasm*, which cannot be formed spontaneously, but is always passed on from the germ-cell in which an organism originates in direct *continuity* to the germ-cells of the succeeding generations. The difference between the body in the narrower sense (soma) and the reproductive cells was also emphasised, and it was maintained that the germ-cells alone transmit the reproductive substance or germ-plasm in uninterrupted succession from one generation to the next while the body (soma) which bears and nourishes the germ-cells, is in a certain sense only an outgrowth from one of them.

In these few sentences is contained the fundamental principle of the *theory* of the mechanism of inheritance. I have emphasised the *theoretical* nature of this description, since although Strasburger's observation of the splitting of the chromosomes in plant cells had been made by 1880, Flemming's similar observation in the case of animal cells had at that time still to be made; and even the relation between cell and nucleus was still in a fluid state. By 1885 Weismann was sufficiently convinced of the seat of the germ-plasm to publish his book *The Continuity of the Germ-Plasm*. Within two years he had so clearly grasped the implications of his theory that he *foretold* " the necessity of *a reduction of the germ-plasm* each time to one half of its bulk as well as a reduction of the number of the ancestral germ-plasms contained in it. The hypothesis of the ' reducing divisions of the germ-cells ' has been thoroughly substantiated by subsequent observations." Not the least interesting aspect of Weismann's comprehensive theory is its exemplification of the *deductive* method in biology. The only one of his contemporaries who envisaged in any degree of detail the mechanism subsequently *observed* was de Vries, to whom he pays generous tribute, but who fell far short of the true picture by mistaking continuity in space (the cytoplasmic fibrils penetrating the cell walls in plants) for continuity in time.

We have not space to follow the argument by which Weismann was led to formulate the scheme finally presented in *The Germ-Plasm*; it must suffice to summarise the principal features. After criticising Galton and others for not realising that heredity has no *essential* connexion with fertilisation, hence with sexual differentiation, he admits that the subject may most easily be approached and most strikingly demonstrated by the phenomena of fertilisation with its accompanying reduction division of the germinal substance. " There can now be no longer any doubt that . . . *the nuclei of the male and those of the female germ cells are essentially similar*, i.e. in any given species *they contain the same specific hereditary substance.*" Now not only do all cells contain the same number of chromosomes, but it is evident that these chromosomes contain the determinants whose existence Darwin and others had postulated. Moreover " it is evident as Wilhelm Roux was the first to point out, that the whole complex but wonderfully exact apparatus for the division of the nucleus exists for the purpose of dividing the chromatin substance . . . not merely quantitatively but also in respect of the different *qualities* which must be contained in it." Once again Weismann shows his insight to be far greater than that of his contemporaries, for he alone realised that in any organism which *grows*, the ultimate determining factors must come into operation in a serial order: there must therefore be what he called an " architecture of the germ-plasm ". For this architecture to fulfil the demands made upon it, it must, he thinks, consist of *biophores*, the actual operative chemical units grouped together into *determinants*, each of the latter corresponding to an " independently varying character " of the organism—for instance all the blood corpuscles might be controlled by a single determinant. Finally the determinants must be grouped in units spatially related to each other—the *ids*. All these are of course " formal " elements—hypothetical only; but from the " beaded " appearance of the chromosomes in the parasitic worm *Ascaris* (upon which much of the early work in animal cytology was done) he hazards the guess that the chromosomes are in fact *idants*, that is strings of *ids*—a remarkable anticipation of the gene theory now holding the field.

We must now return to the experimental work on the behaviour of the *organism* as distinct from that of its cells, the recognition of whose importance was for so long delayed. The practice of selection in the improvement of cultivated seed plants was based on de Vilmorin's discovery that the best results are obtained by selecting those plants with high *average* performance. This was tantamount to the admission that, as it were, a " proportion " of the variation in any one character in a given individual is not inherited—contrary to the Darwinian view. An extension of this idea by de Vilmorin is referred to by Weismann, namely that although certain plants possess the capacity for handing on a desirable character, it is impossible to detect these plants by mere inspection—a fact supporting Weismann in his contention that the *germ-plasm* and not the *soma* is all important. By proceeding in this way de Vilmorin was able to establish seed giving rise to plants which bred true to an average specification: such individuals were subsequently described by Johannsen, who applied statistical analysis to the problem, as " pure lines ": these constitute the analogue to the chemists' " pure chemical individuals " and are of course essential in genetical research.

Of more immediate effect were the observations of de Vries on the strange behaviour of the Evening Primrose (*Oenothera Lamarckiana*). This species breeds true on the whole, as every " good " species is supposed to, but among the many normal progeny appear a few which differ markedly from the average, *and these abnormal forms breed true to their own type.* De Vries interpreted these results as being due to a *sudden* change in the substance of the germ-plasm, and distinguished such " mutations " from the fluctuating variations which are normally observed in all species, due presumably to varying response to minute differences in the environment. It was evident that Darwin's analogy of artificial selection, whereby the fancier develops new strains, was not strictly applicable to the supposed process of " natural selection " operating on *all* variations, the majority of which in the light of the new knowledge are not inheritable. When it was later established that the majority of mutations would certainly be *dis*advantageous in a state of nature, the analogy was seen to be almost wholly false. It is one of the ironies of history that de Vries' mutation theory, which wrecked the naïve Darwinian theory of natural selection, though ultimately confirmed as a true description of a natural process, was itself based on a misinterpretation of the observed facts. For it was later shown that the abnormal forms of *Oenothera* are not due to " mutations " at all, but are the result of a reassortment of Mendelian factors in a particularly unstable complex. This point may be better understood at a later stage.

In 1895 William Bateson published his *Materials for the Study of Variation*—a date which marks the foundation of the science of genetics in Great Britain. Two years later he established the experimental station in the Botanic Garden at Cambridge, where his researches on poultry and flowering plants were undertaken. Before he had time to discover the fundamental laws of genetics however they were re-discovered in 1900, simultaneously by de Vries, Correns, and Tschermak, not by experiment, but by reading the original papers by Mendel.

As Bateson was to say in his Inaugural Lecture in 1908, Mendel's discoveries led biology into a " new world, the very existence of which was unsuspected before "; and this was achieved by applying to biological data the analytical methods of the physical sciences. The accelerating progress of physics since the early years of the seventeenth century derives from the method of isolating in thought the conditions of change and devising experiments by which the results of controlled change of each condition may be determined. The method may be applied to

CONTINUITY AND ORGANISATION 465

physical data because we are there dealing (unconsciously in the early days) with statistically homogeneous masses of non-identical particles. In biology we are primarily concerned with the individual, whose behaviour, just because it is the behaviour of an individual, can lead to no *general* results. The first claim of Mendel to the reverence of posterity was his recognition of the necessity for operating on random groups; the second claim was his recognition for the necessity for isolating the factors of change. Taking then a random group of pea plants all of whose characters save one—say relative tallness—were ignored, he crossed them with a similar group all of which agreed in being relatively dwarf. The unexpected result was that the progeny bred from the seed

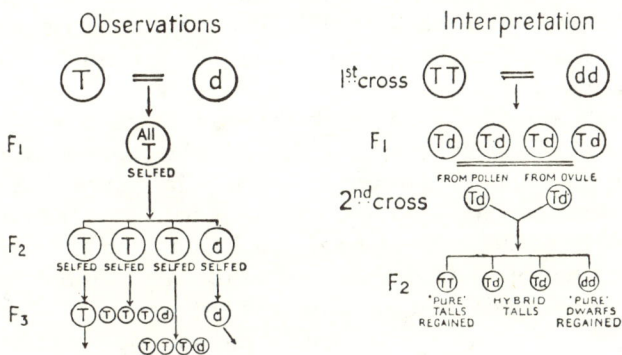

Fig. 35. Mendel's law of segregation.

were *all* tall. When each member of this group was self-pollinated the progeny derived from the seed were found to be mixed tall and dwarf roughly in the ratio of three to one. But whereas the dwarfs of this group bred true in successive generations, the talls continued to throw a mixed population, of which the dwarfs all bred true. Of the talls, some when isolated bred true, others continued to throw mixed populations. Mendel recognised that the implication of these facts is the " atomicity " of heritable characters—or rather of the " factors ", as he called them, which are the material determinants of the characters. In the first hybrid generation dwarfness disappears altogether, but its reappearance *in a definite quantitative measure* proves that it was

" there " all the time, as the atom of the gas, hydrogen, is "there" in the liquid, water. By taking characters in pairs—tall and yellow seed coat with dwarf and green seed coat for instance—Mendel showed that the analogy may be pressed further, since the resulting association of pairs of characters is such as can be mathematically predicted on the assumption of " independent assortment ".

Into the further opening up of this " new world " we cannot enter: suffice it to say that from the year 1900 discoveries have come thick and fast, rival hypotheses have struggled for the mastery and been merged in a higher synthesis. One matter alone concerns us here, namely the discovery by Bateson, Saunders and Punnett in 1904 that Mendel's law of independent assortment is far from being universally valid. Though it holds rigorously for certain pairs of characters—indicating independent assortment of the causative factors—for others the divergence is not only greater than could reasonably be ascribed to the errors inevitably associated with relatively small samples, but the *degree* of divergence, while approximately constant for each pair, varies in a regular manner when one character of the pair is replaced by a different character. The formal conclusion from these results is of course that the " independence " of the factors is only relative; that is the *chance* of their separation in the formation of the germ cells is not absolute, but itself a measurable quantity. The phenomenon was therefore called " linkage "; and in the hands of Sturtevant was ultimately shown to imply a *linear arrangement of the factors* (now called " genes ") on the chromosomes, thus confirming in a remarkable manner the speculation of Weismann as to the pattern of the material basis of heredity.

Between the germinal continuators and the mature soma there stands, or rather emerges, the transitional phase known as the embryo. As in all forms of " becoming " there is no sharp boundary between the embryo and the mature organism; but in all species for a short time, and in many for a long time—almost the whole life of the organism sometimes—the appearance of that which is " becoming " differs so strikingly from that which it will finally become that it is more than mere convention to regard embryology as a separate branch of biology. In the history of the subject we have seen emphasis at first exclusively placed on the cause of the embryo itself—the very nature of life. Later the phases of the embryo came to be regarded as the successive pages of the Family Photograph Album, in which recognisable, if caricatured, pictures of successive generations pass in review. In the Introduction to his Silliman Lectures, *Embryological Development and Induction*, Spemann wrote: " It had been forgotten by many that the changes of form constituting the outer aspect of development must have, besides their hypothetical derivation from adult stages of long ago, their immediate physical causes, the investigation of which might also be a worthy scientific task and, perhaps, the more urgent one. W. His was in conscious opposition to the predominant opinion in attempting to explain mechanically the formation of the chick's embryo by the folding up of the expanded germ layers and in trying to derive these foldings from differential growth of the respective germ-regions." His's book, *Ueber Unsere Körperform* (1874) immediately led to the foundation of a new department of physiology with the suggestive name *Entwicklungsmechanik* (the mechanism of development) which applied the experimental methods of physics and chemistry to the problem of *why* the embryo passes through its successive phases—or why in some cases it does not. In a word, it is ultimately the science of *how the chromosomes work*; though at present, starting from the other end, it has to be content with humbler immediate problems.

Despite Haeckel's placing of His's book on his *Index Expurgatorius* (see p. 422), within ten years work was in progress in many parts of Germany. The first problem—what determines the direction of the first cleavage plane of the fertilised egg cell into two daughter cells or *blastomeres*—was fought out between Pflüger, who backed gravity, and Roux, who regarded the path of the sperm nucleus as all important. The issue was settled by Roux's ingenious adaptation of the klinostat, recently introduced by Sachs, who was at this time founding plant physiology, whereby the action of gravity is neutralised. G. Born and O. Hertwig confirmed the absence of any decisive *external* physical influence in normal development. To clarify the nature of this " self-differentiation " Roux killed one of the blastomeres with a hot needle. The resulting production of a *half-embryo* (the actual "half" depending on the position of the original cleavage plane) seemed to show that the fate of each blastomere is settled from the time of cleavage, and that development proceeds independently of even the *organic* environment. But that such an inference would be theoretically unjustified, seeing that the living blastomere is still in communication with the dead one, was experimentally confirmed by Driesch (1891) and other workers who were able to separate the two blastomeres of the Sea Urchin without damage. In this case each blastomere commences to form a half embryo, but at a certain stage, after many cell divisions, it seems to realise that something is missing, and its subsequent behaviour is different from what it would normally have been. The final result is a mature animal perfect in every respect except of size. Human *identical* twins are known to arise by a similar splitting of the embryo, but at a much later stage.

The implications of this result were profound. For Driesch saw in this behaviour the justification for a revival of vitalism—a vitalism founded on a critical examination of the evidence, but terminating nevertheless in the postulation of the existence in every organism of a determinant of a higher order than that of molecules in physico-chemical relationship. To this determinant Driesch rather unfortunately gave the name of *entelechy*, which thus re-entered biology with a crowd of ambiguous and ill-defined characters associated with its rather sullied past, thereby bringing upon Driesch a good deal of criticism entirely wide of the mark.

The re-direction of the process of development of an isolated blastomere was clearly in contradiction to the extreme interpretation of Weismann's theory of the germ-plasm, namely that development is the ordered unpacking of the determinants whose distribution follows from their supposed serial arrangement on the chromosomes—" evolution " in the literal sense of the word. For on such an interpretation it would be difficult to account for the " change of plan " of the *isolated* blastomere. The situation is however complicated by the division of the cytoplasm. A more direct attack on the theory—one which made its literal acceptance impossible—was the experiment first performed by Loeb, in which a portion of the cytoplasm was isolated from the remainder by a constriction, thus maintaining it without a nucleus. Such a mass may be kept alive while the remainder divides several times. When in the end a nucleus was allowed to slip through the constriction, the latent mass developed in a perfectly normal manner, despite the fact that it was provided with a nucleus very different from the one it would normally have received.

. . . .

At the head of the Second Part of this book we set the question " What is Life ? ", and in the succeeding chapters we have seen in a hasty and perhaps superficial, but I hope not distorting, glance how various lines of attack have been opened up in an attempt to reach nearer the heart of the problem. It is none of the business of this humble view of the growth of these ideas to attempt to see the whole picture as it is revealed to-day; for like the objects of its study it is a constantly changing picture. Not only are obscure and empty patches being clarified or filled in, but others ill-drawn in the past have in a few cases to be erased. One characteristic it has however possessed for centuries, that is an oscillation between what might, to continue the metaphor, be called a negative and positive aspect. For the former (paradoxically known to philosophers as " positivism ") life is " nothing but " the clash and swirl of material systems of outstanding complexity. The latter endows every living unit with a principle of a different *order of being* from that of the lifeless object, however complex this may be after its own kind.

It is important to note that these traditional views have been held in what might seem the most unlikely quarters. To Newton for instance it seemed probable that " God in the beginning formed matter in solid, massy, hard, impenetrable, movable particles ", whose only *innate* characteristic was inertia. So the system which was to form the reasoned basis of the extreme materialism of the French Illumination was created by one who " had an eye upon such principles as might work with considering men for the belief of a Deity . . ." and who, so far as we can judge, regarded even the motion of material particles, let alone the behaviour of living things, as having no *cause* in nature, but only in the will of God. This was no stumbling block to the progress of Newton's mathematical physics, but it left a gap in his natural philosophy which the French materialists filled by endowing matter *with* motion.

It is important also to note that we have to a large extent grown out of this acceptance of unanalysed concepts. " What is the sense ", Whitehead asked, "of talking about a mechanical explanation when you do not know what you mean by mechanics? " " What is the use ", biologists are asking themselves now, " of asking whether life is nothing but the manifestation of physico-chemical relationships of pre-eminent complexity, if you do not know what you mean by life? " The fairly general recognition that the question is so framed as to admit of no answer has come about mainly as the result of attempts to decide whether viruses are " alive ".

Pasteur, as is well known, established the fact that a large number of diseases of man and animals could be contracted only by the entry into the body of small objects, visible under the microscope, and in the majority of cases capable of unlimited multiplication in suitable media from which every trace of " life " had been removed by heat treatment. No one doubted that these objects were alive, the criterion of " life " being the power of unlimited reproduction when provided with suitable nutriment. To remove these microbes from water Pasteur's colleague, Chamberland, made use of unglazed porcelain filters, whose capillary passages were too small to allow any bacteria to pass. It was not long before filters of varying degrees of porosity were being used to separate the larger from the smaller organisms; it thus came about that Iwanowsky passed the juice of a diseased tobacco plant through such a filter and discovered that *the filtrate was just as infectious to healthy plants as the unfiltered juice.* This in itself was of no outstanding theoretical interest, since it might have been due to a chemical toxin generated in the juice. Iwanowsky himself did not follow up his discovery. But by 1897 it had been discovered that not only tobacco mosaic disease but also foot-and-mouth disease could be transmitted *to an unlimited extent* by injecting filtered fluid from a diseased organism. But this " multiplication " of the virus took place *only in the living body*. There can be no doubt about its being " alive " if the power of multiplication be the only criterion; but what are we to make of the fact that no virus can "live" outside another living body? The same *materials* as can maintain it alive *within* the host are present when the tissues are mechanically crushed and the product passed through a filter narrow enough to withold any living cell, wide enough to allow the largest molecules to pass; but the virus will not "grow". How then does it grow within the host? It has been suggested that the virus is only an irritant which promotes activity of the host cells whereby more virus is produced—a fantastic reversal of Bernard's definition of life as the "sum of all those activities which strive to resist death". Fantastic perhaps, but not necessarily false; for no one doubts that the neoplasms of cancer are produced by proliferation of the "host's" own cells run amok, perhaps as the result of an irritant, either foreign or one of its own degradation products. It is clear that the boundary between the "living" and "non-living" matter has broken down. Nor has the attack been entirely one-sided; for Newton's "hard, massy, particles" are, to-day, as dead as the *vis essentialis* of Wolff. The simplest known unit of matter, the electron, is still a complex field of relations, at present, and perhaps for ever, unpicturable except as a mathematical function of abstract terms.

It is perhaps significant that one of the most notable contributions to the growth of the idea of life has come from one of the most notable creators of the mathematical description of the electron, Professor Schrödinger. Before examining his ingenious speculation on the way life "works", let us remind ourselves of the many ways in which the study of life processes has revealed higher and higher levels of *order*. In the new sciences of experimental embryology and genetics we have just seen that the smallest attainable units—chromosomes and fertilised ova—are still highly differentiated complexes, capable of reacting in many different ways. In recent years this characteristic has been further emphasised: in the case of the chromosomes in particular it has been recognised that the Mendelian unit factor corresponding to each character in the organism is over-simplified. Rather is there a potency characteristic of each *locus* on the chromosome, which may be actualised in different ways according to the particular arrangement of the "gene-complex" in which it finds itself. Moreover this complex is continually changing—in some species more than others—by actual exchange of material between homologous chromosomes when they pair at meiosis.

In physiology we have the same story—sufficient " food " gave way to sufficient " calorie-equivalent "; sufficient " protein " to sufficient precursors of appropriate amino-acids; and life may hang by the tenuous thread of a fraction of a milligramme of cobalt ion or riboflavine. It is inconceivable that these minute traces could exert so profound an influence unless they are essential pieces in a jig-saw pattern—perhaps a colloidal enzyme-complex in which certain molecular forms must be scattered according to definite rules, to act as centres from which some specific activity may be maintained. In the study of the nervous system too the greatest advance was signalised by Sir Charles Sherrington's *Integrative Action of the Nervous System* (1906) in which the oversimplified concept of " reflex arcs " is giving way to that of regions of continuous activity, modulated as it were by other impulses superimposed. Moreover the activity demanded is controlled by impulses received from within the organs themselves (by the so-called proprioceptive fibres), not only from the environment. This physiological integration has its counterpart in psychology—a tendency starting from Semon's revaluation of the idea of inheritance of acquired *character* to the dynamical notion of acquired *reaction*, the modification of the material system being called an *engramm*. On the mental side this movement is represented by the Gestalt school of psychology founded by Köhler as a result of his demonstration that the behaviour of apes can be fully understood only in relation to the *pattern* (gestalt) of the presented stimulus. A promising extension into the anthropological field is the " culture-pattern " theory of the late Dr. Ruth Benedict.

Professor Schrödinger's approach to the problem of the nature of life is based on the experimental facts of the *ordered* structure of the chromosomes, and the quantitative data provided by the most recent advances in physics. It would be inappropriate to attempt a detailed exposition of his views, as they cannot yet be said to form a part of the history of the subject; my reason for including any reference to such recent work is that it seems to me to cast a revealing light upon the nature of scientific thinking.

## CONTINUITY AND ORGANISATION

So long as atoms were conceived as solitary existents, the principle of least action, whereby a material system seemed to " know " what path to follow, smacked of the miraculous. A wave-like form however demands smaller draughts on faith; matter is thus feeling its way towards sentience. In every material system also there is a proportion of orderliness, whose amount may be measured by its " negative entropy ". Professor Schrödinger suggests that we may use the two synonymously. Thus the classification of existents is no longer to be regarded as the dichotomy of " living " and " non-living " but of the relatively ordered bodies and the disorderly ones. " The device by which an organism maintains itself stationary at a fairly high level of orderliness . . . really consists in continually sucking orderliness from its environment." For the animal, the material must be the already highly ordered proteins, etc.; for the plant, owing to its ability to extract order from solar radiation, the relatively degraded end-products of animal metabolism suffice. So far this is little more than a formal description of a suggestive kind; but it passes into the realm of articulated hypotheses when it is found that the gene is " *too small* to entail an orderly and lawful behaviour according to statistical physics". The laws of physics do not therefore " break down " in the apparently " unnatural " behaviour of organisms in failing to maximise their entropy; they are merely inapplicable. Nor is this the only link between physics and biology. Here we are dealing with the relation between the size of the gene and the size of the smallest aggregate which can " obey " the (statistical) laws of physics. But what of the extraordinary " stability " of the inherited pattern, and of its even more extraordinary power of very rarely changing abruptly? Here it is field physics which specially illumines the problem; for if, as has in fact been shown, the gene is no bigger than a very large molecule, it is only an occasional local concentration of energy which can bring about the essential *chemical* rupture of the field holding the atoms in their " permanent " pattern.

" In my end is my beginning." In the growth of ideas these words are charged with a new ambiguity; for if the history of science teaches us anything it is that only to those who ask the right questions are any answers vouchsafed. Not even to them are these answers the end of the quest, but only an enlargement and a refining of the question. If we look back to see whence we have come, we may the better set the course on which we are to go. And in frequent converse with those who have brought us to where we are, we enjoy in Whitehead's immortal words " the habitual vision of greatness ", without which the Conquest of Nature must spell the Defeat of Man.

## SOURCES FOR CHAPTER XXXIII

A. Weismann, *The Germ-Plasm* (London, 1893).

J. B. S. Haldane, Art. " Forty Years of Genetics," in the volume *The Background of Science*, ed. Needham and Pagel (Cambridge, 1938).

E. Schrödinger, *What is Life?* (Cambridge, 1944).

H. Driesch, *The History and Theory of Vitalism* (Engl. Tr., London, 1914), also *The Science and Philosophy of the Organism*.

# GENERAL BIBLIOGRAPHY

### General Works

J. G. Crowther, *Social Relations of Science* (London, 1941). Readable but highly informative; extensive references.

G. N. Clark, *Science and Social Welfare in the Age of Newton* (Oxford, 1937). An authoritative study of a critical period. Written from a rather different point of view from that of Crowther.

C. Singer, *A Short History of Science* (Oxford, 1941). Perhaps the best balanced view of the development of science from the earliest time to the end of the nineteenth century. Especially valuable for the routes by which Greek science reached Western Europe.

H. T. Pledge, *Science Since 1500* (H.M.S.O., 2nd ed., 1946). Much more detailed, especially in respect of the modern period.

F. Sherwood Taylor, *A Short History of Science* (London, n.d.). Covers the whole period. Good on the technological side.

A. Wolf, *A History of Science, Technology and Philosophy in the Sixteenth and Seventeenth Centuries* (London, 1935); *A History of Science, etc., in the Eighteenth Century* (London, 1938). Large and detailed works; superbly illustrated.

H. Butterfield, *The Origins of Modern Science* (London, 1949). Of special importance for the light it casts on the gropings towards the the modern concepts of mechanics which took place towards the close of the Middle Ages. The whole book marks an epoch in the development of modern historical thought. It was available to me only after the proofs of the present work had been passed.

# GENERAL BIBLIOGRAPHY
## SPECIAL SCIENCES

*Physics*

F. Cajori, *A History of Physics* (New York, 1899).

H. Buckley, *A Short History of Physics* (London, 1927). Both works of moderate compass.

W. F. Magie, *A Source Book of Physics* (New York, 1935). Long translated excerpts from the great classics, with biographical and explanatory notes.

E. Mach, *The Science of Mechanics* (London, 1893). An epoch making work—historical and critical.

*Astronomy*

H. Shapley and H. E. Howarth, *Source Book in Astronomy* (New York, 1929). A companion to Magie above.

*Chemistry*

J. R. Partington, *A Short History of Chemistry* (London, 1937). A very full treatment considering its compass.

E. von Meyer, *A History of Chemistry* (London, 1898). A large and detailed work still of value for reference.

*Biology*

W. A. Locy, *Biology and its Makers*, 2nd ed. (New York, 1910). Very readable studies of the principal personalities and movements.

C. Singer, *A Short History of Biology* (Oxford, 1931). More detailed than the above.

E. Nordenskiöld, *History of Biology* (London, 1929). A large and detailed work, but eminently readable, well balanced, and fully documented.

# CHRONOLOGY

(1) FROM THE NEOLITHIC REVOLUTION TO THE END OF THE FIFTEENTH CENTURY.

The following Table is intended to give a conspectus of the order in which innovations appeared. Only the earliest recorded appearance is given—diffusion if and when it occurred took as much as a millenium or even more. Names of persons are given in order of birth.

B.C.

8000. Neolithic Revolution—domestication of plants and animals. Introduction of pottery, querns, baking, fermented liquors. Towards the end of the period there appeared copper, wooden wheels and ploughs, sails for boats, and potter's wheel.

3000. Urban Revolution—bronze replaced copper—verbal records of " events " (ritualistic or mythical); numerical tables of priestly property and exchange.

1500. Bellows.

1200. Iron becoming fairly general. ? Fall of " Troy ".

9th Century. Collection of the poems constituting the *Iliad*.

7th Century (late). Thales. Anaximandros.

6th Century. Anaximenes. Pythagoras. Herakleitos. Parmenides.

5th Century. Battle of Marathon, 490. Peloponnesian War broke out. Empedokles. Demokritos. Protagoras. Hippokrates the Physician. Socrates. Plato.

4th Century. Death of Socrates, 399. Battle of Chaeroneia established Philip of Macedon as ruler of Hellas. Eudoxos. Herakleides. Aristotle. Theophrastos. Euclid. Alexander, d. 323. Aristarchos of Samos. Herophilos.

3rd Century. Archimedes. Erasistratos. Apollonius.

2nd Century. Hipparchos.

1st Century. Egypt passed under Roman rule. Lucretius. Crateuas—the earliest recorded drawings of plants. Battle of Actium established Augustus.

A.D.

1st Century. Vitruvius. Celsus. Hero.

2nd Century. Galen. Ptolemy's *Syntaxis*, 150. Diophantos—algebra.

4th Century. Constantinople capital of Roman Empire. Zozimos.

5th Century. Sack of Rome, 410. St. Augustine's *Civitas Dei*, 413.

6th Century. Boethius (d. 525) translated Aristotle's logical works into Latin.

8th Century. Modern method of harness (bridle). Western incursion of Islam halted at Poitiers, 732.

9th Century. Charlemagne crowned Holy Roman Emperor, 800. Al-Khuarizimi, *Al-Gebr.* Geber (Jābir).
10th Century. Avicenna, *Canon of Medicine.* Gerbert reintroduced abacus.
11th Century. Anselm of Canterbury—rise of scholasticism. Infiltration of Greek learning through Salerno.
12th Century. University of Paris. Averroes introduced Aristotelian thought (at first pronounced heretical). "Arabic" numerals in the West.
13th Century. *Book of the Abacus,* 1202 (Leonardo of Pisa.) Roger Bacon—gunpowder. William of Moerbeke translated Archimedes, Michael the Scot, Aristotle's biological works. Magnetic compass in West. St. Albert. St. Thomas's christian interpretation of Aristotle. Frederick II—"modern" interest in experiment. Dante. Edward I.
14th Century. William of Occam—withering of scholasticism. Rise of Florentine Banking Houses. Petrarch. Chaucer. Naturalism in Italian painters. Regiomontanus—high water mark of Ptolemaic astronomy.
15th Century. Henry the Navigator. Sack of Constantinople, 1453. Nicholas of Cusa anticipated many modern ideas. Printing from movable type. Aldine Greek text of Aristotle. America made accessible to Europeans.

(2) FOR THE "MODERN PERIOD" AN ATTEMPT IS MADE TO CORRELATE ADVANCES IN THE SEPARATE SCIENCES AND TO PLACE THIS AGAINST THE BACKGROUND OF EUROPEAN HISTORY.

16th Century. Luther's 95 theses of reform, 1517. Leonardo da Vinci, d. 1519. Battle of Mohacs (1526); Advance of Ottoman Turks to Vienna. Dissolution of Monasteries in England. *De Revolutionibus Orbium Coelestium,* 1543. *De Humani Corporis Fabrica,* 1543. *De Re Metallica,* 1546. Decimal Notation. Mercator's World Map, 1569. Euclid in English, 1570. Gresham College, 1575. This century saw the rapid rise of the great financial House of the Fuggers of Augsburg and a catastrophic fall in the value of money.
17th Century. *De Magnete,* 1600. *Advancement of Learning,* 1605. *Astronomia Nova,* 1609. Invention of telescope and microscope. Authorised translation of Bible, 1611. Logarithms, 1614. Harvey taught circulation of the blood. *Dialogues on the Two Chief Systems of the World,* 1630. *Dialogues on Two New Sciences,* 1638. *Discourse on Method,* 1636. French Academy, 1637. *Leviathan,* 1651. Atomism reintroduced by Gassendi. Air pump. Pendulum clock patented, 1657. *Accademia del Cimento,* 1657. Royal Society's Charter, 1662. Colbert's new system of state economy. Graunt on Mortality Tables and Petty's political arithmetic—basis of social statistics. *Sceptical Chymist,* 1661. Spinoza's *Tractatus Theologico-Politicus,* 1670. Greenwich Observatory, 1675. Finite velocity of light, 1676. Spermatozoa observed, 1679. *De Motu Animalium,* 1680. Halley's tables of world magnetism, 1683. Newton's *Principia,* Revocation of Edict of Nantes, 1687. Locke's Chief Works, 1685-90. Bank of England, 1694.
18th Century. First Daily Newspaper (*Daily Courant,* 1702). Act of Union England-Scotland, 1707. Coke used in iron smelting, 1713. Newcomen—

# CHRONOLOGY

Savery steam driven pump, 1712. Foundation of accurate thermometry, 1724. Compensated pendulum and chronometer, 1726. Death of Newton, 1727. First formulation of Principle of Least Action, 1747. Reform of Julian calendar, 1752. *Experiments on Magnesia Alba*, 1756. *Systema Naturae*, 1758. Watt's First Patent, 1765. Ramsden's Escapement, 1774. *Critique of Pure Reason*, 1771. *Wealth of Nations*, 1776. Declaration of Independence, 1776. Cavendish's Work on Water, 1784. Torsion Balance, 1785. Roy's base line for Ordnance Survey, 1787. Hutton's first *Theory of the Earth*, 1788. Storming of Bastille, 1789. Smith's geological map, 1790. Foundation of current electricity, 1790-1800.

19th Century. Charter of Royal Institution, 1800. Young's work on interference, 1801. Davy starts work on electro chemistry. Dalton's atomic theory, 1807. Geological Society founded, 1811. Amici's improved microscope, 1812. Congress of Vienna, 1814. *Règne d'Animal*, 1819. Ampère founded electrodynamics 1820-3. Ohm's law; Foundation of University College, London, 1826. Lyell's *Principles of Geology*, 1830. Electromagnetic induction, British Association founded, 1831. Reform Bill, 1832. Liebig's *Agricultural Chemistry*, 1840. Repeal of Corn Laws, 1846. First Medical Officer of Health (Liverpool); Joule and Helmholtz—conservation of energy, 1847. Communist Manifesto, 1848. Great Exhibition, 1851. Meteorological Office, 1855. Bunsen and Kirchhoff on Spectrum Analysis, *Origin of Species*, 1859. Maxwell's *Dynamical Theory of the Electro-magnetic Field*, 1864. Edison founded first commercial research laboratory, 1869. Cavendish Laboratory founded 1874. Telephone invented 1876. Hertzian waves 1887. X-rays, 1895.

# INDEX OF NAMES

Under each author's name are listed all the works mentioned in the text or to whose subject matter explicit reference is made. 'In general, the title is given in the language of publication.

Adams, J.C. (1819-92), 108
Agassiz (1807-73), 416
Agricola (1490-1555), 166
   *De Re Metallica*
Albert, St. (1206-80), 291, 367, 396
Alembic Club, 185
Alexandria, 37, 44, 160 f.
Al-Mamun, 322
Alphonso X, 42
Amici (1784-1860), 385, 425
Ampère (1775-1836), 247 f.
*Anatomia Cophonis*, 335
Anaxagoras (c. 488-428), 332 n.
Anaximandros (c. 611-547), 13
Anaximenes (b. 570 B.C.), 14, 16
Apollonius (fl. 220 B.C.), 38, 85
Arabs, 162
Arago (1786-1853), 151, 247
Archimedes (287-212), 26, 38, 56, 82-3, 96, 97, 269
Aristarchos (310-230), 38, 45
Aristillus, 41
Aristotle (384-22), 7, 11, 16, 27, 34, 44, 45, 49, 54, 58, 62, 68, 104, 116, 126, 129, 130, 154, 160, 269, 322 f., 340, 342, 344, 353, 357, 365, 396
   *de Anima*, 326
   *de Generatione*, 322 f.
   *de Partibus Animalium*, 322 f., 365
   *Metaphysica*, Ch. II. *passim*
Aristoxenos, 21
Arkwright, 202
Arnold de Villeneuve (1240-1311), 165
Arrhenius (1859-1927), 293
Asklepiads, 12, 331
Athenaeus, 331
Athens, 37

Atomists (see Atom and Atomism)
Avicenna (980-1037), 165, 335
   *Canon of Medicine* (Arabic)
Avogadro (1776-1856), 237
Azyr, Vic d' (b. 1748), 452

Babylonia, 6, 20, 21, 31
Bacon, F. (1561-1626), 60 f.
   *Advancement of Learning*
   *Novum Organum*
   *New Atlantis*
Bacon, R. (1214-94), 164, 207
Baer, von (1792-1876), 381, 390, 417, 419, 420
Balfour, F. M. (1851-82), 424
   *Comparative Embryology*
Balfour, Stewart (1828-87), 251
Barrow (1630-77), 88, 374
Bartholin (1655-1738), 142
Bassi (1773-1856), 448
Bastian, 447
Bateson (1861-1926), 464
   *Materials for the Study of Genetics*
Beaumont (1785-1853), 434 n.
Becher (1635-82), 197
   *Physica Subterranea*
Bell, C. (1774-1842), 454
   *Idea of a New Anatomy of the Brain*
Belon (1517-64), 348
Benedict, R. (d. 1948), 472
   *The Culture Pattern Theory*
Bennet, 212, 223 n.
Bentham (1748-1832), 413
Bergman (1735-84), 291, 398
Bernard, C. (1813-78), 442 f., 451, 471
   *An Introduction to Experimental Medicine* (Fr.)
Bernoulli, D. (1700-82), 276

## INDEX OF NAMES

Bernoulli, John (1667-1748), 274
Berthelot (1827-1907), 291
Berthollet (1748-1822), 195, 229-30, 291, 440
   *Essai d'une Statique Chimique*
Berzelius (1779 1848), 240, 434, 435, 446
Bichat (1711-1802), 381 f.
   *Traité des Membranes*
Biot (1774-1862), 245-6
Black (1728-99), 166, 184 f., 199-201
   *Experiments on Magnesia Alba*
Boerhaave (1668-1738), 184, 432
Boltzmann (1844-1906), 288 f., 297
Bonnet (1720-93), 362
Borelli (1608-79), 348 f., 432
   *De Motu Animalium*
Born, G., 467
Boscovich (chief work 1759), 287
Boulton, M., 201
Boyle (1627-91), 65 n., 132, 168 f., 181 f., 190, 197, 210, 234, 291, 349 n.
   *New Experiments Physico Mechanicall touching the Spring of the Air*
   *The Sceptical Chymist*
   *The Origin of Forms and Qualities*
Brahe, Tycho (1546-1601), 45 f.
Brethren of Purity, 163, 177
Brewster (1781-1868), 257, 265
Bridgman, P. W., 118
Brogniart (1770-1847), 377, 416
Brougham, 149
Brouncker (1620-84), 179
Brown, R. (1773-1858), 289, 389
Brunfels (1489-1534), 368
Bruno (1548-1600), 49, 396
Buchner (1860-1917), 451
Buffon (1707-88), 361, 397
   *Histoire Naturelle*
Bullialdus, 76
Bunsen (1811-99), 258 f.
Buridan (fourteenth century), 271
Byzantium, 44

Caesalpinus (1519-1603), 369
Calcar, 336-7
Cannizzaro (1826-1910), 240
Canton (1718-72), 212-3
Cardan (1501-76), 184, 208, 209
Carlisle (1768-1840), 223

Carnot (1796-1832), 202, 277 f.
   *Réflexions sur la Puissance Motrice du Feu*
Cartwright, 202
Cavalieri (1598-1647), 84, 88
Cavallo, 212 n.
Cavendish (1731-1810), 192 f., 221, 252
Challis, 108
Chamberland, 470
Charles (1746-1823), 193
Chevreul (1786-1889), 434
Childe, G., 4
Chladni, 153
Clapeyron (1799-1864), 282
Clark, Alvan (1808-87), 110
Clausius (1822-88), 281 f., 289
   *Ueber die Art der Bewegung welche wir Wärme nennen*
   *Ueber die bewegende Kraft der Wärme*
*Codex Aniciae Julianae*, 367
Coiter (1534-76), 348
Collingwood, R. G., 8
Columbus, C., 208
Columbus (*de Re Anat.*), 344
*Communist Manifesto*, 413
Comte (1798-1857), 413
   *Cours de Philosophie Positive*
Conant, J. B., 8, 458
Condillac (1715-1780), 442
   *Essai sur l'Origine des Connaissances Humaines*
Constantine the African (1017-87), 355
Copernicus (1473-1543), 44, 49, 50, 58
   *De Revolutionibus Orbium Coelestium*
Cornford, F. M., 69
   *The Laws of Motion in Ancient Thought*
Correns, 464
Cotes (1682-1716), 105
Coulomb (1738-1806), 245-6
Crookes (1832-1919), 258
Cullen (1710-90), 185
Cuvier (1769-1832), 377, 395, 403, 405, 407, 416
   *Règne d'Animal*

D'Alembert (1717-1803), 273
Dalton (1766-1844), 231 f., 277
   *A New System of Chemical Philosophy*

# INDEX OF NAMES 483

Darwin, C. (1809-82), 395 f., 401, 408 f., 415 f., 459 f., 463 f.
  *The Origin of Species*
  *The Variation of Animals and Plants under Domestication*
Davaine, 449
Davy (1778-1829), 195, 225 f., 276, 437
Debray, 296
Dedekind (1831-1916), 26
Delafond, 449
Delsaux, 290
Demokritos (c. 460-370), 27, 82, 322, 395
Descartes (1596-1650), 85 f., 122, 177, 234, 272, 345-6, 347
  *Discours de la Méthode*
  *Le Monde*
Deville (1818-81), 296
Digges, 49 n.
Diogenes Laertius (second cent. A.D.), 9
Dioskurides (first cent. A.D.), 367-8
Dobell, 364
Dogmatists, 331
Dollond (1706-61), 146, 385
Doppler (1803-53), 263
Draper, 257-62
Driesch (b. 1867), 468
Dufay (1698-1739), 211 f.
Dujardin (1801-62), 391, 424
Dulong (1785-1838), 277
Dumas (1800-84), 437
Dutrochet (1776-1847), 442

Ecole Polytechnique, 150 n., 247, 277
Eddington, A. S. (1882-1944), 117
*Edinburgh Review*, 149
Egypt, 6, 12, 21, 31, 47
Ehrenberg, 424
Einstein (b. 1870), 117 f., 290
Empedokles (c. 490-430), 15, 17
Empiricists, 331
*Entwicklungsmechanik*, 467
Epikuros (340-270), 26, 322
Erasistratos (*fl.* 280 B.C.), 328 f., 331
Estienne, 337
Euclid (*fl.* 300 B.C.), 22, 56, 87, 129
Eudoxos (c. 408-355), 26, 34, 35, 38, 82

Fabricius (ab Aquapendente) (1537-1619), 341, 348, 353 f.
Faraday (1791-1867), 242, 251, 277
  *Experimental Researches in Electricity*
Fermat (1601-65), 84, 88
Fernel (1497-1558), 338, 339, 352
Fichte (1762-1814), 386
Flamsteed (1646-1719), 107
Flemming (1843-1915), 394, 461
Forbes (1809-68), 262
Foster, M., 432
Foucault (1819-68), 156, 257
Fourcroy (1755-1809), 195
Fourier (1768-1830), 36, 202, 277
  *Théorie Analytique de la Chaleur*
Fracastoro (1484-1553), 396, 448 and n.
Franklin (1706-90), 214 f.
Fraunhofer (1787-1826), 255 f.
Fresnel (1788-1827), 150 f., 149 n.

Galen (c. 130-200), 165, 330 f., 347, 452
Galileo (1564-1642), 55 f., 62 f., 73 f., 100, 109, 130, 132, 134, 152, 272
  *Dialogo dei due massimi sistemi del mondo*
  *Dialoghi delle nuove scienze*
  *Sidereus Nuncius*
Galle, 108
Galton (1822-1911), 460, 462
Galvani (1737-98), 219 f., 453
  *De Viribus Electricitatis*
Gassendi (1592-1655), 174, 234
Gauss (1777-1855), 286
Geber (c. A.D. 850), 162
Gegenbaur (1826-1903), 422, 430
Geminus, 38
Geoffroy (1772-1844), 291
Gerhardt (1816-56), 435
Gesner (1516-65), 369
Gibbs (1839-1903), 294 f.
  *On the Equilibrium of Heterogeneous Substances*
  *Elementary Principles in Statistical Mechanics*
Gilbert (1540-1603), 209 f.
  *De Magnete*
Gilson, E., 346 and n.
Goethe (1749-1832), 377, 381 f., 386, 394, 455

21

# INDEX OF NAMES

Golgi (1884-1926), 394
Gouy, 290
Graaf, de (1641-73), 359, 390, 431
Graham, T. (1805-69), 439
Grassmann (1809-77), 297, 317
Gray (1726), 210
Green (1793-1841), 316
Gregory, J. (1638-75), 88, 96
Gresham College, 61
Grew (1641-1712), 348, 350, 375, 380
Grimaldi (1618-63), 143, 149 n.
  *Physico-Mathesis de Lumine, etc.*
Guericke (1602-86), 132, 210
Guldberg (1836-1902), 291

Haeckel (1834-1919), 393, 419 f., 424
  *Die Welträtsel*
Hahn, 451
Haldane, J. B. S., 411
Hales, S., 184, 349
  *Vegetable Statics*
Hall (1761-1832), 400
Haller (1708-77), 154, 432, 452
  *Elementa Physiologiae*
Halley (1656-1742), 76, 79, 105
Hamilton, W. R. (1805-65), 297, 317
Hanno, 37
Harun-al-Rashd (eighth cent. A.D.), 322
Harriot (1560-1621), 84
Harvey (1578-1657), 340 f., 348, 353 f., 358, etc.
  *Exercitatio Anatomica de Motu Cordis*
  *De Generatione Animalium*
Hauksbee (d. 1713), 210, 212
Heath Sir T., 26
Heiberg, 83
Helmholtz (1821-94), 280, 292, 293
  *Die Erhaltung der Kraft*
Helmont, van (1577?-1644), 166, 168, 341, 355, 431
Henderson, L., 445
Henle (1809-85), 448
Henly, 212
Henry the Navigator (1394-1460), 208
Henslow, 408
Herakleides (c. 388-315), 37
Herakleitos (c. 540-475), 16 f., 27, 47, 395
Herapath, 285
Herodotos (c. 484-425), 396

Herophilos (*fl.* 280 B.C.), 328 f., 331
Herschel, J. F. W. (1792-1871), 265
Herschel, W. (1738-1822), 107, 109, 111, 257, 400 n.
Hertwig, O. (1849-1922), 421, 467
Hertwig, R. (*b.* 1850), 421
Hesiod, 10, 11
Hipparchos (c. 130 B.C.), 38 f, 49
Hippasos, 23
Hippokrates (c. 460-370), 12, 322, 328, 331
His (1831-1904), 422, 467
*History of the Royal Society* (Birch), 170
*History of the Royal Society* (Sprat), 60, 179
Hobbes (1588-1679), 65 n.
Hofmeister (1824-77), 425 f.
Homer, 10, 11, 17
Hooke (1635-1703), 76, 78, 130, 147, 168, 177 f., 190, 197, 349 n., 380
  *Micrographia*
Hooker, J. D. (1817-1911), 145
Horrocks, 207
Horrox (1619-41), 77 n.
Horstmann (1842-1929), 292
Huggins (1824-1910), 264
Hume (1711-76), 452
Hunter, J. (1728-93), 434
Hutton (1726-97), 398 f.
  *Theory of the Earth*
Huxley, J. S., 414
Huxley, T. H. (1825-95), 419, 420
Huygens (1629-95), 71, 75, 135 f., 142 f., 147-149 n., 274 n.
  *Horologium Oscillatorium*
  *Traité de la Lumière*

Ibn-Sina (Avicenna), 165, 335
  *Canon of Medicine*
Ingenhousz (1730-99), 188
Ionians, 4
Iwanowsky, 470

Jabir, see Geber
Jaillard, 449
Janssen, 259
Johannsen (*b.* 1857), 411, 463
Joule (1818-89), 279 f.
Jung (1587-1657), 370
Jussieu, A. L. de (1748-1836), 376
Jussieu, B. de (1699-1777), 376

INDEX OF NAMES 485

Kalippos, 36
Kant (1742-1804), 385 f., 400, 413
  *Kritik der Reinen Vernunft*
  *Kritik der Urteilskraft*
  *Theorie des Himmels*
Kelsen, 14 n.
Kepler (1571-1630), 47, 49 f., 58, 67, 88
  *Mysterium Cosmographicum*
  *Astronomia Nova*
  *Harmonices Mundi*
Kircher (1602-80), 357
Kirchhoff (1824-87), 258 f., 268
Kirwan (1733-1812), 298
Koch (1843-1910), 449 f.
Kolbe (1818-84), 436
Kowalewsky (1840-1901), 421
Krateuas, 367
Kühne (1837-1900), 451
Kundt (1839-94), 289

Lack, D., 414
Lacaille, 400 n.
Lagrange (1736-1813), 204, 274, 316, 317
  *Mécanique Analytique*
Lamarck (1744-1829), 378, 402 f. 422, 452 f.
  *Histoire des Animaux sans Vertèbres*
  *Philosophie Zoologique*
Lankester, R. (1847-1929), 421
Laplace (1749-1827), 107, 112, 134, 196, 203, 286, 316, 407, 440
  *Mécanique Céleste*
Large, E. C., 458
Latour, C. de (1777-1859), 277, 446
Lavoisier (1743-94), 188 f., 197-8, 203, 225 f., 229
  *Traité Elémentaire de Chimie*
Le Chatelier (1850-1936), 292
Leeuwenhoek (1632-1723), 358 f.
Leibniz (1646-1716), 85, 91 f., 272, 360 f.
  *Monadologie*
Leonardo da Vinci (1452-1519), 62, 271, 396
Leplat, 449
Leuckhart (1822-1898), 423
Leukippos (*fl.* 475 B.C.), 25, 26
Leverrier (1811-77), 108
Lewis, G. N., 296 (1875-1946)

Liebig (1803-73), 435, 436 f., 446, 450
  *Chemistry in its Application to Agriculture and Physiology*
Linnaeus (1707-78), 370, 373, 374 f., 381, 395
  *Species Plantarum*
  *Genera Plantarum*
  *Systema Naturae*
Lipperhey, 55
Locke (1632-1704), 65 n., 331 n.
  *Essay on the Human Understanding*
Lockyer (1836-1920), 259
Loeb (1859-1924), 468
Lower (1631-91), 181
  *Tractatus de Corde*
Luc, de (1728-1817), 399
Lucretius (*c.* 95-55), 204, 232, 241, 322, 395
  *De Rerum Natura*
Lyell (1797-1875), 405 f., 415 f.
  *Principles of Geology*
  *The Antiquity of Man*

Mach (1838-1916), 274
  *The Science of Mechanics* (Ger.)
McKie, D., 435
Maclean, 204
Magendie (1783-1855), 442
Malpighi (1628-94), 344, 350, 358, 380
Malthus (1766-1834), 410
  *Essay on Population*
Malus (1775-1812), 150
Marée, 124
Maricourt, de (thirteenth century), 207
Marx (1818-83), 413
Maxwell (1831-79), 11, 254, 286 f., 295, 297
  *Treatise on Electricity and Magnetism*
Mayer (1814-87), 279-80
Mayow (1645-79), 179, 181 f., 349 n.
Meckel (1761-1833), 417
Melloni (1798-1854), 251-2
Melsens, 436
Mendel (1822-84), 411, 460 f., 464 f.
Mersenne (1588-1648), 130
Messier, 400 n.
Methodists, 331
Mettrie, de la (1709-51), 387

# INDEX OF NAMES

Michael Angelo (1475-1564), 336
Michael the Scot, 332
Michelson (1852-1931), 117, 135
Mill, J., 413
Miller, H. (1802-56), 416
Miller, W. A., 257
Milne-Edwards (1800-85), 423
Mohl (1805-72), 391, 426
Monte Cassino, 335
Morley (1838-1923), 117
Morveau, G. de (1737-1816), 195
Mouffet (1553-1604), 373
 *Theatrum Insectorum*
Müller, F. (1821-97), 419
 *Für Darwin*
Müller, Johannes (1801-58), 325, 420, 454 f.
 *Handbuch der Physiologie des Menschen*
Murdoch, 258
Musschenbroek (1692-1761), 214
*Mustelus*, 324

Nägeli (1817-91), 391, 393
*Naturphilosophie*, 377, 381, 386, 387, 454, 455
Neckam, 207
Needham, J. T. (1713-81), 361
Needham, W., 371
Neptunists, 397
Nernst (b. 1864), 293
Newcomen (1663-1729), 200, 201
Newton (1642-1727), 72 f., 84 f., 100 f., 114 f., 120 f., 132 f., 175, 210, 234, 241, 272, 273, 350, 454 n., 469, 471
 *Philosophiae Naturalis Principia Mathematica*
 *Opticks*
Nicholas of Cusa (1401-64), 271
Nicholson (1753-1816), 223
Nidd, 371
Nobile (1784-1835), 223, 251
Norman (fl. 1590), 208

Obel de'l (1538-1616), 369
Oersted (1777-1851), 243 f.
Ohm (1787-1854), 262
Oken (1779-1851), 381, 387 f.
Oldenburg, 359
Olympiodoros, 162
Ostwald, Wilhelm (1853-1932), 293

Paine, T., 413
Palissy (sixteenth century), 167
Pallas (1741-1811), 392
Pander (1794-1865), 420
Paracelsus (c. 1493-1541), 165 f.
Parmenides (b. c. 510 B.C.), 5, 16 n., 26
Pascal (1623-62), 88, 132-286
Pasteur (1822-95), 440, 445 f., 470
Peregrinus, 207
Perkin, W. H. (1838-1907), 394
Perrin (b. 1870), 290
Peters, 110
Petit (1791-1820), 277
Pflüger, 467
Pharos, 55
Pictet (1752-1825), 260
Planck (b. 1858), 265
Plato (429-348), 5, 25, 30, 31 f., 37, 44, 49, 52, 56, 129, 365, 395, 416
 *Theaetetus* and *Timaeus* and *Sophist*
Plattner, 259
Playfair (1748-1819), 400
Pledge, T., 277
Pliny the Elder (A.D. 23-79), 20, 367
Pneumatists, 331
Poisson (1781-1840), 286, 316
Pouchet, 446
Prévost, B., 448
Prévost, P. (1751-1839), 261
Priestley (1733-1804), 187, 204, 225
 *Experiments and Observations on Different Kinds of Air*
Protagoras (b. c. 490 B.C.), 29
Proust (1755-1826), 230
*ps*-Geber, 177
Ptolemy, Claudius (fl. 127-51)
Ptolemy Philadelphos, 37
Punnett, 466
Purkinje (1787-1869), 391
Pythagoras and Pythagoreans, 19, 22, 27, 30, 37, 39, 52, 66, 129
Pytheas, 37

Quetelet (1796-1874), 286

Radcliffe, 202
Ramsay (1852-1916), 259
Raoult (1830-1901), 293
Rathke (1793-1860), 417, 420
Raven, C. E., 364, 370 f.

# INDEX OF NAMES 487

Ray (1627-1705), 370 f., 375, 396-7
Rayleigh (1842-1919), 285
Réaumur (1683-1757), 433-4
Redi (1627-97), 355 f.
Remak (1815-65), 420
Rey (early seventeenth century), 179
Rhazes, 335
Richter (1762-1807), 229
Ritter (1776-1810), 257
Robison, 199
Roebuck, 201
Roemer (1644-1701), 135
Rondelet (1507-66), 348
Rouelle (1703-70), 189
Rousseau (1712-72), 413
Roux (1850-1924), 462, 467
Royal Institution, 225
Royal Society, 60, 78, 168
Rudolphi (1771-1832), 392, 455
Ruini, 348
Russell, B., 30
Rutherford (1749-1819), 186

Sachs (1832-97), 467
St. Gilles, 291
Salisbury, Lord, 156
Sars, 390
Saunders, 466
Saussure, H. B. de (1740-99), 399
Saussure, Th. de (1767-1845), 437
Savart (1791-1841), 245-6
Savery (1650-1715), 200
Scheele (1742-1806), 227, 363, 434
Schelling (1775-1854), 386
Schleiden (1804-81), 389, 426
Schrödinger (b. 1887), 471 f.
Schultze (1825-74), 393
Schwann (1810-82), 389, 390, 446
Sedgwick, 408
Seleukos, 38
Servetus (1509-53), 344
Sherrington, Sir C. (b. 1859), 472
Sidgwick, A., 415
Siebold (1804-84), 390, 392, 393, 423
Simplicius, 35
Smith, W. (1769-1839), 401
Snell (1591-1626), 56
Socrates (c. 469-399), 17, 30, 129
Sophists, 30, 129
Spagyrists, 172
Spallanzani (1729-99), 362 f., 434

Spemann (b. 1869), 467
Spencer, H. (1820-1903), 459
Spratt, 60
Stahl (1660-1734), 166, 180 f., 197, 346, 383, 432
Steenstrup (1813-97), 392, 422
Steno (1638-86), 398
Stevinus (1548-1620), 271, 342
Stokes (1819-1903), 265-7
Strasburger (1844-1912), 394, 426, 461
Swammerdamm (1637-80), 358
Sylvius (1614-72), 338, 431

Talbot, Fox, 257
Thales (seventh century B.C.), 6, 9 f., 20
Theban, Papyrus, 160
Thenard (1777-1857), 225
Theodoros, 25
Theophrastos (380-287), 328 n., 366-7
Thomas Aquinas, St. (1225-74), 258, 269, 271, 396
Thomson, B (Count Rumford) (1753-1814), 275-6
Thomson, T., 232
Thomson, W. (Kelvin) (1824-1907), 283 f.
Tillet, 448
Timocharis, 41
Torricelli (1608-47), 132
Tournefort (1656-1708), 374
Treviranus (1779-1864), 404
Tschermak, 464
Turner, 368
Tyndall (1820-93), 447

Ubaldo, 271
Usher, 360

Van't Hoff (1852-1911), 293
Varignon, 274
Vesalius (1514-64), 335 f.
 *Tabulae Anatomicae Sex*
 *De Humani Corporis Fabrica*
Vilmorin, de, 411, 460, 463
Virchow (1821-1902), 389, 391, 443
Vitruvius (*fl. c.* A.D. 10), 130
Vogel, 264
Volta (1745-1827), 212, 221 f.
Vries, de (b. 1848), 462, 463-4

# INDEX OF NAMES

Waage (1833-1900), 291
Waddington, C. H., 460
Waldeyer (1836-1921), 394
Wallace, A. R. (1823-1913), 409, 415
Wallis (1616-1703), 84, 144
Warburg, 289
Waterston (1811-83), 285
Watson, 214, 217
Watt (1736-1819), 200 f., 268
Weiditz, 368
Weierstrass (1815-98), 26, 93
Weismann (1834-1919), 460 f., 466, 468
　*The Continuity of the Germ Plasm*
Wenzel (1740-93), 291
Whitehead (1861-1947), 156, 201, 384, 389, 470, 474
Wilcke (1732-96), 200
Wilhelmj (1812-64), 291

Wheler, 212
Werner (1750-1817), 397 f.
Wilkins, 371
William of Moerbeke (*d.* 1286)
William of Occam (*d.* 1347), 276
Willis (1621-75), 347, 452
　*Anatome Cerebri*
Willughby (1635-72), 371
Wöhler (1800-82), 435
Wolff, C. F. (1733-94), 139 f., 416, 471
Wollaston (1766-1828), 232, 256, 257
Wren (1632-1723), 76

Young, T. (1773-1829), 146 f., 275

Zaluzian (1558-1613), 369
Zeno (fifth century B.C.), 24, 26, 27, 82
Zozimos (*fl. c.* A.D. 300), 162

# INDEX OF SUBJECTS

Abstraction, 5, 17, 22, 86
Achromatic combination, 146, 385
Acids, 167, 193, 227
Action, 103
   at a distance, 143
   Principle of least, 308 f.
Activity, 296
ἀήρ, 14, 15 n.
Affinity, chemical, 291 f.
Agriculture and chemistry, 225, 350
Air, dephlogisticated, 185, 187, 190, 352
   element, 14
   fixed, 186, 191
   germs in, 446 f.
   and plant life, 187
   kinds of, 184
   inflammable, 192
αἰθήρ, 15 n., 326
Alchemy, 167, 177
Alcohol, 167
Alkali, nature of, 185 f.
   decomposition of, 225-6
Alternation of generations, 422-3, 428
Analogy, 129, 386, 411, 416
Analysis, chemical, 203
   harmonic, 35
   spectrum, 257 f.
Anatomy, 328, 335 f., 348, 382
   comparative, 348
Animalcules, 359 f., 424
Anthrax, 449-50
ἄπειρον, 14
Apparatus, chemical, 161
Apperception, synthetic unity of, 385
   *Archeus*, 166
Arithmetic, 21
Art and medicine, 336-7
   and illustration, 368
Arterial loop, 417-8
Arteries, 330, 341
Assimilation, 431 f., 438 f.
Astronomy, Greek, 30

Atom and atomism, 26, 102, 174, 177-8, 231 f., 321, 330
Atomic weight, 232, 239, 240
Atomicity, 246
   of energy, 265
   of inheritance, 465

Bacteria, 359 f., 361
Bacteriology, 450
Balance, torsion, 246
Balloon, 193
Being, 16 n.
Bile, 432-3
Blood, motion of, 330, 333-4, 335, 340 f., 355, 358
Body, black, 262 f.
Brachistochrone, 309, 311 n.
Brain, 325
Breathing, 181, 348-9

Caesium, 259
Calculus, 22, 96, 252
   fluxional, 82
   of variations, 311
   vector, 312, 317
Caloric, 198, 202, 277, 281
Calorimeter, 203
Calx, 181, 190
Capacity (electric), 217, 222
   specific inductive, 300
Catastrophism, 405
Causality and retribution, 14
Cause, 63, 68, 74, 104, 156, 180, 414, 469
   four kinds of, 324-327, 385
Cell and cell nucleus, 379, 389, 390, 391, 393, 394, 426, 459, 461, 468
Ceramics, 167
Chemistry, and agriculture, 437 f.
   and biology, 363, 431 f.
   organic, 431-7
   and medicine, 164, 166, 176, 332, 341

490 INDEX OF SUBJECTS

Chlorine, 227
Chromosomes, 461, 467
 reduction of number of, 462
Classification, 363, 403, 422 f., 430-1
 logical, 365
 natural, 367, 376, 377
Cleavage (embryonic), 390, 420, 467 f.
Clock, pendulum, 69-71, 274 n.
 water, 65
Codex, 366 n.
 *Codex Aniciae Julianae*, 367
Coelome, 421
Colour, nature of, 125 f.
 of thin plates, 130, 137 f.
Colloidal state, 439
Combustion, 163, 176 f., 203
Comet, 105
Concept, 5
Concrete, 173, 184, 186
Condensation, 15
Condenser, 214, 222
Continuity, 24
 principle of, 106
Co-ordinates, cartesian, 84
 celestial, 41
Creation, 360, 397
 special, 375, 405
Crustacea, 419
Culture-pattern, 472
Cycle, reversible, 264, 278

Decay, 321, 358, 440 f., 446 f., 450
Declination, 208
Deferent, 40
Determininsm, 112
Differential, 91-3
Digestion, 431 f., 444
Dip, 208
Discharge, of point, 215
 oscillatory, 307, 314
Discovery, accidental, 110
Disease, germ theory of, 448 f.
Dissection, 325, 329, 330, 338, 353
Documents, 6

Earth, annual revolution of, 38, 45
 diurnal motion of, 37, 45
 size of, 54
Elasticity, 133, 136, 144
Electricity, 192, 210 f., 220
 animal, 220, 223

Electricity, contact, 323
 current, 224
 frictional, 248
 galvanic, 248
 two fluid theory of, 212, 245, 248
 induction of, 214
 one fluid theory of, 215, 221
 wave of, 223
Electrodynamics, 250 f., 301
Electrolysis, 293, 304
Electromagnetism, 244 f.
Electroscope (versorium), 210, 213, 222
 condensing, 222
Electrotonic state, 301, 305
Elements, primordial, 10, 16, 126, 158, 167 f., 195
 four, 15
Embryology, 321. 332, 352 f., 380-1, 416 f., 423-4, 466 f.
 plant, 425
Encasement, 360
Energy, conservation of, 280
 distribution of, 265, 268, 275
 free, 292
 kinetic, 274, 280, 298
 partition of, 288
 potential, 274
 total, 292
Engine, steam, 200 f., 277 f., 281
Entelechy, 355, 360, 385, 468
Entropy, 294, 383, 473
Enzyme, 451, 472
Epicycle, 38, 40, 49
Epigenesis, 381
Ether, 117, 136, 140, 154, 214-5, 265, 303, 314
Evolution, organic, 14, 360, 377, 395 f., 468
 of earth, 398 f.
 of societies, 413
 stellar, 400
Exchanges, theory of, 260 f.

Factor, highest common, 22
 isolation of, 124
Facts, 29
 absolute, 32
Fermentation. 203, 440 f., 450
Field, electromagnetic, 299 f.
 simplified, 65

## INDEX OF SUBJECTS

"Fits", theory of, 139-40
Fluents, 89
Fluids, imponderable, 183, 220, 222 n., 243, 247, 305, 454
Fluorescence, 265-7
Fluxions, 88, 89, 94
Food, preservation of, 363, 446
Force, 67, 73, 272
  and moment, 269-70
  and mass, 115, 118
  and weight, 272
  central, 79, 82, 96
  electromotive, 248
  lines of, 299 f.
Form, 216, 269, 305, 312-3, 416, 430, 457
Formulae, chemical, 240
Fossils, 377, 396, 398, 399, 401, 405-6, 410
Frame, spinning, 202
Freedom, degrees of, 288, 311
Fugacity, 296
Function, 85, 91, 93, 97
  Hamilton's, 317
  Lagrange's, 311

Galvanometer, 248
Gas (see "Airs"), 341
  laughing (nitrous oxide), 225
Gastrulation, 420
Gemmules, 459
Gene, 466
Generalisation, 86, 97, 152, 308
Generation, spontaneous, 325 f., 354, 355 f., 358, 361 f., 388, 392, 403, 446 f.
Genetics, 411, 461 f.
Genus, 369, 372, 375
Geometry, 21
  analytical, 84
  Gene-complex, 471
Germinal layers, 419-20, 467
Germ-plasm, 461 f.
Gestalt, 472
Glass manufacture, 159
Glycogen, 444-5
Gravitation, 20, 99, 101, 144, 454 n.
  Cavendish's demonstration, 99 n.
  Einstein's law of, 118
Gravity, centre of, 104, 109, 242, 274 n.

Growth, 321, 332, 379, 462, 471
  of plants, 328 n., 349 f., 366, 390
  and differentiation, 354
Gunpowder, 164, 190 n.

Harmony, 21
Heart, 333 f., 341 f., 346, 352
Heat, 195
  and electricity, 279
  and matter, 197
  and motion, 197, 275-6
  and work, 279 f.
  animal, 203, 354
  atomic, 277
  capacity for, 199
  enclosure, 262 f.
  flow of, 202
  kinetic theory of, 285 f.
  latent, 199
  motive power of, 202
  specific, 289
Helium, 259
Herbals, 367 f.
Histology, 383
History, nature of, 112
Homology, 416
Humours, four, 165, 334
Humus, 439 f.
Hydrogen, 193
Hylozoism, 12
Hypothesis, 65, 140-1, 152, 183, 185, 194, 197, 198, 242
  Avogadro's, 237

Induction, 301
  electromagnetic, 300 f.
  self-, 301 n.
Inertia, 75, 469
Infinitesimal, 26, 82, 84, 91, 93, 96
Infusoria, 389-9, 393
Inheritance of acquired characters, 403, 460
Insects, 355 f., 362, 373, 433
Instruments, 338
Integration, 94
Intervals, musical, 19
Invariant, 33
Invertebrate, 376, 402
Iodine, 328
Irritability, 452

## INDEX OF SUBJECTS

Jupiter (satellites), 58, 135

Kidney, 334

Lacteal, 330
Lamp, miners', 225
Law, absolute, 51
  biogenetic, 419
  Lenz's, 292
  of constant composition, 229 f.
  of inverse squares, 75, 246
  of mass action, 230
  of multiple proportions, 231
  of partial pressures, 231
  of reaction, 103
  of thermodynamics, 282-92
  of volumes (Gay-Lussac), 236
  parallelogram, 252
  statistical, 286, 290, 473
Leyden Jar, 214, 222
Liberalism, 413, 460
Light, 195
  and magnetism, 303
  diffraction of, 143, 149, 150
  double refraction of, 142
  electromagnetic theory of, 299, 304
  interference of, 147 f., 153
  Newton's queries on, 141
  polarisation of, 143, 150 f.
  propagation of, 56, 129, 141
  reflection of, 56
  refraction of, 55-7
  velocity of, 134-5, 307
Limit, 24, 93
Linkage, 466
Lodestone (see Magnets)
Loom, power, 202

Magnets, 200-8
Magnetism, terrestrial, 209-10
Maps, 37
  geological, 401
Mass, 100, 114
  active, 292
  and force, 115, 117
  and weight, 292
Materialism, 26
Mathematics, Babylonian, 4, 20
  Egyptian, 21
  nature of, 23
Matter, 101-2, 144, 268, 269

Matter, and electricity, 303
  and heat, 197
  conservation of, 203
  states of, 277
  transmutation of, 160
Meaning, 111
Mechanism, 326, 346, 347, 450, 469
Medicine, Alexandrian, 328
  and chemistry, 164, 166, 176, 332, 341
  and religion, 12
  Greek, 12, 322 f.
  schools of, 331
Mercury, motion of, 117
Metals, extraction of, 101, 159, 166, 226
  transmutation of, 160
Metamorphosis, principle of, 382
Meteorology, 189
Method,
  Archimedes', 83, 97
  Aristotelian, 323 f., 340, 353
  concrete, 327
  controlled experimental, 356-7, 363, 456
  deductive, 63, 133, 155
  empirical, 187
  historical, 6, 8
  of exhaustion, 82
  of fluxions, 88
  of residues, 108
  of tangents, 84
  quantitative in biology, 344
  scientific, 3 f., 64 f.
  statistical, 290, 296, 464
Microscope, 358 f., 385
Microscopic technique, 358, 391 f., 394
Models, mechanical, 287, 298, 305
Molecular chaos, 289
Molecule, 238, 290
  weight of, 240
Momentum, 103, 272
Moon, gravitation of, 72 f.
  motion of, 31, 77
  surface of, 57
Morphology, 382
Motion, 24, 62, 69, 87
  circular, 33, 49, 50, 75
  laws of, 68, 75, 115
  perpetual, 207, 272

# INDEX OF SUBJECTS

Motion, quantity of, 103, 273
   relative, 115
   wave, 130
Movement, Brownian, 289-90
   muscular, 348
Mover, prime, 37 n., 54, 268
Muscle, 348
Mutation, 412, 463-4

Navigation, 135 n., 207
Nebulae, 57, 400
Nerves, action of, 154, 220, 329 f., 349, 452 f., 472
   cranial, 347
   roots of, 454
Nitrogen, 186
Nomenclature, 195, 363, 367, 369, 374, 375
Notation, chemical, 195
   mathematical, 22
Notochord, 421
Numbers, 86
   and things, 19
   incommensurable, 25
   Pythagorean 22-3
   rational, 23, 26
Nutrition, 332, 350

Operator, 93
   Hamilton's, 317
Operations, physical, 118
Optics, formal, 56, 122, 129
Order, " natural ", 372
Organism and organisation, 11, 323 361, 364, 383-4, 387 f., 390, 443, 451, 469,
   and history, 413
Osmosis, 293, 442
Ovum, 360, 361, 379, 390, 417
Oxygen, 178-9

Pancreatic secretion, 431 f.
Pangenesis, 459
Paradox, Zeno's, 24, 27
Parasitology, 391 f., 423
Path, mean free, 288
Percept, 5
Phenomena, saving of, 35, 37, 52 and noumena (things-in-themselves), 335
Philosophers' stone, 164

Phlogiston, 180 f., 197, 203
Phosphorescence, 267
Photography, 257
Physiology, 329, 338, 335 f., 339, 442
   plant, 350 f.
Pile, voltaic, 223-4
Planet, motion of, 31 f.
   light of, 256
   satellite of, 287
Planetary motion, theory of, viz.:
   Eudoxos's, 34; Copernicus's, 45;
   Kepler's, 50-2, 78 f., 94, 99
   Ptolemy's, 40
   Tycho's, 47
Polarisation, electromagnetic, 304
   of light, 143, 150 f.
Pollination, 375
$\pi\nu\epsilon\tilde{\upsilon}\mu\alpha$, 326 f., 329 f., 331
Potential, 17, 222-3, 248, 295, 305, 307, 316 (Appendix)
Precession of equinoxes, 41, 105
Preformation, theory of, 360, 381
Principle, d'Alembert's, 309
   Euler's, 310
   Fermat's, 309
   Gauss's, 309
   Hamilton's, 312, 317
   Hero's, 309
   Huygens', 136, 142
   Maupertuis's, 310
   of least action, 310, 473
   " stationary ", 317
" Principles ", 163, 170, 182, 186, 189
   the three, 166
Probability, calculus of, 286
Progression, organic, 415 f.
Protista, 393, 424
Protoplasm, 391, 393
Protozoa, 424
Pump, 132, 342
Pure line, 463
Pyramid, Great, 4

Qualities, 125, 158
   emergent, 173
   secondary, 65 n.
Quaternion, 317

Radiation, 255 f.

## INDEX OF SUBJECTS

Radiation, absorptive and emissive power of, 262
  full, 262
  infra-red, 257
  selective absorption of, 266
  ultra-violet, 257
Radioactivity, 267
Rainbow, 122
Ratio, 26
  limiting, 88
Ray (light), 56, 121
Reaction,
  between currents, 247
  law of, 103
  magnetic, 246
Recapitulation, 418
Reference, frame of, 41, 115
Refraction (atmospheric), 56-57
Regeneration, 362
Relativity, 118
Resonance, 266
Respiration, 187, 203
Revolution, industrial, 268
Rubidium, 259
Rule, phase, 295

Saros, 106
Scholium, Newton's general, 115
Science, a connected system, 111
" Seeds ", 361, 448
Selection, artificial, 412
  natural, 411
Sensation, nature of, 455 f.
Simultaneity, 118
Solenoid, 250
" Souls ", 326 f, 357, 360, 385, 452
Sound, 130 f., 146
  velocity of, 133
Space, absolute and relative, 116, 118
Species, 375, 377, 406, 408 f, 412, 457
  endemic, 409-10
Spectroscope, 121, 264
Spectroscopy, 255
Spectrum, absorption, 257, 259
  and temperature, 257
  continuous, 258
  emission, 257, 258
  solar, 256
Sphere, celestial, 31-2, 49
  sublunary, 47
Spiral, Archimedean, 88

Spirit, animal, 332 f, 349
  natural, 355
  vital, 329, 332 f, 341
Standards, electrical, 283
Star, discovery of new, 47
  double, 108
  fixed, 30
  in milky way, 57
  parallax of, 108
Sterilisation, 363, 446
Sugar, blood, 444
Sun, motion of, 31, 45
  size of, 38
Surgery, Egyptian, 12
Symbols, 22
  algebraic, 64, 84, 85
System, conservative, 310
  isolated, 103
  Ptolemaic, 40, 58
  stability of solar, 107 n.

Telegraphy, 250
Teleology, 335, 385
Temperature, absolute scale of, 284
  equilibrium of, 260
Tendency, escaping, 296
Tension, electric, 248, 249
Terminology, 195, 369-70
Tests, chemical, 171, 186
Text, Aristotelian, 7, 322 f.
  chemical, 162
  medical, 335-6
Thallium, 259
Thermochemistry, 196
Thermodynamics, 278, 292 f.
  statistical, 297
Thermoelectricity, 262
Thermopile, 261
Tides, 105
Time, nature of, 24
  absolute and relative, 116
Tissues, 384
Transmutation, organic, 415 f.

Uniformitarianism, 400, 405-6
Uranus, 107
Urea, synthesis of, 435-6
Urinogenital system, 418

Valves, arterial, 343
  venous, 341

# INDEX OF SUBJECTS 495

Variations, calculus of, 311
Variations, organic, 411, 459 f.
Vector, 249
  algebra, 297, 317
Velocity, distribution of in gas, 288
  root mean square, 286, 288
Venus, phases of, 58
*Verae causae*, 10, 414
Virus, 441, 450, 470-1
Viscosity, 288
*Vis essentialis*, 380-1, 385, 471
*Vis formativa*, 363, 396
Vitalism, 326, 335, 347, 434, 435, 438, 441, 456, 468
Vortex filament, 305

Water, element, 10, 169
  electrolysis of, 224
  soda, 187 n.
  synthesis of, 192-3
Wave, electric, 314 f.
  longitudinal, 133-4, 143
  stationary, 147
  transverse, 134, 151 f.
Weight, 272
Work, 251, 272 f.
  and heat, 262, 275, 279 f., 289
  virtual, 274, 309
Worms (and larvae), 323, 355 f, 373

Zero, 26